International Conference on Natural Products and Bioresource Science (ICONPROBIOS 2017)

IOP Conference Series: Earth and Environmental Science
Volume 101

Jakarta, Indonesia
23 – 24 October 2017

ISBN: 978-1-5108-5492-5
ISSN: 1755-1307

Printed from e-media with permission by:

Curran Associates, Inc.
57 Morehouse Lane
Red Hook, NY 12571

CURRAN ASSOCIATES INC. proceedings.com

Some format issues inherent in the e-media version may also appear in this print version.

Printed by Curran Associates, Inc. (2018)

Institute of Physics
Dirac House, Temple Back
Bristol BS1 6BE UK

Phone: 44 1 17 929 7481
Fax: 44 1 17 920 0979

techtracking@iop.org

Additional copies of this publication are available from:

Curran Associates, Inc.
57 Morehouse Lane
Red Hook, NY 12571 USA
Phone: 845-758-0400
Fax: 845-758-2633
Email: curran@proceedings.com
Web: www.proceedings.com

Table of contents

Volume 101

The International Conference on Natural Products and Bioresource Science 2017 (ICONPROBIOS 2017)

23–24 October 2017, Balai Kartini Convention Center, Jakarta, Indonesia

Accepted papers received: 13 December 2017
Published online: 30 December 2017

Preface

Papers

PREFACE

The International Conference on Natural Products and Bioresource Science (ICONPROBIOS) 2017 is a conference organized by Research Unit for Natural Product Technology, Indonesian Institute of Sciences (BPTBA-LIPI), which is conducted in conjunction with the Indonesian Science Expo (ISE) arranged by the Indonesian Institute of Sciences (LIPI) to commemorate its 50th anniversary. ICONPROBIOS 2017 was held on October 23-24, 2017 at Balai Kartini (Convention Center), Jakarta, Indonesia.

By carrying the theme "Natural Products for Sustainable Life", ICONPROBIOS 2017 brought together more than 50 participants and presenters from several background such as university lecturer, researcher, student, and industrial practitioners. ICONPROBIOS 2017 provides a forum for exchange of information on natural products within all of the related topics as well as aims to build and strengthen scientific cooperation among the research institutions. The topics covered include biologically active natural products and drugs, marine natural products, bioactive metabolites from microbes, extraction and separation science, food science and technology, food processing and food engineering, food packaging, feed science and technology, ethno-veterinary, bio-resource for biofuels, bio-resource (including waste) recovery and recycling, bio-refinery and biotechnology, bio-resources for bio-based chemicals and products, and bioremediation. We hope all of us get lots of benefit from this conference thus it will give us inspiration how to enhance natural product to reach sustainable life.

Sincerely,

Dr. Anastasia Wheni Indrianingsih, M.Sc.Eng
ICONPROBIOS 2017 General Chair

IOP Conf. Series: Earth and Environmental Science **101** (2017) 011002 doi:10.1088/1755-1315/101/1/011002

Committee of ICONPROBIOS 2017

Host and Organizer: Research Unit for Natural Product Technology, Indonesian Institute of Science (BPTBA LIPI)

Advisory Board:

1. Prof. Dr. Ir. Iskandar Zulkarnaen (Chairman of Indonesian Institute of Sciences)
2. Dr. Laksana Tri Handoko (Deputy Chairman for Engineering Sciences of the Indonesian Institute of Sciences)
3. Hardi Julendra, S.Pt, M.Sc. (Director of the Research Unit for Natural Product Technology, Indonesian Institute of Sciences)

Organizing Committee:

1. Chair person: Dr. Anastasia Wheni Indrianingsih, M. Sc. Eng.
2. Co-Chairperson: Hernawan, M. Sc.
3. General Secretary: Muslih Anwar, M. Sc.
4. Secretary: Septi Nur Hayati, M.Sc., Apt.
5. Program Coordinator: Dr. Khoirun Nisa, M. Sc.

Member of Organizing Committee:

1. Lusty Istiqomah, M. Biotech.
2. Miftakhussolikhah, M. Sc.
3. Wiku Ismu Dalpito
4. Wahyu S Prabowo, S. Kom.
5. Wahyu Anggo Rizal, S.T.

Scientific and Editorial Board of ICONPROBIOS 2017

Sciences-Indonesia)

30. Dr. Asep Nurhikmat (Research Unit for Natural Product Technology, Indonesian Institute of Sciences-Indonesia)
31. Dr. Khoirun Nisa, M. Sc (Research Unit for Natural Product Technology, Indonesian Institute of Sciences-Indonesia)
32. Dr. Ahmad Sofyan, M.Sc (Research Unit for Natural Product Technology, Indonesian Institute of Sciences-Indonesia)
33. Dr. Roni Maryana, M.Sc.Eng (Research Unit for Natural Product Technology, Indonesian Institute of Sciences-Indonesia)
34. Dr. Anastasia Wheni Indrianingsih (Research Unit for Natural Product Technology, Indonesian Institute of Sciences-Indonesia)

IOP Conf. Series: Earth and Environmental Science **101** (2017) 011003 doi:10.1088/1755-1315/101/1/011003

Peer review statement

All papers published in this volume of *IOP Conference Series: Earth and Environmental Science* have been peer reviewed through processes administered by the proceedings Editors. Reviews were conducted by expert referees to the professional and scientific standards expected of a proceedings journal published by IOP Publishing.

Distribution and occurrence of polyisoprenoids in rambutan (*Nephellium lappaceum*)

M Basyuni* **and R Wati**

Department of Forestry, Faculty of Forestry, University of Sumatera Utara, Medan, North Sumatera 20155, Indonesia

E-mail address: m.basyuni@usu.ac.id

Abstract. Rambutan (*Nephellium lappaceum* L., Sapindaceae) is a tropical plant having some biological activities. All parts of *N. lappaceum* either edible or non-edible part have shown some components beneficial to human health. Plant polyisoprenoids have been reported to have biological and pharmaceutical activities. Using a two-dimensional thin layer chromatography (2D-TLC), we analyzed composition and occurrence of polyisoprenoid alcohols (polyprenols and dolichols) in *N. lappaceum* consisting of leaves, roots, fruits, flowers, seeds, and rinds. The distribution of polyprenols and dolichols were detected and categorized into three types. Type-I, displaying a predominance of dolichols over polyprenols, was observed in the roots and flowers. Type-II, having the presence of both polyprenols and dolichols, was found in the fruits and seeds. Type-III, having the predominance of polyprenols over dolichols, was observed in the leaves. The diversity of polyisoprenoids in various tissues of *N. lappaceum* is the first report, suggested the chemotaxonomic importance of polyisoprenoids in this tropical fruit plant.

1. Introduction

Rambutan (*N. lappaceum* L.) is a native tropical fruit plant to Malaysia and Indonesia and grown throughout Southeast Asia, Australia, South America, and Africa [1]. The *N. lappaceum* trees distributed in the lower or middle story in different kinds of primary and secondary forests [2]. In Indonesia, there are several cultivars of rambutan with different characteristics, namely Rapiah, Narmada, Sinyonya, Binjai, Garuda, Kapulasan, Lebak bulus, Tangkue Lebak, Bahrang, and Sibongkok. Rambutan fruit is ovoid with red or yellow pericarp. It is covered with soft spines and varied in color from green, yellow, and red [1].

Rambutan has several biological activities derived from various tissues of rambutan either edible or non-edible part have been reported to contain some components beneficial to human health [3]. For example, fruits have astringent, stomachic, and anthelmintic; the leaves are used in poultices for headaches. Furthermore, rambutan fruit peels/rinds had as an antihyperglycemic activity, antidiabetic, anti-hypercholesterolemia activities, anti-inflammation, anticancer, antioxidant, and antibacterial [3-5].

Several studies have reported that plant polyisoprenoids play a significant role in controlling cholesterol level [6]. Polyprenols from *Ginkgo biloba* leaves to have antibacterial and antifungal activity [7]. Given the critical biological and pharmaceutical activities of rambutan as well as plant polysioprenoids. Nonetheless, the distribution and occurrence of polyisoprenoid in *N. lappaceum* have not been studied yet. To get more insight into the biological activities of polyisoprenoids, it is important to obtain the information of distribution and occurrence of polyisoprenoid in tropical fruit plants especially rambutan. Here we report the distribution and occurrence of polysioprenoids from edible and non-edible parts of *N. lappaceum* for the first time.

2. Materials and Methods

2.1 Chemicals
A mixture of dolichols (C_{90}-C_{95}) and polyprenols (C_{90}-C_{100}) standard compounds as previously described [8] was used to identify the polyisoprenoids in this study. The identification of the family corresponding to polyprenols or dolichols was performed in at least three experiments. Silica gel 60 TLC plates and reversed-phase silica RP-18 HPTLC plates were purchased from Merck (Darmstadt, Germany). All of the other chemicals and solvents were of reagent grade.

2.2 Plant materials
The leaves, roots, fruits, flowers, seeds, and rinds of rambutan (*N. lappaceum*) were collected from Binjai, North Sumatra, Indonesia, in September 2016. Rambutan is naturally growing in direct sun light. The average temperature in the month of the collection was 29 °C with an average humidity of 74%. All of the fresh samples were kept at -20 °C until used.

2.3 Isolation of polyisoprenoid alcohols
A procedure for the isolation of polyisoprenoids as previously described [9]. The leaves, roots, fruits, seeds, flowers, and rinds were dried at 70°C for two days. The dried tissue (5 g each) was crushed into fine powder and immersed in 30 ml of chloroform/methanol (2/1, v/v) solvent for 48 h. The lipid extract of leaves, roots, fruits, flowers, seeds, and rinds was saponified at 65°C for 24 h in 50% ethanol containing 2 M KOH. The non-saponifiable lipids of each tissue sample extracted with hexane and the organic solvent were evaporated and redissolved in hexane.

2.4 Analysis by two-dimensional thin layer chromatography (2D-TLC)
First-dimension TLC was carried out for about 45 min on a silica gel glass plate (20 × 3 cm) with a solvent system of toluene-ethyl acetate (9:1) as previously described [9]. The longitudinal edge of the first dimension TLC plate was 1 cm in width, and the concentration zone of a reversed-phase C-18 TLC was clamped using two magnetic bars (4.0 × 1.1 × 0.8 cm) facing each gel phase. The bound TLC plate was then developed vertically to the first-dimension to transfer polyprenol or dolichol to the concentration zone of the reversed-phase TLC plate.

The second-dimension reversed-phase C-18 silica gel TLC was performed with acetone as the solvent for approximately 30 min. The spots of plants extract mixtures, and standard mixtures are separated and being developed by 2D-TLC was then identified and visualized with iodine vapor. The improved chromatographic images were obtained and digitally scanned with a Canon E-400 series printer. The polyisoprenoid family was determined by the comparison of mobility on TLC with that of authentic standards of dolichol or polyprenol that were applied in the second-dimension development. The polyprenols and dolichols that were detected on RP-18 HPTLC plates were quantified using ImageJ version 1.46r [10], with dolichol and polyprenol standards as references.

3. Results and Discussion
The results discussed in two subsections; they are the occurrence of polyisoprenoid and analysis polyisoprenoid by two-dimensional thin layer chromatography (2D-TLC).

3.1. Occurrence of polyisoprenoid
Table 1 shows the total lipid content of vegetative organs of *N. lappaceum* ranged from 509.5 to 786.5 mg g^{-1} dry weight with the lowest and highest pressures in the rinds and leaves of *N. lappaceum*,

respectively. The total lipid content of generative organs ranged from 696.7 to 987.8 mg g^{-1} dry weight with the lowest and highest contents in the fruits and flowers of *N. lappaceum*, respectively. The number of polyisoprenoids was least in the leaves (42.3 mg g^{-1} dry weight), and the highest was in the seeds (110.2 mg g^{-1} dry weight). The lowest content of polyisoprenoids was in the rinds (509.7 mg g^{-1} dry weight) and the fruits (696.7 mg g^{-1} dry weight) (Table 1). Tables 1 and 2 summarize the occurrence and distribution polyprenols and dolichols with the carbon-chain lengths given for each family.

Present results indicate that distribution of polyisoprenoid in rambutan varies depend on each tissue. In *N. lappaceum* leaves, polyprenols were abundantly detected, this finding supported the previous report on the dominance of polyprenols over dolichols in plant leaves [11-13], minor mangrove species, such as *Excoecaria agallocha, Heritiera littoralis,* and *Hibiscus tiliaceus* [8]. Shorter chain of polyprenols has been characterized by the young and old rubber leaves [13], soybean leaves [14], young spinach leaves [15], and in the families of Lauraceae, Tiliaceae, and Magnoliaceae [16]. The occurrence of ficaprenols in this study and other plants above-mentioned suggested the localization of ficaprenols in the chloroplast [14-15]. The presence of phytol may be biosynthesized ficaprenols in this organelle [15].

3.2 Analysis polyisoprenoid by two-dimensional thin layer chromatography (2D-TLC)

The structural groups of polyprenols and dolichols in the *N. lappaceum* tissues were classified as previously described [8-9] into three types (I, II, and III). In the tissues of roots, rinds, and leaves, type I, II and III was detected, respectively and on the other hand, in flowers, fruits, and seeds only type I and type II were found. In type-I, the predominance of dolichols over polyprenols (nine-fold) was observed in root and flowers of *N. lappaceum* (Figure 1A and B).

Dolichol has known as dominating polyisoprenoids in the roots of plants, for example, young rubber roots [13], soybean roots [14], and the majority of mangrove plants [8-9]. A typical feature of polyisoprenoids is their occurrence in leaf tissues as a mixture of homologous, more complicated polyprenols (such as ficaprenols; medium and longer prenols). By contrast, dolichols in the root family are enough "small" (6-8 dolichols) when accumulated in this tissue [17]. However, as has been reported recently [8-9] and in *N. lappaceum* seeds of the present study, dolichols also occurred as longer-chains in several mangrove leaves. For example, *Avicennia marina* (C_{65-130}), *Lumnitzera racemosa* (C_{60-140}), *Phempis acidula* (C_{50-140}), *Sonneratia alba* (C_{65-130}), *Acanthus ilicifolius* (C_{60-125}), *Aegiceras corniculatum* (C_{60-140}), and *S. caseolaris* (C_{50-120}).

Table 1. Occurrence and distribution of polyisoprenoids in *N. lappaceum*

Species	Tissue	TL (mg/g dw)	PI (mg/g dw)	Pol (mg/g)	Dol (mg/g)	% in total lipid			% in polyisoprenoid		Type
						PI	Pol	Dol	Pol	Dol	
N. lappaceum	leaves	786.5	42.3	42.3	nd	5.4	5.4	nd	100	nd	III
N. lappaceum	roots	761.5	25.3	nd	25.3	3.3	nd	3.3	nd	100	I
N. lappaceum	rinds	509.7	22.0	9.1	12.9	4.3	1.8	2.5	41.4	58.6	II
N. lappaceum	fruits	696.7	65.4	42.0	23.4	9.4	6.0	3.4	64.2	35.8	II
N. lappaceum	flowers	987.8	8.2	nd	8.2	0.8	nd	0.8	nd	100	I
N. lappaceum	seeds	808.2	110.2	45.9	64.2	13.6	5.7	7.9	41.7	58.3	II

nd= not detected, TL = Total lipids, PI = Polyisoprenoids, Pol = Polyprenols, Dol = Dolichols. Data are expressed as mean of triplicate analyses.

In seeds, a trace amount of polyprenols with chain-lengths similar to those of dolichols was detected (Figure 1D). In type-II, the presence of both polyprenols and dolichols was observed in rinds, fruits, and seeds of *N. lappaeceum* (Table 2, Figure 1). In the fruits and rinds polyprenols detected (ficaprenols and longer polyprenols) with a chain-length similar to that of dolichols and also polyprenols much longer than dolichols (>C_{100} and more) in chain-length were also detected, as shown in Figure 1C and 1E. Similar results of type-II have reported in seeds of rubber plant, ginkgo, and pine [13]. The occurrence of both polyprenols and dolichols was also detected in *Coluria geoides* seeds [17]. In contrast to these observations, several species of gymnosperms contained only polyprenols no dolichols identified [18], however, soybean seeds contained only dolichols [14].

In a case of *N. lappaceum* flowers, the occurrence dolichols only were found. This result did not agree with previous reports on occurrence polyprenols only in rubber flowers [13], *Philesia magellanica* and *Fuchsia magellanica* flowers [19]. It is yet known, the accumulated dolichols detected only in *N. lappaceum* flowers and roots.

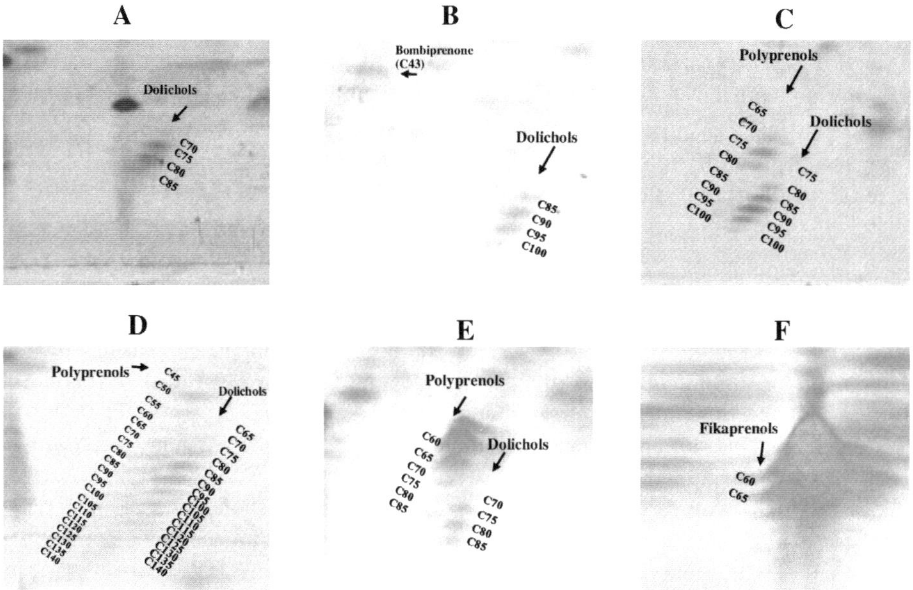

Figure 1. 2D-TLC chromatograms of polyisoprenoid extract from *N. lappaceum* roots (A), flowers (B), fruits (C), seeds (D), rinds (E), and leaves (F).

Table 2. Carbon-chain lengths of polyprenol and dolichol of Rambutan

Species	Tissue	(C43)	Polyprenol	Dolichol
N. lappaceum	leaves		60 65	
N. lappaceum	roots			70 75 80 85
N. lappaceum	rinds		60 65 70 75 80 85	70 75 80 85
N. lappaceum	fruits		65 70 75 80 85 90 95 100	75 80 85 90 95 100
N. lappaceum	flowers	O		85 90 95 100
N. lappaceum	seeds		45 50 55 60 65 70 75 80 85 90 95 100 105 110 115 120 125 130 135 140 and more	65 70 75 80 85 90 95 100 105 110 115 120 125 130 135 140 and more

For type-III, the occurrence of polyprenols over dolichols (more than nine-fold) was observed in leaves of *N. lappaceum* as shown in Figure 1D, in these species contained shorter-chain polyprenols only, ficaprenol-type (C_{60}-C_{65}), and that dolichols and longer polyprenols are in the detectable level.

The predominance dolichol over polyprenol in some plant tissue, on the other hand, polyprenols dominated on polyprenols in this study and other reports remain unclear. It has been suggested that it might be the result of changing the balance of specific compounds as biosynthesis end-products [12]. Furthermore, the predominance dolichols over polyprenols in mangrove leaves may be the presence active enzyme of polyprenol reductase [8,9], as consequence of tropical or sub tropical conditions. In this context, *N. lappaceum* is a tropical plant may have a similar pattern as mangrove plants.

4. Conclusion

The diversity of polyisoprenoids in various tissues of *N. lappaceum* is the first report, suggested the chemotaxonomic importance of polyisoprenoids in this tropical fruit plant. Present findings indicate that distribution of polyisoprenoids in rambutan varies depend on each tissue.

Acknowledgment

We thank the Directorate for Research and Community Service, Ministry of Research, Technology and Higher Education, the Republic of Indonesia for an Excellent Research of Higher Education Grant 2017 (No. 003/SP2H/LT/DRPM/IV/2017).

References

[1] Tindall H D 1994 *Hortic. Rev* **16** 143–196
[2] Mei C S W, Ismail A, Esa M N, Akowuah G A, Wai C H, Seng H Y 2014 *Antioxidants* **3** 371–386
[3] Muhtadi M, Haryoto H, Sujono A T, Suhendi A 2016 *J. Appl. Pharm. Sci.* **6** 190–194
[4] Rohman A 2017 *Res. J. Phytochem* **11** 66–73
[5] Thitilerdecha N, Teerawutgulrag A, Rakariyatham N 2008 *Food Sci. Technol.* **41** 2029-2035
[6] Pronin A V, Danilov L L, Narovlyansky A N, Sanin A V 2014 *Arch. Immunol. Ther. Exo.* **62** 31–39
[7] Tao R, Wang C Z, Kong Z W 2013 *Molecules* **18** 2166–2182
[8] Basyuni M, Sagami H, Baba S, Iwasaki H, Oku H 2016 *Dendrobiology* **75** 167–175
[9] Basyuni M, Sagami H, Baba S, Oku H 2017 *Dendrobiology* **78** 18–31
[10] Schneider, Rasband W S, Eliceiri K W 2012 *Nature Methods* **9** 671–675
[11] Sagami H, Kurisaki A, Ogura K, Chojnacki T 1992 *J. Lipid Res.* **33** 1857–1862
[12] Swiezewska E and Danikiewicz W 2005 *Prog. Lipid Res.* **44** 235–258
[13] Tateyama S, Wititsuwannakul R, Wititsuwannakul D, Sagami H, Ogura K 1999 *Phytochemistry* **51** 11–15
[14] Kurisaki A, Sagami H, Ogura K 1997 *Phytochemistry* **44** 45–50
[15] Sakaihara T, Honda A, Tateyama S, Sagami H 2000 *J. Biochem.* **128** 1073–1078
[16] Roslinska M, Walinska K, Swiezewska E, Chojnacki T 2002 *Dendrobiology* **47** 41–50
[17] Skorupińska-Tudek K, Bieńkowski T, Olszowska O, Furmanowa M, Chojnacki T, Danikiewicz W, Swiezewska E 2003 *Lipids* **38** 981–990
[18] Chouda M and Jankowski W 2005 *Acta Biochim. Pol.* **52** 243–253
[19] Strzałka K A, Szymańska R E, Świeżewska E, Skorupińska-Tudek K A, Suwalsky M A 2009 *Acta Biol. Cracoviensia Ser. Bot.* **51** 39–44

Investigation of Total Phenolic and Flavonoid Contents, and Evaluation of Antimicrobial and Antioxidant Activities from *Baeckea frutescens* Extracts

K Nisa[1], S Nurhayati[1], W Apriyana[1] and A W Indrianingsih[1]

[1] Research Unit for Natural Products Technology, Indonesian Institute of Sciences, Yogyakarta, Indonesia

E-mail: nisa.khoirun @ yahoo.com

Abstract. *Baeckea frutescens* L. is a medicinal plant endemic to the tropical area and it has been used by locals for topical and oral ailments. This study investigated total phenolic and flavonoid contents and also evaluated *in vitro* antimicrobial and antioxidant activities of of *Baeckea frutescens* crude extracts. These extracts were assessed for their antibacterial activities against strains of *Escherichia coli*, *Staphylococcus aureus*, *Salmonella thypii*, and *Pseudomonas aureginosa* by the broth micro-dilution methods using a modified tetrazolium-based colorimetric assay (3-(4, 5-dimethylthiazol)-2, 5-diphenyl tetrazolium bromide (MTT) assay). *Baeckea frutescens* crude extracts were also tested against the stable DPPH (2,2-diphenyl-1-picryl-hydrazyl-hydrate) free-radical. The results indicated that *Baeckea frutescens* water and ethanol extracts possesed remarkable antibacterial activity with the minimum inhibitory concentration less than 100 μg/ml against *Escherichia coli* and *Salmonella thypi*. On the evaluation of the antioxidant activity *via* DPPH assay, *Baeckea frutescens* ethanol extracts exhibited a good antioxidant activity with IC_{50} less than 50 μg/ml and *Baeckea frutescens* water extracts showed a moderate antioxidant activity with IC_{50} less than 100 μg/ml.

1. Introduction

Plants have the ability to produce secondary metabolites like proteins, flavonoids, alkaloids, and phenolic substances which responsible to their pharmacological properties [1,2]. Phenolic and flavonoid existed in fruits, seeds and leaves are the most important phytochemical constituents. Phenolic components in natural compounds put a notable source of antioxidant and antifungal metabolites. The source of potent antioxidants is also supplied by flavonoids content from vegetables and fruits which reported could reduce incidence of chronic inflammatory diseases or cancer. The important tool to understand the therapeutic effect of medicinal plants is through investigation of the polyphenols and flavonoids as antioxidant compounds in herbs [3,4].

Baeckea frutescens L., a coastal shrub belongs to the family Myrtaceae. This plant is mostly present throughout Southern China, Australia, and South East Asia including Indonesia, Malaysia, and Vietnam. The leaves have been used to treat fever, headache, snake bites, and rheumatism [5-7]. Previous investigation on its main constituents suggested that the terpenoids, flavonoids, and phloroglucinols possessed the biological effects such as anti-inflammatory, anti-bacterial, and cytotoxic activities. Prior phytochemical studies of the leaves of this species led to the isolation of flavanone and phloroglucinol derivatives [8,9].

Therefore, the aim of this study was to determine total phenolic and total flavonoid contents in water, ethanol and chloroform extracts of *Baeckea frutescens* and also establish their potential effects on health promotion as antioxidant and antimicrobial agents.

2. Materials and Methods

2.1. Collection of plant materials
The leaves of *Baeckea frutescens* were purchased from Pasar Beringharjo, Yogyakarta, Indonesia.

2.2. Preparation of Baeckea frutescens crude extracts
Baeckea frutescens leaves were air-dried and powdered. Approximately 20 g of materials were exhaustively macerated with water, ethanol and, chloroform solvents (3 times, with ratio 1:5 w/v) in an ultrasonic bath, for 60 min each at room temperature.

2.3. Determination of the total phenolic contents
The amount of total phenolic content in *Baeckea frutescens* crude extracts was determined according to Folin-Ciocalteu reaction method [10] with slightly modification. Briefly, 0.5 mL of extracts from the stock solution was mixed with 7.5 ml distilled water and 0.5 mL of Folin-Ciocalteu reagent. After 8 min of incubation, 1.5 mL of 20% Na_2CO_3 solution was added and further incubated for 2 hours. The absorbance was measured at 765 nm using a UV-Vis spectrophotometer (UV Vis Spectrophotometer Hitachi HALO RB-10) against blank sample. The results were expressed as gallic acid equivalents (% GAE) of dry weight and the experiment was conducted in triplicate analysis.

2.4. Determination of the total flavonoid contents
Total flavonoid contents in the *Baeckea frutescens* extracts were measured according to a colorimetric assay (UV Vis Spectrophotometer Hitachi HALO RB-10) [11] by taking 10 µL of each extract and mixed with 60 µL methanol, 10 µL of 10% $AlCl_3$ solution, 10 µL of 1 M KCH_3COO, and 120 µL distilled water. After 30 min of incubation, the absorbance was measured at 415 nm. The total flavonoids were determined as quercetin equivalents (% QE) dry weight. The experiment was conducted in triplicate analysis.

2.5. Assay for antimicrobial activity
Antimicrobial activity was evaluated through inhibition of bacterial strains growth by the standard MTT assay according to a previous procedure [12] with slight modifications. A stock solution of samples was prepared at 10 mg/mL in DMSO and varied into different concentrations in 96-well plates that contained microbial strains incubated in Potato Dextro Agar (Merck, Germany) and Nutrient Broth (Merck, Germany) medium. Bacterial strains of *Escherichia coli*, *Staphylococcus aureus*, *Salmonella thypii*, and *Pseudomonas aeruginosa* were used for this assay. The plate containing bacteria strains and samples was further incubated at 37 °C overnight, and ampicillin (Merck, Germany) was used as a positive control for the bacterial strains. Finally, the 50 mL of the 3-(4,5-dimethylthiazol-2-yl)-2,5-dimethyltetrazolium bromide (MTT) (Sigma Aldrich, USA) solution (0.5 mg mL^{-1} isopropyl-HCl) was then added to each well and incubated further for 1 h.

2.6. Assay for antioxidant activity
The ability of the plant extracts to scavenge DPPH free radicals was assessed using the microplate assay for radical scavenging activity DPPH standard method [13]. Briefly, in a 96-well plate, successively sample dilutions (standard stocks of different samples 10 mM), in triplicate, received DPPH solution (40 µM in methanol) in a total volume of 0.2 mL. After incubation period of 30 min in darkness at room temperature, the absorbance was recorded with a microplate reader (Thermo-

scientific). Absolute methanol was used as a blank and gallic acid (Sigma Aldrich, USA) as a positive control.

3. Results and Discussion

3.1. Chemical composition of Baeckea frutescens

The ethanol extract of *Baeckea frutescens* has the highest total phenol content as 5.796 % GAE dw. The total phenol content of ethanol extract was significantly higher compared to water extract, whereas the chloroform extract of *Baeckea frutescens* showed the absence of phenolic content. The efficiency of the solvents to the phenolic compounds was in the order of ethanol > water > chloroform. Total flavonoid content of the *Baeckea frutescens* extracts were recorded in least quantities in quercetin equivalents (QE). The ehanol extract of *Baeckea frutescens* showed the highest total flavonoid content as 0.192 % QE dw, in contrast the phenolic content was not detected in chloroform extract of *Baeckea frutescens*. The efficiency of the solvents to the flavonoid compounds was in the order of ethanol > water > chloroform.

3.2. Antimicrobial Activity of Baeckea frutescens

The antibacterial activity of different *Baeckea frutescens* extracts has been demonstrated against *E. coli, S. aureus, S. thypi,* and *P. aeruginosa* bacterial strains. The ethanol and water extracts of *Baeckea frutescens* showed antibacterial activity even against *E. coli* and *S. thypi*, but none of the extracts was active against *S. aureus* and *P. aeruginosa*. The result showed that the ethanol extract of *Baeckea frutescens* exhibited strong antibacterial activity against *S. thypi*, with an MIC value less than 50 µg/mL. Both ethanol and water extracts remarkably inhibited the growth of *E. coli* and *S. thypi*, with MIC values less than 50 µg/mL. Each bacterial strain demonstrated a significant degree of sensitivity to the *Baeckea frutescens* extracts. It has been reported that antibacterial activity of the extracts is due to the flavonoids, aromatic acids, terpenoids, and its esters contents. Flavonoids, one of the major constituents existed in *Baeckea frutescens*, are found to be very useful as an antimicrobial agent, a mitochondrial adhesion inhibitor, an antiulcer agent, an antiarthritic agent, an antiangiogenic agent, and an anticancer agent [14].

Table 1

Antibacterial activities of *Baeckea frustescens* L.extracts

Extracts	MIC (µg/mL)			
	E. coli	*S. aureus*	*S. thypi*	*P. aeruginosa*
Ethanol	< 50	> 200	< 50	> 200
Chloroform	> 200	> 200	> 200	> 200
Water	< 50	> 200	< 50	> 200
Ampicillin[a]	1.0	1.0	1.0	1.0

[a] Ampicillin: Positive control

3.3. Antioxidant Capacity

The antioxidant activity of the extracts was assessed on the basis of the radical scavenging effect using stable 1,1-diphenyl-2-picrylhydrazyl (DPPH). The antioxidant activity may occur via various mechanisms such as scavenging the radicals and chelating metal ions. The good DPPH scavenging activity was showed by *Baeckea frutescens* ethanol extract (IC_{50} 41.96±6.74 µg/mL) and moderate DPPH scavenging activity by *Baeckea frutescens* water extract (IC_{50} 93.3±3.41 µg/mL). In contrast, chloroform extract was not active as compared to gallic acid (IC_{50} 3.81±0.18 µg/mL). Phenolic compounds are secondary metabolites that have ability to donate hydrogen, quench singlet oxygen and act as metal chelators, which can act as antioxidant agents. Flavonoids, as one of major components in *Baeckea frutescens* consist of a large group of polyphenolic compounds provide benefits in multiple ways to the plant producing them [15]. The phenolic compounds presence in *Baeckea frutescens* leaves are attributed many of the antioxidant properties.

Table 2
Antioxidant activities of *Baeckea frustescens* L.extracts

Extracts	IC_{50} (μg/mL)
Ethanol	41.96 ± 6.74
Chloroform	> 200
Water	93.30 ± 3.41
Gallic acid[a]	3.81 ± 0.18

[a] Gallic acid: Positive control

4. Conclusion

The objective of the present work was to find the medicinal value of the plant *Baeckea frutescens*. The phytochemical properties of *Baeckea frutescens* was studied as not much work done earlier. The ethanol extract of *Baeckea frutescens* showed higher antioxidant activity than water extract. In addition, the ethanol and water extracts were also active to inhibit some bacterial strains. The highest antibacterial activity of the *Baeckea frutescens* extracts was observed against bacterial strains of *E. coli* and *S. thypi*, and is thought to be due mainly to the presence the phenolic compounds. The antioxidant and antimicrobial activities of these extracts were positively correlated with their total phenolic and flavonoid contents. It has been shown that *Baeckea frutescens* possesses significant antioxidant and antimicrobial properties as a source of natural antioxidants for use in pharmaceutical applications.

Acknowledgements
This work was supported by a grant from Indonesian Institute of Sciences, Indonesia

References

[1] Thatte U M, Dahanukar S A 1986 Ayurveda and contempory scientific thought Trends *in Pharmacological Sciences* **7** 247–249
[2] Chakraborthy G S, Patil V, Kaushik K N 2009 Evaluation of immunomodulatory activity of *Aesculus indica Journal of Pharmacy Research* **2** 385-387
[3] Gul M Z, Bhakshu L M, Ahmad F, Kondapi A K, Qureshi I A, Ghazi I A 2011 Evaluation of Abelmoschus moschatus extracts for antioxidant, free radical scavenging, antimicrobial and antiproliferative activities using in vitro assays *BMC Complementary and Alternative Medicine* **11** 64
[4] Coutinho I D, Coelho R G, Kataoka V M F, Honda N K, Silva J R M, Vilegas W, Cardoso C A L 2008 Determination of phenolic compounds and evaluation of antioxidant capacity of *Campomanesia adamantium* leaves *Ecl. Quím., São Paulo* **33** 53–60
[5] Bean A R 1997 A Revision of *Baeckea* (Myrtaceae) in eastern Australia, Malesia and South East Asia *Telopia* **7** 245–268
[6] Mardisiswojo S 1985 Cabe Puyang Warisan Nenek Moyang *PN Jakarta* (Balai Pustaka) pp 95
[7] Navanesan S, Wahab N A, Manickam S, Sim K S 2015 Evaluation of selected biological capacities of *Baeckea frutescens BMC Complem. Altern. Med.* **15** 186
[8] Makino M, Fujimoto Y 1998 Flavanones from *Baeckea frutescens Phytochemistry* **50** 273–277
[9] Nisa K, Ito T, Kodama T, Tanaka M, Okamoto Y, Asakawa Y, Imagawa H, Morita H 2016 New cytotoxic phloroglucinols, baeckenones D–F, from the leaves of Indonesian *Baeckea frutescens Fitoterapia* **109** 236-240
[10] Nowak A, Czyzowska A, Efenberger M, Krala L 2016 Polyphenolic extracts of cherry (*Prunus cerasus* L.) and blackcurrant (*Ribes nigrum* L.) leaves as natural preservatives in meat products *Food Microbiology* **59** 142-149
[11] Muanda F N, Soulimani R, Diop B, Dicko A 2011 A Study on chemical composition and biological activities of essential oil and extracts from Stevia rebaudiana Bertoni leaves *LWT -*

Food Science and Technology **44** 1865-1872

[12] Nisa K, Ito T, Subehan, Matsui T, Kodama T, Morita H 2016 New acylphloroglucinol derivatives from the leaves of *Baeckea frutescens Phytochem Lett.* **15** 42–45

[13] Chatatikun M and Chiabchalard A 2013 Phytochemical screening and free radical scavenging activities of orange baby carrot and carrot (*Daucus carota* Linn.) root crude extracts *Journal of Chemical and Pharmaceutical Research* **5** 97-102

[14] Choi Y M, Noh D O, Cho H J S, Kim K M, Kim J M 2006 Antioxidant and antimicrobial activities of propolis from several regions of Korea *LWT – Food Science and Technology* **39** 756-761

[15] Mensor L L, Menezes F S, Leita G G, Reis A S, Santos T C D, Coube C S and Leita S G 2001 Screening of Brazilian Plant Extracts for Antioxidant Activity by the Use of DPPH Free Radical Method *Phytother. Res.* **15** 127–130

IOP Conf. Series: Earth and Environmental Science **101** (2017) 012003 doi:10.1088/1755-1315/101/1/012003

Antibacterial properties of parasitic mistletoe - *Scurrula ferruginea* (Jack) Danser of Brunei Darussalam

Sheba R David, Amira Amni Adam and Rajan Rajabalaya

PAPRSB Institute of Health Sciences, Universiti Brunei Darussalam, Jalan Tungku Link, BE1410, Brunei Darussalam

E-mail address: sdsheba@gmail.com

Abstract. Natural products especially from plants have become subject of much interest in drug discovery. *Scurrula ferruginea* is parasitic mistletoe used traditionally to treat various illnesses. The aim of this study is to investigate the potential antibacterial effect of *Scurrula ferruginea* native to Brunei Darussalam. The plants were collected from Brunei-Muara district air-dried, pulverized into powder, extracted by water maceration and freeze dried. Later, the extract was screened by disc-diffusion assay for two bacterial strains. The positive controls were Gentamycin and Ampicillin for *Staphylococcus aureus* while Sulphamethoxazole was used for *Escherichia coli*. Dimethyl sulfoxide was the negative control for both strains. 100, 200 and 500 mg/ml of the extract were used for the assays. The extracts did not show any inhibition activity against *Escherichia coli*, while sulphamethoxazole, showed inhibition of 24 mm. Ampicillin, Gentamycin, 100 and 500 mg/ml extract exhibited 29, 20, 8.3 and 10.7 mm inhibition zone, respectively against *Staphylococcus aureus*. There was an increase in zone of inhibition on *Staphylococcus aureus* strain with increase in extract concentration. Further investigations including Minimum Inhibition Concentration and Minimum Bactericidal Concentration methods as well as investigation on additional bacterial strains will be conducted for confirmation of antibacterial effect of the extract.

1. Introduction

Antibiotics have enormously helped in treating infectious diseases caused by bacteria and consequently aid the quality of health of the infected individual. However, these benefits are under threat because many of the common antibiotics were known to produce adverse reactions as well as the increase in emergence of the multiple drug resistance strains in microorganisms [1]. This calls for development of new drugs that produces lesser side effect and resistance.

There has been an increase of interest on developing drugs from natural product especially from traditional medicinal plants as they are recognized as important sources of therapeutically active compounds. Hence, research on medicinal plants is becoming more popular especially for the identification and isolation of bioactive compounds from natural sources. [2]

Scurrula ferruginea is parasitic mistletoe plant that comes under *Scurrula* genus from *Loranthaceae* family and is believed to have healing properties such as anticancer, antimicrobial, antioxidant as well as antihypertensive activities. [3]

Ameer *et al.* research group verified the effect of active hydrophilic extract of *Scurrula ferruginea* plant in lowering blood pressure and treating gastrointestinal ailments by experiments based on pharmacological investigations. [4]

However, antibacterial study on this plant is still in its early stage. Hence, the aim of this study is to investigate the potential antibacterial effect of *Scurrula ferruginea* plant native to Brunei Darussalam.

2. Materials and methods

2.1. *Plant materials*

The plant *Scurrula ferruginea* was freshly collected from the host plant, *Tabeuia palleida* from Bandar Seri Begawan, Brunei Darussalam (Latitude N 4°54'57.0", Longitude E 114°56'53.8"). The plant material was taxonomically identified and deposited in the Universiti Brunei Darussalam Herbarium.

2.2. *Preparation of aqueous extract*

The leaves were cleaned thoroughly with distilled water, dried with paper towel and dried under the shade (30 ± 5°C). Then, the dried leaves were pulverized into powder and stored in an airtight container until further use. Aqueous extract was prepared by maceration of the powdered leaves (20 grams) with distilled water (100 ml) and sonicated for two hours. The extract was filtered using Whatman filter paper and stored overnight in -80°C. On the following day, the extract was freeze dried using lyophiliser. The percentage yield was 3.9%. Finally, the extracts were stored in -80°C until the next experiment. Three different concentrations of aqueous extract (100, 200 and 500 mg/ml) were investigated in this study.

2.3. *Growth and maintenance of Bacterial strains*

The antibacterial effect of the aqueous extract was investigated on ATCC 25923 *Staphylococcus aureus* (gram positive) strain and clinical strain of *Escherichia coli* (gram negative). Both of the strains were cultured in nutrient broth and the bacterial isolates were preserved on Mueller Hinton agar at 4°C until further analysis.

2.4. *Preliminary Screening: Disc diffusion method*

Disc diffusion assay was carried out as a preliminary screening for antibacterial effect of the aqueous extract. Agar plates, prepared broth and other utensils were autoclaved to sterilize them prior to the test. First, the bacterial stock suspension was removed from -80°C freezer and thawed at room temperature. Inoculation from the bacterial suspension was done on a Mueller Hinton agar and incubated at 37°C overnight in the inverted position. This procedure was done to ensure the bacteria was still alive and was not contaminated by any other organisms. After 18 – 24 hours, a few colonies of the bacterial growth were cultured in nutrient broth and incubated for approximately two hours to reach the standardized concentration of 10^8 CFU/ml (0.5 McFarland turbidity). Then, 100μl of the bacterial suspension was spread gently onto a Mueller Hinton agar plate. Finally, sterilized Whatman filter paper discs containing different concentrations of the aqueous extract were placed on the agar plate. These procedures were repeated three times and all the agar plates were incubated at 37°C overnight in the inverted position. Ampicillin/Penicillin and Dimethyl Sulfoxide (DMSO) were used as the positive and negative control of the assay respectively. The diameter of the zone of inhibition was measured to determine the antibacterial effect of the aqueous extract.

2.5. *Statistical analysis*

All the data obtained were presented as the mean ± standard deviation. One- way ANOVA test was done and subjected to Scheffe post hoc test where values $p < 0.05$ was considered as significant. All statistical analysis was processed using SPSS, version 21, IBM Corporation, NY, USA.

3. Results

Table 1. Anti bacterial effect of aqueous extract of *Scurrula ferruginea* leaves on *Staphylococcus aureus* and *Escherichia coli*

Bacterial strains	**Zone of inhibition (mm) Mean ± standard deviation**						
	DMSO	Amp	Gen	SXT	*Scurrula ferruginea* Aqueous Extract (mg/ml)		
		10 µg	10 µg	25 µg	500	200	100
***Staphylococcous aureus* ATCC 25923**	0 ± 0.00	29.33 ± 3.79	20 ± 1.41	NT	10.67 ± 0.58***	9.33 ± 0.58***	8.33 ± 0.58***
Escherichia coli	0 ± 0.00	NT	NT	24.0 ± 1.00	0 ± 0.00	0 ± 0.00	0 ± 0.00

Values are represented as mean of inhibition (mm) ± standard deviation of triplicates; NT = Not Tested; DMSO = Dimethyl Sulfoxide; Amp = Ampicillin; Gen = Gentamycin; SXT = Sulphamethaxazole * = P <0.05; ** = P < 0.01 and *** = P < 0.001 significantly lower when compared with Ampicillin.

Table 1 shows the results of the antibacterial effect of aqueous extract of *Scurrula ferruginea* leaves on *Staphylococcus aureus* and *Escherichia coli*. Dimethyl Sulfoxide was used as the negative control for both strains which did not exhibit any zone of inhibition. Ampicillin and Gentamycin (10 µg each) were used as positive control and exhibited 29.33 ± 3.79 and 20 ± 1.41 mm zone of inhibition respectively against *Staphylococcus aureus*. Additionally, 25 µg of Sulphamethaxazole was used as positive control and exhibited 24.0 mm zone of inhibition against *Escherichia coli*. The concentrations of the aqueous extract tested were 100, 200 and 500 mg/ml.

The aqueous extracts suggest minimal antibacterial action against *Staphylococcus aureus* because the zones of inhibitions were significantly lower when compared with Ampicillin ($P < 0.001$) and Gentamycin ($P < 0.01$). Nonetheless, the study has shown that when the aqueous extract was tested against *Staphylococcus aureus* strain, the zone of inhibition increases from 8.33 to 10.67 mm as the concentration of aqueous extract increase from 100 mg/ml to 500 mg/ml. However, the aqueous extract did not show any inhibition activity against *Escherichia coli*.

4. Discussion

There has been an increase in reports of antimicrobial research on medicinal plants from different parts of the world. About 80% of the world population used plant extract or their active compounds as traditional medicine as estimated by the World Health Organization. [5]

Few studies particularly on antibacterial activity has been done on mistletoe *Scurrula ferruginea* which belongs to *Loranthaceae* family. Marvibaigi *et al.*, investigated the antibacterial study of different parts such as leaf, stem and flower of *Scurrula ferruginea* found in Malaysia against *Staphylococcus aureus*, *Escherichia coli, Pseudomonas putida* and *Bacillus subtilis*. The difference is that the plant materials were macerated using Acetone and water (8:2 v/v) although the specific concentration of extracts being used were not mentioned. The study demonstrated extract from stem possessed highest antibacterial activity on *Pseudomonas putida*. The other crude extracts also capable of inhibiting the growth of *Staphylococcus aureus*, *Bacillus subtilis* and *Escherichia coli* at minimal effect. [6]

Staphylococcus aureus belongs to gram positive bacteria whereas *Escherichia coli* to gram negative bacteria. The main difference between gram positive and negative bacteria is the existing peptidoglycan cell wall in gram positive bacteria and unique outer membrane in gram negative bacteria. [7]

In this study, aqueous extract of leaves of *Scurrula ferruginea* was found to possess antibacterial effect on *Staphylococcus aureus*. However, the aqueous extract did not show any antibacterial effect on *Escherichia coli*. As a preliminary explanation, this could be because of the presence of the phytochemical constituents present in the aqueous extract of *Scurrula ferruginea* leaves which may inhibit only the growth of gram positive bacteria instead of gram negative bacteria by targeting the peptidoglycan cell wall.

Quorum sensing is a system of cell-to-cell communication between bacteria involved in regulation of genes expression of disease causing traits, such as motility, biofilm information, sporulation, and expression of many virulence determinants. The signalling molecules known as autoinducers involved in quorum sensing found in gram positive and gram negative bacteria are different. For instance, gram negative bacteria use N-acyl homoserine lactones (AHLs) while gram positive use the post-translationally modified oligopeptides as signalling molecules. [8] Numerous researches have published the ability of plant extracts and phytochemicals to interfere the quorum sensing system. The difference of signalling molecules involved between gram positive and gram negative bacteria could be one of the reasons why the aqueous extract of *Scurrula ferruginea* only inhibit the gram positive bacteria. The most common inhibition mechanisms are due to the similarity of their chemical structure to the signalling molecules or the ability to degrade the signal receptors[8]. However, further *in vivo* studies need to be done to validate if the results are truly based on quorum sensing inhibition.

Moreover, another factor that is often reported is the higher resistance of gram negative bacteria to the plant- based extracts and essential oils. This is because gram negative bacteria have hydrophilic cell wall structure which constitutes mainly of lipo-polysaccharides (LPS). The LPS structure will prevent any hydrophobic agent from penetrating and accumulating in the target cell membrane of the bacteria. [9] Hence, this could be another possible factor for inhibition of gram positive bacteria only by the aqueous extract of *Scurrula ferruginea.*

Further studies which include investigation of methanolic extract of *Scurrula ferruginea* extracted using Soxhlet apparatus is still ongoing in order to validate the overall potencies of antibacterial effect of *Scurrula ferruginea* and the outcome of the studies will be published at a future date when the results are available.

As reported previously, the phytochemical compounds that have been isolated so far from *Scurrula ferruginea* were quercitrin, quercetin and 4''-O-acetylquercitrin which are common flavanols. [10] Further in 2010, Terpenoids were discovered by Ameer *et al.* [11]. Several molecules of the secondary metabolites from phytochemical constituents such as alkaloids, flavonoids, saponins and tannins were found to be active against pathogenic microorganisms. [12]

Moreover, many researchers have consistently mentioned inhibition potential of these metabolites which significantly reduce the growth of bacteria; thus, flavonoids and terpenoids have been found to be an effective antimicrobial substance against various microorganisms. [13]

This study has also exhibited the proportionate increase of the zone of inhibition with increase in the concentration of the aqueous extract. This could also mean a higher concentration of extract contains more of the active phytochemical compounds therefore displaying larger zone of inhibition.

In another study, Tripathi *et al.*, investigated antibacterial effect of *Scurrula atropurpurea* found in South West Bengal which comes from the same *Scurrula* genus against *Bacillus subtilis*, *Escherichia coli*, *Klebsiella pneumoniae* and *Vibrio Cholerae.* They have prepared two types of macerated extracts namely, 70% crude methanolic extract and aqueous extract. The study demonstrated the effectiveness of crude methanolic extract possessing antibacterial effect against all of the bacterial strains mentioned previously. However the aqueous extract did not exhibit antibacterial effect against any of the bacterial strains including *Escherichia coli.* [14] From their study, it is possible that aqueous extract from *Scurrula* genus does not have any antibacterial effect on *Escherichia coli*. However, to confirm this statement, further investigation needs to be done specifically on the other *Scurrula* genus.

It could be also the aqueous extract of *Scurrula ferruginea* does not have any antibacterial effect on gram negative bacteria. This also needs to be confirmed by screening more bacterial strains especially on gram negative bacteria.

5. Conclusion

This is the first study reported on investigation of antibacterial effect of aqueous extract from *Scurrula ferruginea* leaves found in Brunei against *Staphylococcus aureus* and *Escherichia coli*. In summary, this study has shown that aqueous extract of *Scurrula ferruginea* leaves found in Brunei has a potential of antibacterial effect on *Staphylococcus aureus* ATCC 25923 strains but shows no inhibition against *Escherichia coli*. Further investigation which includes Minimum Inhibition Concentration (MIC) and Minimum Bactericidal Concentration (MBC) has been initiated to confirm the antibacterial effect. In addition, investigation of *Scurrula ferruginea* against more bacterial strains with different types of extracts will also be preferred.

References

[1] Nayan R B and Shukla V J 2011 Antibacterial and antifungal activities from leaf extracts of Cassia fistula l.: An ethnomedicinal plant. *J Adv Pharm Technol Res* **2** 104–9.

[2] Adebayo S A, Dzoyem J P, Shai L J and Eloff J N 2015 The anti-inflammatory and antioxidant activity of 25 plant species used traditionally to treat pain in southern African. *BMC Complement Altern Med* **15** 159.

[3] Lim Y, Rajabalaya R, Lee S, Tennakoon K, Le Q-V, Idris A, et al 2016 Parasitic Mistletoes of the Genera Scurrula and Viscum: From Bench to Bedside. *Molecules* **21** 1048.

[4] Ameer O Z, Salman I M, Quek K J and Asmawi M Z 2015 Loranthus ferrugineus: a Mistletoe from traditional uses to laboratory bench. *J Pharmacopuncture* **18** 7–18.

[5] Nayan R B and Shukla V J 2011 Antibacterial and antifungal activities from leaf extracts of Cassia fistula l.: An ethnomedicinal plant. *J Adv Pharm Technol Res* **2** 104–9.

[6] Marvibaigi M, Amini N, Supriyanto E, Jamil S, Abdul Majid F A and Khangholi S 2014 Total phenolic content, antioxidant and antibacterial properties of Scurrula ferruginea extracts. *J Teknol* **5** 65–72.

[7] Poh-Hwa T, Yoke-Kqueen C, Indu Bala J and Son R 2011 Bioprotective properties of three Malaysia Phyllanthus species: an investigation of the antioxidant and antimicrobial activities. *Int Food Res J* **18** 887–93.

[8] Abd Hamid M 2016 A new promising target for plant extracts: inhibition of bacterial quorum sensing. *J Mol Biol Biotechnol* **1** 3–5.

[9] Al Laham S A and Al Fadel F M 2014 Antibacterial activity of various plants extracts against antibiotic-resistant Aeromonas hydrophila. Jundishapur *J Microbiol* **7**:e11370.

[10] Lohezic-Le Devehat F, Tomasi S, Fontanel D and Boustie J 2002 Flavonols from Scurrula ferruginea Danser (Loranthaceae). Zeitschrift Fur Naturforsch - Sect C *J Biosci* **57** 1092–5.

[11] Ameer O Z, Salman I M, Siddiqui M J A, Yam M F, Sriramaneni R N, Sadikun A, et al. 2010 Cardiovascular activity of the n-butanol fraction of the methanol extract of Loranthus ferrugineus Roxb. *Brazilian J Med Biol Res* **43** 186–94.

[12] Dzotam J K, Touani F K, Kuete V 2015 Antibacterial and antibiotic-modifying activities of three food plants (Xanthosoma mafaffa Lam., Moringa oleifera (L.) Schott and Passiflora edulis Sims) against multidrug-resistant (MDR) gram-negative bacteria. *BMC Complement Altern Med* **16** 9.

[13] Ukwueze S E, Osadebe P O and Ezenobi NO 2013 Bioassay-guided evaluation of the antibacterial activity of Loranthus species of the African mistletoe. *Int J Pharm Biomed Res* **4** 79–82.

[14] Tripathi S, Ray S, Das P K, Mondal A K and Verma NK 2013 Antimicrobial activities of some rare aerial hemi parasitic taxa of south West Bengal, India. *Int J Phytopharm* **4** 106–12.

Antiviral Effect of Sub Fraction *Cassia alata* Leaves Extract to Dengue Virus Serotype-2 strain New Guinea C in Human Cell Line Huh-7 it-1

Marissa Angelina[1,2], Muhammad Hanafi[2] , Franciscus D. Suyatna[4] T. Mirawati S.[3] , Shirley Ratnasari[3] ,Beti Ernawati Dewi[3]

[1]Doctoral Programme in Biomedical Science Faculty of Medicine, University of Indonesia, Jl. Salemba Raya 6, Jakarta 10430, Indonesia
[2]Research Centre for Chemistry LIPI , Kompleks Puspiptek Serpong Indonesia 15416
[3]Departement of Microbiology Faculty of Medicine, Universitas Indonesia-Cipto Mangukusumo Hospital, Jalan Pengangsaan Timur No. 16 Jakarta Indonesia 10320
[4]Department of Pharmacology and Therapeutics Faculty of Medicine- RSCM Universitas Indonesia, Jl. Salemba Raya 6, Jakarta 10430, Indonesia

E-mail: marissa_angelina@yahoo.com

Abstract. Dengue virus (DENV) is one of the most common viral infections found Indonesia and tropical regions, and no specific antiviral for DENV. Indonesia has several of herbal medicine that were not explored of their potency as antiviral DENV. This study was done to evaluate the activity and toxicity of 4 derived fractions: Hexane (CA1), ethyl acetate (CA2), buthanol (CA3) and water (CA4) of *Cassia alata* leaf extract (CA) as an antiviral drug to DENV. The DENV was treated with various concentration of extract and added to Huh-7 it-1. The decrease of virus titer was determined by Focus assay. The toxicity of extract was measured by MTT assay. In our previous study ,we found that CA on Huh-7 cells showed IC_{50}, CC_{50} and SI values of <10 μg/mL, 323.45 μg/mL, and more than 32.3, respectively. For the fractions, CA3 showed best antiviral activity among other, with IC_{50}, CC_{50} and SI of <10 μg/mL, 645.8 μg/mL, and more than 64.5, respectively. CA and CA3 were proven to possess antiviral activity that is potent when tested against DENV-2. Future study was needed to explore the inhibition mechanism and compound of CA that have potency as antiviral drug to DENV.

1. Introduction

Dengue fever (DF) is one of the most prominent diseases in countries lying in near the equator, including Indonesia. According to the data from Ministry of Health Republic Indonesia, more than 190 000 people were diagnosed with dengue fever in the 2016 and were found to be fatal in several cases. The dengue virus had infected more than 80% of the cities in Indonesia and still continues to spread.[1] The virus has four serotypes (DENV-1, DENV-2, DENV-3, DENV-4), where the serotype DENV-2 has an approximately 75% chance of progressing the dengue fever to dengue hemorrhagic fever (DHF).[2] This serotype has the highest progression rate, where only up to 65% of the disease progresses in other dengue virus serotypes[3]

Currently, treatment for dengue revolves around supportive intravenous fluids to improve hematocrit result. Researches have been made, aiming at the inhibition of replication and development of DENV. However, to date, there are no specific antiviral for the DENV of any serotype that can be safely used in medical centers.[3] Screening for alternative medicine in plant extracts have been done many times due to their low toxicity, low risk of side effects, and low costs. Since then, they have been successful in producing therapeutic result in a form of antibacterial, antifungal, and also antiviral.[4] One of the known plants reported for these qualities are of the Fabaceae family and Caselpiniaceae subfamily, that is *Senna alata* or also known as *Cassia alata*. *Cassia alata* is a plant that can grow to a height of 2-3 m with the number of compound leaves. This plant is native to Indonesia and also grows in America, India, Malaysia, Brazil and Africa. Plants contain biomolecules that have varying bioactivity that can be used as a source of medicine for several different diseases. Cassia genus plants are known to contain secondary metabolites, ie antraquinon glycosides, polyphenols, flavanoids and polysaccharides.[4] Aloe-emodin, emodin, rein, crysopanol are compounds that have been isolated from *C. alata* leaves.[5] It is reported that *C. alata* has biological activity as a medicinal scab, malaria, anti oxidant, anti-bacterial wound healing against gram-positive and gram-negative bacteria, antirotavirus, and anti-fungi. [5-8]

Focusing on the leaf extracts alone, the plant contains chemical materials that is probable for developing antivirals, namely the phytosterol derivatives such as stigmasterol and β-Sitosterol, kaempferol and its derivatives, anthraquinone derivatives like aloe-emodin and chrysophanol derivatives, also rhein, ellagitinnin, phenolic acids, adenine, and xanthone.[5] The phytosterol groups for example, are known capable to inhibit absorption of cholesterol, inhibiting growth of carcinogenic cells, angiogenesis, and invasion and metastasis with an addition of trypanocidal, mosquito larvicidal as well as neutralizing agents for snake venoms. The phytosterol group also suggests have strong antioxidant abilities.[5-10] In addition, stigmasterol has a proven antiviral activity against HSC-1 while β-Sitosterol has been used as herbal medicine for its chemo preventive and neuro protective ability against BPH as well as prostate cancer through its oxidative stress mechanism.[9,11,13] Extracts mentioned above also delivers a potential antiviral property ranging from treatment against Japanese Encephalitis virus (JEV), influenza virus, enteroviruses, herpes simplex, even hepatitis virus, with some of them have also been used in antimicrobial and antifungal medicine. [9]

In this study, we screened subfraction of *Cassia alata* leaves ethanol extract and its againts DENV-2 using human cell line Huh-7 it-1 in vitro. Future study was needed to explore the inhibition mechanism and compound of CA that have potency as antiviral drug to DENV.

2. Materials and methods

2.1. Preparation of Extract

The botanical identities of *Cassia alata* leaves extract was determined and authenticated by the Botanical in Research Center for Biology LIPI. After the taxonomy identification, the plants were washed and cleaned prior to air drying at room temperature. Dried and ground leaves (1000 g) were submitted to extraction at room temperature with ethanol 70%. The solvent was removed used vaccum evaporator yielding 225.7 g of the thick extract. The extract (60 mg) was initially partitioned with n-hexane, ethyl acetate, butanol and water. The solvent was removed used vacuum evaporator yielding 4.2 g; 7.7 g; 7.1 g; 31.5 g respectively.

All of natural extracts were diluted at concentration of 100 mg/ml in dimethyl sulfoxide (DMSO) (Sigma Aldrich, USA). Centrifugation was done to remove unsolved material. The stock solution was further diluted with culture medium to the desired concentration for the assays. Stock was stored at -20°C, until used. The subtitle for each extract/fraction used in this work is defined below: • CA– Crude ethanol extract Leaves • CA1 – Hexane fraction • CA2 – Ethyl acetate fraction • CA3 – Buthanol fraction • CA4 – Water fraction

2.2. Standardization of extract

Standardization of extracts and subfraction were carried out based on the total flavonoids content (TFC) of the extracts measured using aluminium chloride colorimetric assay, and also total phenolic content (TPC) measured using follin-ciocalteu colorimetric assay. For TFC determination: Extract/ subfraction solution (100 μL) and standard (quercetin) solution (50, 100, 150, 200 and 250 μL) mixed with 2 mL distilled water. After 5 minutes, the solution was then mixed with 150 μL $NaNO_2$ 5% and 150 μL $AlCl_3$ 10%. Incubate for 6 minutes, then added with 2 mL NaOH 1 N. The absorbance reading of the reaction mixture was measured at 510 nm using spectrophotometer (UV-Vis Spectrophotometer,

Japan) with methanol as blank. A standard curve was constructed using Quercetin (Sigma-Aldrich, USA). Total flavonoids content of the extracts was then compared with the standard curve, expressed in mg quercetin equivalents per 100 g dry weight (mg QE/100 g) and further calculated as % of total flavonoids content. For TPC determination: Extract/ subfraction solution (100 μL) and standard (gallic acid) solution (50, 100, 150, 200 and 250 ul) mixed with 7 mL distilled water and 0.5 mL Follin-Ciocalteau. After incubate for 8 minutes, the solution was then added with 1.5 mL Na_2CO_3 20% ,keep in room temperature for 2 hours. The absorbance reading of the reaction mixture was measured at 765 nm using spectrophotometer (UV-Vis Spectrophotometer, Japan) with methanol as blank. A standard curve was constructed using gallic acid (Sigma-Aldrich, USA). Total phenolic content of the extracts was then compared with the standard curve, expressed in mg gallic acid equivalents per 100 g dry weight (mg GA/100 g) and further calculated as % of total phenolic content.[10]

2.3. Preparation of DENV-2 and Huh 7it cells.

Dulbecco's Modified Eagle's Medium (DMEM) High glucose cat no 11965- 092 (GIBCO, UK) was used to maintain of Huh-7 it-1 cell line in this study. DMEM was added with sodium bicarbonate to maintain of the pH of the medium and 10% of Fetal Bovine Serum (FBS) (GIBCO). Cells were then incubated at 37°C with 5% CO_2 for 4-5 days until confluent. At 80-100% confluence, the cells were sub-cultured. The sub-culturing process was initiated by removing the used medium followed by rinsing the cells twice with Phosphate buffer solution (PBS) (Gibco) cat no 70011-044. Then, 1 mL of 0.25% trypsin-EDTA (GIBCO, Canada) cat no 05200-056 was added and incubated for 5 to 10 minutes at 37°C. After addition with trypsin, 1 mL of fresh medium with 10% FBS was added and mixed. The mixture was then centrifuged for 10 minutes at 1500 rpm. The supernatant was discarded while the pellet was resuspended with 2 mL of fresh medium and re-distributed into new tissue culture flask for further maintenance.

We used DENV serotype 2 strain NGC adapted in human cell line of Huh-7 it-1 cell (Dr. Y. Shimidzu, Kobe University). A monolayer of Huh-7 it-1cell in T-75 flasks were infected with DENV-2 NGC with moi of 0.5 PFU/cell and incubate at 37oC with 5% CO_2 for 7 days. During the time of virus propagation, the FBS concentration of the cell culture medium was reduced to 2%. Supernatant was harvested and centrifuged at 1000 g for 5 minutes. Subsequently, filtered using a syringe driven 0.22 mm (Millipore, Co. Bedford MA USA). Culture supernatant was stored at -80°C and checked for the titer of dengue virus by Focus assay. [12]

2.4. Determination of cytotoxicity (CC_{50})

In vitro cytotoxicity (CC_{50}) was determined by MTT assay based on viability of Huh-7 cells after treated with extract. In 96 well flat-bottom plates (Corning, USA), we added 2×10^4 cells/well and incubated at 37°C, 5% CO_2 for 24 hours. After 24 hours, the cells were treated with various concentration of extract from 0.1 to 80 μg/mL and incubated at 37°C, 5% CO_2. After 48 hours, 20 μL of 3-(4,5-Dimethylthiazol-2-yl)- 2,5-diphenyltetrazolium bromide (MTT) (Promega) salt solution was added into each well and incubated for 4 hours according to the manufacturer's instruction. The absorbance reading of each well was measured using micro plate reader at 490 nm. The percentage of cell viability and toxicity was further determined based on the absorbance readings. First, we calculate the theoretical percentage toxicity of the samples by dividing the mean blanked sample ODs by the mean blanked control ODs for each sample. The resulting percentage toxicity was provided as the "Data In" to curve fit, with the requested interpolation set to 50. The concentrations of the samples were calculated from the curve and the interpolated 50% value was provided. The absorbance was measured at 620 nm. Background absorbance of the micro plate was removed by subtraction of blank wells. We used the viability data correspond to the cytotoxic effect. The viability of the cells was determined by this following equation: The percentage viability cell, we used to determine CC_{50}. The CC_{50} was obtained from nonlinear regression analysis of concentration-effect curves by the Graph and represented the means ± standard deviation

2.5. Determination of antiviral activity (IC_{50})
A total of 2×104 cells/well were seeded into 96-wellplate and incubated at 37°C with 5% CO_2. After 24 hours, the cells were infected with DENV-2 at a moi of 1 FFU/cell which contained various concentration. After 2 hours infection, added 100 ul of DMEM + 2% FBS contained various concentration of natural extract. Plates were further incubated at 37°C for 3 days. Then we harvest and determine virus titer by focus assay (Igarashi, 1995). Briefly, a 10-fold serial dilution of supernatant was inoculated onto Huh-7 it-1 cell monolayer in duplicate wells. Absorption was carried out at 37°C in 5% CO_2 for 2 hours with agitation at 30 minutes interval. Methylcellulose 1.5% overlay medium was added to the cell and incubated at 35°C in 5% CO_2 for 3 days. The infected cells were stained according to previous publication (Payne AF et al, 2006) with slight modification. First, infected cells were fix with 10% formaldehide in PBS and incubate at room temperature for 1 hour. Wash the cells with PBS for 3 times. To permeabilize the cells, add 100 ul /well of 1% of P40 (P40 and incubate at room temperature for 30 minutes. Block the cell with 5% skim milk in PBS and incubate at room temperature for 1 hour. After wash, add the cell with 1/1000 of human IgG anti dengue and incubate at room temperature for 1 hour. We use 1/1000 antihuman IgG label HRP as a secondary antibody. After wash, add substrate and observed infected cell with brown color. The result from focus assay was used to determine IC50. [13]

3. Results

3.1 Standardization of Cassia alata extracts and subfraction

Table 3.1 Total flavonoid and phenolic content of *C. alata* leaves ethanol extract and subfraction

Sample	Total Flavonoid Content[1]	Total Phenolic Content[2] (2)
CA	4.87±0.46	5.39±0.03
CA1	9.43±0.08	1.96±0.02
CA2	11.22±0.3	14.79±0.04
CA3	3.54±0,92	9.65±0.06
CA4	2.01±0.12	4.46±0.03

• CA– Crude of leaves ethanol extract • CA1 – Hexane fraction • CA2 – Ethyl acetate fraction • CA3 – Buthanol fraction • CA4 – Water fraction
[1]Measured with regreesion equation y=0.0088x+0.041
[2]Measured with regression equation y= 0.1174x-0.1432

Ethyl acetate fraction showed the highest result for Total flavonoid content (TFC) 11.22±0.3 mg/g euqivalence with quercetin as a standard. And also for Total phenolic content (TPC) 14.79±0.04 mg/g equivalence with gallic acid as a standard.

3.2 Activity on inhibition of DENV by C. alata
Determination of the adequate concentration needed for the extract to produce the inhibition needed to be a potential antivirus for DENV-2 is done by introducing the *C. alata* leaves extract of six different levels of concentrations to infected hepatocarcinoma cells (Huh7 it-1), with DMSO 0.1% as a negative control. To establish a more authentic result, triplicates are made for each of the concentration and also the DMSO. After the process of focus assay and immuno staining, brownish spots will be presented when looked through the microscope. These are the brown focuses that will be counted to figure out the virus titers.

IOP Conf. Series: Earth and Environmental Science **101**(2017) 012004 doi:10.1088/1755-1315/101/1/012004

Table 3.2 Virus titer achieved after given *C. alata* extract and its subfractions

Concentration	Average of Virus Titer (FFU/mL)				
μg/mL	CA*	CA1	CA2	CA3	CA4
320	0	ND	ND	ND	ND
160	0	ND	ND	ND	ND
80	0	0	0	0.3x 10^4±0.03	0
40	0.02 x 10^4±0.01	0	0	4.2 x 10^4±0.1	0.73x 10^4±0.3
20	0.13 x 10^4±0.01	0	0	2.5x 10^4±0.3	2.03 x 10^4±0.1
10	0.20 x 10^4±0.04	0	0.2 x 10^4±0.01	5.67 x 10^4±0.1	1.5x 10^4±0.3
5	ND	0.35 x 10^4±0.1	0.2 x 10^4±0.01	8.3x 10^4±0.33	2.67 x 10^4 ±0.1
2.5	ND	1.3 x 10^4±0.33	0.83x 10^4±0.3	5.13 x 10^4± 1.5	2.93x 10^4 ±0.33
DMSO 0.1%	1.60 x 10^4±0.27	14.4 x 10^4± 1.5	7.77x 10^4± 1.5	14.4 x 10^4± 1.5	7.77 x 10^4± 1.5

*= Results of CA from our previous study

Table 3.3 Percentage of Inhibition DENV-2 after Given *C. alata* Extract and its subfractions

Concentration	Percentage of Inhibition (%)				
μg/mL	CA*	CA1	CA2	CA3	CA4
320	100	ND	ND	ND	ND
160	100	ND	ND	ND	ND
80	100	100	100	97.9	100
40	98.7	100	100	70.04	90.06
20	92.0	100	100	82.7	73.8
10	87.5	100	97.4	62.6	80.07
5	ND	96.8	97.4	42.7	65.7
2.5	ND	90.08	89.3	64.4	62.2
DMSO 0.1%	0	11.6± 0.56	0	11.6± 0.56	0

*= Results of CA from our previous study

Consequently, from the virus titer we will have gotten the percentage number of infectivity, and also percentage of inhibtion to gain the final half inhibitory concentration (IC_{50}). It is shown that the higher the concentration of extract given for the viral treatment, the percentage virus titer will also decreases. This reveals a strong viral inhibition from the extract, as it only needs a modest amount to deliver its response. Thus, explaining why extracts taken from *C. alata* is a potent inhibitor of DENV-2. IC_{50} result was 0.026 μg/mL, with logarthmic regreesion $y = 6.3628 \ln(x) + 73.277$ and thus it will be taken to consideration that concentration needed to achieve a 50% inhibition needs to be much lower than the dose that were tested. Therefore, it will be concluded that the IC_{50} is lower than the lowest concentration tested which is <10 μg/mL. The result for hexane subfraction and ethyl acetate sub fraction The IC_{50} result is < 2.5 μg/mL. with logarthmic regreesion $y = -6.665 \ln(x) + 14.885$ and thus it will be taken to consideration that concentration needed to achieve a 50% inhibition needs to be much lower than the dose that were tested. Study on butanol sufraction activity shown that the higher the concentration of extract given for the viral treatment, the inhibitory effect also increases. The IC_{50} result is 6.47 μg/mL. with linear regreesion y = -2.5707 x + 14.885. Resutl for water sub fraction, shown that the higher the concentration of extract given for the viral treatment, the inhibitory effect also increases also same result with other subfraction. The IC_{50} result is < 2.5 μg/mL. with logarthmic regreesion y = -10.85 ln(x) + 51.038 and thus it will be taken to consideration that concentration needed to achieve a 50% inhibition needs to be much lower than the dose that were tested.

3.3 Toxicity of C. alata extract on the Cells

To be a potential antivirus, the extract not only has to be able to inhibit growth of the virus, but also remain harmless to the cells. Thus, half cytotoxic concentration (CC_{50}) is evaluated using MTT assay.

Table 3.4 Viability cells after treatment with *C. alata* leaves extract and its sub fractions

Concentration	Viability Cells (%)				
μg/mL	CA*	CA1	CA2	CA3	CA4
320	29.4	ND	ND	ND	ND
160	54	ND	ND	ND	ND
80	70.7	17.1	24.2	104.3	110.9
40	91.2	39.5	68.5	108.6	111.7
20	103.1	83.2	90.9	109.9	109.3
10	103.3	105.2	110.5	111.1	110.7
5	ND	112.4	111.1	108.0	112.6
2.5	ND	108.6	108.4	109.6	111.2

*= Results of CA from our previous study

Calculation use Polynomial Regression Model of the Average Cell Viability Percentage against the Extract of *C. alata*, proven that the cells become more viable in lower extract concentrations. The result of equation, $y = -0{,}0003x^2 - 0{,}2871x + 174.25$, which the CC_{50} is determined to be 323.45 μg/mL. The result for subfraction shown the same with the exatract that the cells become more viable in lower extract concentrations. The CC_{50} value of hexane, ethyl acetate, butanol and water was 47.46, 22.17, 645.8 and 311.33 μg/mL, respectively.

3.4 Selectivity Index (SI)

Selectivity index is calculated from dividing CC_{50} by IC_{50}. As a result, the final SI number is more than 32.3 for *C. alata* extract. Selectivity index for fraction can't measured clearly but in average > 18.98, >8.868, 98.29, >124.53, respectively for CA1, CA2, CA3, CA4.

4. Discussion

From this study above about standarized the extract and has been concluded that ethyl acetate extract has an high contain of total flavonoid and total phenol. From the study performed, results of IC_{50}, CC_{50}, and selectivity index has been gathered for treatment of DENV-2 by *C. alata* extract, in order to establish whether it has a potential antiviral inhibitory activity that is also non-toxic to the Huh7 cells..

Our previous study showed that the viability cells after treated with C. *alata* extract, were significant differences among each concentrations. Higer concentrations such as 320 μg/mL and 160 μg/mL showed low viability compare to control. Cells treated with 20 ug/mL or less did not show any viability difference with control. The half-cytotoxic concentration of *C. alata extract* was 323.45 µg/mL (Prasetyo et al not publish yet). The CC_{50} value of hexane, ethyl acetate, butanol and water was 47.46, 22.17, 645.8 and 311.33 µg/mL, respectively. When we compared with the crude extract, we found that CC_{50} value of hexane and ethyl acetate were lower. In other word that hexane and ethyl acetate subfraction more toxic that crude extract. In contrary in buthanol and water subfraction we found the CC_{50} value higer that crude extract of *C. Alata*. Based on the standarized data, we found that hexane and ethyl acetat subfraction of *C alata* contain more flavonoid compound compare with crude extract, buthanol and water fraction. Report in previous report that flavonoid has cytoxicity activity.[14] The other report, showed that ethanol extracts of *C. alata* leaves exhibiting significant cytotoxic activity using the brine shrimp lethality bioassay.[15] Compounds from C alata have also exhibited anti-angiogenic activity and cytotoxic activity in breast cancer cell lines as well as protective effects against pancreatic cancer. [16,17]

The half-inhibitory concentration of *Cassia alata* was below 10 µg/mL as the lowest concentration tested yielded more than 50% inhibition. This result is quite similar with a study that was also searching for an antiviral inhibitory effects from several essential plant oils, although against yellow fever virus. Nevertheless, the trend for virus titer percentage still follows as the concentration of extract gets lower. In addition, the result shows that the extract is very potent even in the highest concentration tested, as it already produces 0% virus titer percentage. Comparing this result with the hexane and ethyl acetate subfraction, a study analyzing showed until concentration 10 ug/mL still produce 0 % virus titer percentage, and because the lowest concentration of testing was 2.5 ug/mL , we concluded the $IC_{50} < 2.5$ ug/mL, also either for water subfration. The result for buthanol subfraction showed the virus titer percentage increased from the highest to the lowest concentration with the IC50 < 6.47 ug/mL. Comparing the virus titer percentage wtih the capability of polyphenol-derived compounds from plants as an antiviral for DENV-2, the outcome is alike, with the virus titer percentage trend also going up as the concentration gets lower , confirming that it is potential for the extract or subfraction to be used to inhibit dengue virus growth.[18] Moreover, phenolic acids are one of the components that can be found within the *C. alata* extract and also subfraction , as explained in the above result. Though, the virus titer result is stronger correlated with the activity of antiviral extract and subfractions.

Meanwhile, the viability percentage shows that the cell toxicity became more apparent in higher concentrations as the viable cells percentage are lesser, which is true to the theory and are similar in terms of trend with other correlated researches. The number of CC_{50} is also greater in this study (323.45 µg/mL) compared with other similar studies with a CC_{50} of 252.6 µg/mL and lesser. This also might suggest that some polyphenol compounds (included in the bioflavonoids category) may differ in producing cytotoxicity effect. Having quite a high concentration to produce cytotoxicity effect means that it will be less harmful to the cell. The result for subfraction showed the buthanol and water fraction less toxic than ethyl acetate and hexane fraction. The citotoxiciy of fraction strong correlated with the major compound in the fraction. [19]

Therefore, it can be concluded that with the IC_{50} value that is <10 µg/mL and the CC_{50} value that is >320 µg/mL, it is determined that the *C. alata* extract is a suitable antiviral for DENV-2, because of its ability to produce viral inhibition even in a relatively small concentration and an agreeable cell toxicity in a concentration higher than the ones that were studied. For the sbufraction can be concluded that buthanol fraction is more suitable antiviral for DENV-2 than other fraction because IC_{50} value that is < 10 ug/mL and the CC_{50} value that is 645.8 ug/mL.

Besides that, the selectivity index received from dividing CC_{50} (323.45 µg/mL) by the IC_{50} calculated using the logarithmic regression equation (0.026 was specifically 12,440.38, which is very high. Although taking into consideration that the IC_{50} was actually less than the concentration experimented (<10 µg/mL), the SI value became > 32.3, and > 100 for buthanol extract which are still quite high as other similar studies generally have a much lower selectivity index of around the range of 1-5, suggesting that the compound in this study is most selective towards the virus and not damaging to the host.

Conclussion

Cassia alata leaves extract and buthanol subfraction are potent candidate antiviral againt DENV-2 with the IC_{50} 0.0256 and 6.47 ug/mL and CC_{50} 323.45 and 645.8 ug/mL, respectively.

Acknowledgments

This study was supported by grant of Publikasi International Terindeks untuk Tugas Akhir Mahasiswa (PITTA) 2017 Universitas Indonesia, The Development of Medical Education & Research Center and also Two University Hospital Project no IND-0137& IND-0138 and JICA-SATREP 2010-2014. Project " Kompetitif LIPI" Molecular Farming 2014-205. Thank you for Lia Meilawati, with her assintance in extraction process.

References

[1] Kementerian Kesehatan Republik Indonesia 2015 *Profil Kesehatan Indonesia* Tahun 2014 Kementerian Kesehatan RI Jakarta

[2] FriedJR, Gibbons RV, Kalayanarooj S, Thomas SJ, Srikiatkhachorn A, Yoon I-K, et al. 2010 Serotype-Specific Differences in the Risk of Dengue Hemorrhagic Fever: An Analysis of Data Collected in Bangkok Thailand from 1994 to 2006 *PLoS Negl Trop Dis.* **2:4** (3):e617

[3] Halstead SB 2012 *Handbook for clinical management of dengue*. Geneva, Switzerland: World Health Organization

[4] Panichayupakaranant P, Sakunpak A, Sakunphueak A. 2009 Quantitative HPLC determination and extraction of anthraquinones in Senna alata leaves *J Chromatogr Sci.* **47**(3):197–200.

[5] Zige D V., Ohimain EI, Nengimonyo B.2014 Antimicrobial Activity of Ethanol Extract of Senna Alata Leaves against Some Selected Microorganism in Bayelsa State, Nigeria. Greener *J Microbiol Antimicrob.* **2**(2):26–31.

[6] Wikaningtyas P, Sukandar EY, Pharmacy C. 2015 The antibacterial activity study of senna alata leaf extract and fraction toward MRSA (methicilin-resistant staphylococcus aureus) and its mode of action **4**(4):126–33.

[7] Ehiowemwenguan G, Inetianbor J., Yakubu JM. 2014 Antimicrobial Qualities of Senna Alata. *IOSR J Pharm Biol Sci.* **9**(2):47–52.

[8] Kazeem MI, Azeez GA, Ashafa AOT. 2015 Effect of senna alata (L) roxb (fabaceae) leaf extracts on alpha-amylase, alpha-glucosidase and postprandial hyperglycemia in rats. *Trop J Pharm Res.* **14**(10):1843–8.

[9] Chang SJ, Huang SH, Lin YJ, Tsou YY, Lin CW. 2014 Antiviral activity of Rheum palmatum methanol extract and chrysophanol against Japanese encephalitis virus. *Arch Pharm Res.* **37**(9):1117–23.

[10] Artanti N, Triana Dewi R, Maryani F. 2014 The effect of location and extraction solvent on phytochemical content and antioxidant activity of Pegagan extract (Centela asiatica L. Urb) *JKTI* vol. **16** No. 2

[11] [11] Igarashi A. 1997 Impact of dengue virus infection and its control. *J. Pathogen and Disease* Vol.**18** Issue 4

[12] Payne AF, Binduga G., Kauftman E G, Kramer E D. 2006 Quantification of flavivirus by pluresecent focus assay. *J. Virol. Methods* **134** (1-2)

[13] Ridhanya A. 2015 *An In Vitro Study of Curcuma longa Activity As Antiviral Against Dengue Virus*. Jakarta: Universitas Indonesia.

[14] Srivastava J et al. 1996 Medicinal plants: an expanding role in development: *World Bank World Bank Agriculture and Forestry Systems*.Washington 320.

[15] Aburjai T, Natsheh FM. 2003 Plants used in Cosmetics. *Phototherapy Research*, **17**: 98-1000

[16] Dutta, S., Chatterjee, S., Chatterjee, S. 2012 Overview on the ethnophytopathological studies of Cassia alata- an important medicinal plant and the effect of VAM on its growth and productivity. *International Journal of Research in Botany*, **2**(4): 13-19

[17] Igwe, O. U., Onwu, F. K. 2015 Leaf essential oil of Senna alata Linn from South East Nigeria and its Antimicrobial activity. *International Journal of Research in Pharmacy and Chemistry*, **5**(1): 27-33

[18] Abd Kadir, Siti Latifah, Harisun Yaakob, and Razauden Mohamed Zulkifli. 2013 Potential Anti-Dengue Medicinal Plants: A Review *J. Nat. Med.* **67**.4 : 677-689

[19] Anjoo Kamboj , Ajay K Saluja , Munish Kumar , Pooja Atri. 2012 Antiviral activity of plant polyphenols *J.of Pham. Res.* **5**(5) 2402-241

[20] Zandi K, Teoh BT, Sam SS, Wong PF, Mustafa MR, Abubakar S 2011 Antiviral activity of four types of bioflavonoid against dengue virus type-2 *J.virol* **28**;8:560

IOP Conf. Series: Earth and Environmental Science **101** (2017) 012005 doi:10.1088/1755-1315/101/1/012005

Antiviral effects of *Curcuma longa* L. against dengue virus in vitro and in vivo

M Ichsyani[1,3], A Ridhanya[2], M Risanti[1], H Desti[1], R Ceria[1], D H Putri[1], T M Sudiro[1] and B E Dewi[1]

[1] Department *of Microbiology, Faculty of Medicine, Universitas Indonesia- Dr Cipto Mangun Kusumo Hospital, JalanPegangsaanTimur no 16, Jakarta, 10320, Indonesia.*

[2] *Undergraduate student of Faculty of Medicine, Universitas Indonesia- Dr Cipto Mangun Kusumo Hospital, Jl. Salemba no 6, Jakarta, Indonesia.*

E-mail address: betied@yahoo.com

Abstract. Dengue is the most common infective disease caused by dengue virus (DENV) and endemic diseases in tropical and subtropical areas. Until now, there is no specific antiviral for dengue infection. It is known that viral load is related to disease severity. *Curcuma longa* L. (turmeric) with curcumin as major active compound has been identified for its antiviral effect. This study to determine antiviral effect of *C. longa* extract on DENV-2 in vitro and in vivo along with its toxicity in liver and kidney of ddY mice. Antiviral activity (IC_{50}) and toxicity (CC_{50}) in vitro was examined on Huh7it-1 cells by focus assay and a MTT assay, respectively. To determine the selectivity index (SI), we used CC_{50} and IC_{50} value. The safe doses obtained were used for toxicity tests of liver and kidney with histopathological and biochemical observations. The *C. longa* extracts was given orally with dose of 0.147 mg/mL for each mice at 2 hours after injected with DENV-2 infected Huh7it-1 cells. Serum was collected from intraorbital at 6 hours and 24 hours after infection and focus assay was used to determine viral load. In this study, the acquired value of IC_{50} was 17,91 µg/mL whereas the value of CC_{50} was 85,4 µg/mL. The value of SI of *C. longa* was 4.8. In vivo, we found that *C. longa* remarkable reduced of viral load after 24 hour. Histopathological examination showed no specific abnormalities in liver and kidney. There was no significant increase in levels of SGPT, SGOT, urea, and creatinine. From this study it can be concluded that *C. longa* could potentially be used as antiviral against DENV with low cytotoxicity and effective inhibition.

1. Introduction

In the recent years, it is evident that dengue, which is largely found in the tropical and subtropical areas, is the most common systemic viral disease infecting humans [1]. According to WHO, about 390 million people of the world has infected with DENV where is 500.000 with dengue cases with more than 22.000 deaths every year [2]. Based on the number of dengue cases reported to WHO in 2004-2010, Indonesia ranks second after Brazil [3]. In Indonesia, Bali and DKI Jakarta have the highest number of DHF cases in 2013 up to 168.5/100.000 population [4]. In addition, dengue has always been a major health problem in Indonesia, which became the leading cause of the hospitalization and death of children [4].

DENV is a family of *Flaviviridae* viruses with 4 serotypes (DENV-1, DENV-2, DENV-3, and DENV-4) distinguished by genetic and antigenic properties. The infection of one serotype will not provide immunity to other serotypes [5]. Clinical manifestations of DENV infection range from asymptomatic, Dengue Fever (DF), Dengue Hemorrhagic Fever (DHF), up to Dengue Shock Syndrome (DSS) [3]. DHF will be developed into DSS in case of shock due to plasma leakage and may be fatal [6]. Decreased viremia in infected patients was expected to prevent the development of manifestation into DHF/DSS [7-8]. Currently, there are no specific treatments for DENV. Antiviral therapy and vaccines are also unavailable at this moment. Treatment for DENV is only by maintaining the body fluid of the patient, as it is critical in managing the severe form of DENV.

In Asia, *Curcuma longa* L. (Sinonim *C. domestica* Val.) or turmeric has long been used not only as spice but also as remedies [9]. Researchers have discovered that *C. longa* extract and its major active compound curcumin has many benefits in health aspect. Moreover, recent experimental studies also indicate that *C. longa* extract has properties that act as antioxidant, anti-inflammatory, antiviral, antimicrobial, etc [10]. With its component, it is expected that *C. longa* extract has an antiviral effect on the replication of DENV [9-10]. The objective of this study to evaluate the effectiveness of *C. Longa* extract as antiviral to dengue virus infection. The specific objectives were: i) to optimize the antiviral assay for dengue virus, ii) to measure the CC_{50} and IC_{50} of *C. longa* in vitro and iii) to determine the toxicity and antiviral activity in vivo.

2. Materials and Methods

2.1. *Ethics statement and experimental animals*

Male and female mice strain ddY 8-12 weeks old (30-40 g) and male mice strain ddY 2-4 weeks old (20-30 g), from Institut Pertanian Bogor (IPB) were used in this study. The mice were acclimatized for a week before start of the experiments. The animals were housed under standard laboratory condition at temperature 22 ± 2°C with relative humidity at 65 ± 10%. Standard pellet rodent diet and water were provided to the animals ad libitum. The experimental protocols were approved by the Institutional Animal Ethics Committee of Universitas Indonesia (no. protocol 16-10-337).

2.2. *Curcuma longa L. extract preparation*

Curcuma longa L. CDS-13 extract contained 19% of curcuminoid was obtained from PT Konimex, Solo. Extraction was performed by PT Konimex, Indonesia with 90% alcohol solvent. The extract is then thickened for further fractionation with methanol. The extract was dissolved in dimethyl sulfoxide (DMSO) (Sigma, Singapore) or in 0.5% Carboxymethyl cellulose (CMC) for in vitro and in vivo assays respectively.

2.3. *Preparation of DENV-2 and Huh7it-1 cells.*

Dulbecco's Modified Eagle's Medium (DMEM) High glucose (GIBCO, UK) with 10% of Fetal Bovine Serum (FBS) (GIBCO) was used to maintain of Huh-7 it-1 cell line in this study. DMEM was added with sodium bicarbonate to maintain of the pH of the medium. Cells were then incubated at 37°C with 5% CO_2 for 4-5 days until confluent. We used DENV serotype 2 strain NGC adapted in human cell line of Huh-7 it-1 cell. A monolayer of Huh7it-1 cell in T-75 flasks were infected with DENV-2 NGC at an MOI of 0.5 FFU/cell and incubate at 37°C with 5% CO_2 for 7 days. During the time of virus propagation, the FBS concentration of the cell culture medium was reduced to 2%. Supernatant was harvested and centrifuged at 1000 g for 5 minutes. Subsequently, it was filtered using a syringe driven 0.22 mm (Millipore, Co. Bedford MA USA). Supernatant was stored at -80°C and checked for the titer of DENV by Focus assay that was performed previously described by Igarashi *et al*, 1999 [11].

2.4. *Determination of cytotoxicity (CC_{50}) in vitro*

In vitro cytotoxicity (CC_{50}) was determined by MTT assay based on viability of Huh-7 cells after treated with extract. In 96 well flat-bottom plates (Corning, USA), 2×10^4 cells/well was added and incubated at 37°C, with 5% of CO_2. After 24 hours, the cells were treated with various concentrations of extract from 2.5 to 80 μg/mL and incubated at 37°C with 5% of CO_2. After 48 hours, 20 μL of 3-(4,5-Dimethylthiazol-2-yl)- 2,5-diphenyltetrazolium bromide (MTT) (Promega) salt solution was added into each well and incubated for 4 hours according to the manufaturer's instruction. The absorbance reading of each well was measured using micro-plate reader at 490 nm. First, the theoretical percentage toxicity of the samples were calculated by dividing the mean blanked sample ODs by the mean blanked control ODs for each sample. The resulting percentage toxicity was provided as the "Data In" to curve fit, with the requested interpolation set to 50. The concentrations of the samples were calculated from the curve and the interpolated 50% value was provided. The absorbance was measured at 620 nm. The viability data was used to represent the cytotoxic effect. The viability of the cells was determined by this following equation: The percentage viability cell, we able to determine CC_{50}. The CC_{50} was obtained from nonlinear regression analysis of concentration-effect curves by the Graph and represented the means ± standard deviation experiments.

2.5. *Determination of antiviral activity (IC_{50}) in vitro*

A total of 2×10^4 cells/well were seeded into 96-wellplate and incubated at 37°C with 5% CO_2. After 24 hours, the cells were infected with DENV-2 at a MOI of 1 FFU/cell which contained various concentration of natural extract from 80 μg/mL, 40 μg/mL, 20 μg/mL, 10 μg/mL, 5 μg/mL and 2.5 μg/mL. After 2 hours infection, 100 μL of DMEM+2% FBS contained various concentration of natural extract was added. Plates were further incubated at 37°C for 3 days. Virus was harvested and virus titer was determined by Focus assay [11]. Briefly, a 10-fold serial dilution of supernatant was inoculated onto Huh-7 it-1 cell monolayer in duplicate wells. Absorption was carried out at 37°C in 5% CO_2 for 2 hours with agitation at 30 minutes interval. Methylcellulose 0.5% overlay medium was added to the cell and incubated at 35°C in 5% CO_2 for 3 days. The infected cells were stained according to previous publication by Payne AF *et al*, 2006 with slight modification [12]. First, infected cells were fixed with 10% formaldehyde in PBS and incubated at room temperature for 1 hour. Cells were rinsed with PBS 3 times. To permeabilize the cells, 100 μL/well of 1% of Nonidet P40 was added and incubated at room temperature for 30 minutes. Then blocking solution (5% skim milk in PBS) was added and incubated at room temperature for 1 hour. After wash, the cell was added with 1/1000 of human IgG anti-dengue at room temperature for 1 hour. 1/1000 anti-human IgG label HRP was added as a secondary antibody. After wash, substrate was added and observed infected cell with brown colour. The result from Focus assay was used to determine IC_{50} with following equation: Based on percentage inhibition, we able to determine IC_{50}. The IC_{50} was obtained from nonlinear regression analysis of concentration-effect curves by the graph and represented the means ± standard deviation experiments.

2.6. *Determination of toxicity in vivo in mice liver and kidney*

Toxicity test was performed on the basis of non-clinical toxicity test guidelines in vivo by BPOM-RI [13]. The ddY mice was divided into 3 groups with 10 mice (5 male and 5 female mice) each group. Group 1, mice were administered with 500 mg/kgBW of extract in 0.5% CMC. Group 2, mice were administered with 1000 mg/kgBW of extract in 0.5% CMC. Group 3, mice were administered with 0.5% CMC only. Extract was admitted orally admitted orally once day for 14 days. Animals were weighed regularly during the experiment.

At the end of the experiment (i.e. on 14th day), the animals were anesthetized with ketamine intraperitoneally (i.p) injection and sacrificed by cervical decapitation. Blood was collected intraorbitally from all the experimental mice and serum was separated. Serum samples were stored at -80°C and assayed for Serum Glutamic Pyruvate Transaminase (SGPT) and Serum Glutamic Oxaloacetic Transaminase (SGOT), ureum and creatinine using standard DiaSys kit that were purchased from PT DiaSys Diagnostic System GmbH (Holzheim, Germany). The liver and kidney were isolated and stored in 10% formalin for Histopathological analysis using Haematoxylin Eosin staining.

2.7. Preparation of Huh7it-1 infected DENV-2 NGC cells for in vivo experiment
A total 5x10^6 Huh7it-1 cells were infected with 800 μL DENV-2 NGC at an MOI of 0.5 FFU/cell and 200 μL DMEM with 2% FBS was added. The infected cells were incubated for 2 hours at 37°C with 5% CO_2 and homogenized every 20 minutes. After 2 hours of incubation, the infected cells were transferred to the T-75 flasks and 9 mL DMEM with 2% FBS was added. Then, cells were incubated at 37°C with 5% CO_2 for 2 days.

After incubation, medium was removed and rinsed with 4 mL PBS 1x. Then, 1 mL of 0.25% trypsin-EDTA was added and incubated for 5 to 10 minutes at 37°C with 5% CO_2. After addition with trypsin, 1 mL of fresh medium with 10% FBS was added and mixed. The mixture was then centrifuged at 4°C for 10 minutes at 1500 rpm. The supernatant was discarded while cells were re-suspended with PBS 1x 2 mL before being injected into mice. The DENV infected Huh7it-1 cells was stained by immunofluoresence assay (IFA) and observed under fluorescence microscope. The infected cell were used when the percentage of infected cells above 80%.

2.8. Determination of antiviral activity in vivo
ddY mice were injected intraperitoneally (i.p) with DENV infected Huh7it-1 to produce DENV viremia in mice. This method was based on the development of Yamanaka A. and Konishi E. method in 2009 in vaccine testing [14].The dose was determined by calculating the conversion value of IC_{50} *Curcuma longa* L. extract [15].

The mice were divided into 2 groups containing 3 male mice each group. Group 1, mice were injected with 1 mL DENV infected cells (5.5 x 10^6 Huh7it-1 in 1 mL PBS 1x) (i.p) and administered once orally with 0.147 mg/mL *C. longa* extract in 0.5% CMC at 2 hours after infection. Group 2 (control), mice were injected with 1 mL DENV infected cells (5.5 x 10^6 Huh7it-1 cells in 1 mL PBS 1x) and administered once orally only with 0.5% CMC at 2 hours after infection.

Blood was collected intraorbitally (i.o) on 6th and 24th hours after infection. The animals were anesthetized with ketamine (i.p) injection. Then, blood was added 10 μl 3.8% sodium citrate and centrifuged at 200 rpm for 5 minutes. Serum samples were collected and stored at -80°C. The titer of DENV in serum was determined by Focus assay [11].

2.9. Statistical Analysis
The data from in vitro assay was analysed with unpaired t-test on GraphPad Prism 6. The homogeneous and normally distributed data were analysed by ANOVA followed by Tukey test as Post Hoc analysis. Data with no normality and homogeneity distribution will be analysed using Wilcoxon test. All the statistical was analysis using SPSS 21 software.

3. Results

3.1. DENV Inhibition by Curcuma longa L.in vitro
DENV were treated with various concentration of extract before infect to the Huh7it-1 cells. The concentrations were 80 μg/mL, 40 μg/mL, 20 μg/mL, 10 μg/mL, 5 μg/mL and 2.5 μg/mL. CyclosporinA 2 μg/mL was used as positive control whereas DMSO 0.1% was used as negative control. The data distribution of the virus titer from each concentration was normal, with p-value more than 0.05. The Table 1 illustrated the average titer of the virus with its standard deviation.

Table 1. Titer of DENV-2 upon treatment with various concentration of *C. longa* extract.

Treatment *C. longa*	Average titer (FFU/mL)	SD
80 μg/mL	0	0
40 μg/mL	0.06 x 10^2	3.8
20 μg/mL	0.75 x 10^2	16.5

10 μg/mL	1.21×10^2	7.9
5 μg/mL	1.80×10^2	11.5
2.5 μg/mL	1.87×10^2	18.7
CyA 2 μg/mL (+ control)	1.73×10^2	3.6
DMSO 0.1% (- control)	2.04×10^2	14.3

The average titer of the virus was then further calculated to find the percentage of the virus that was inhibited upon treatment. The equation from the linear regression curve of the percent inhibition was defined to determine the half inhibitory concentration (IC_{50}).

Table 2. Inhibited DENV-2 upon treatment with *C. longa* extract

Treatment *C. longa*	Inhibition (%)
80 μg/mL	100
40 μg/mL	97.1
20 μg/mL	63.2
10 μg/mL	40.7
5 μg/mL	11.8
2.5 μg/mL	8.3
CyA 2 μg/mL (+ control)	15.2
DMSO 0.1% (- control)	0

The viral inhibition increased as the *C. longa* extract concentration increased (Table 2., Figure 3.). This means that there was an antiviral activity of *C. longa* to DENV. From the equation, the IC_{50} of *C. longa* extract was 17.91 μg/mL (Figure 1.).

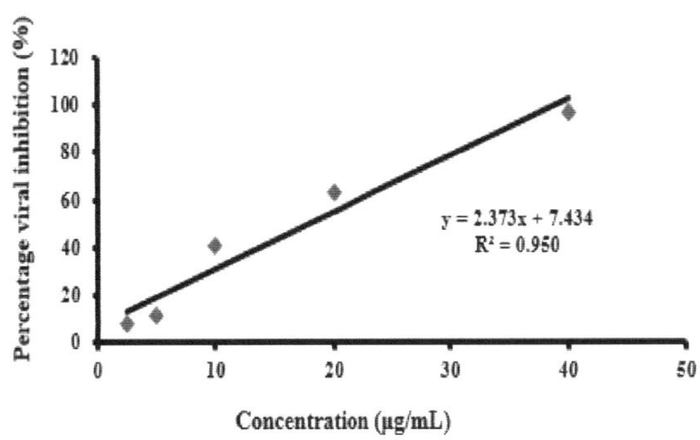

Figure1. Linear regression curve of the inhibited DENV-2 percentage upon treatment with *C. Longa*extract.

3.2. Cytotoxic Effect of C. longa

To ensure that extract was not toxic to the cell, the half cytotoxic concentration (CC_{50}) was measured. This was achieved from the result of MTT assay. The cell viability still showed high level after treated with *C. longa* extract at concentration up to 40 μg/mL (Table 3.).

Table 3. The viability cells upon treatment with various concentration of *C. Longa* extract.

Concentration (μg/mL)	Viability (%)	SD
80	51.3	2.39
40	86.5	11.94
20	99.1	3.14
10	97.4	4.01
5	101.8	2.63
2.5	111.7	4.53

The CC_{50} was analyzed from the linear regression equation of the percent viability. It was discovered that the CC_{50} of turmeric was 85.4 μg/mL (Figure 2.)

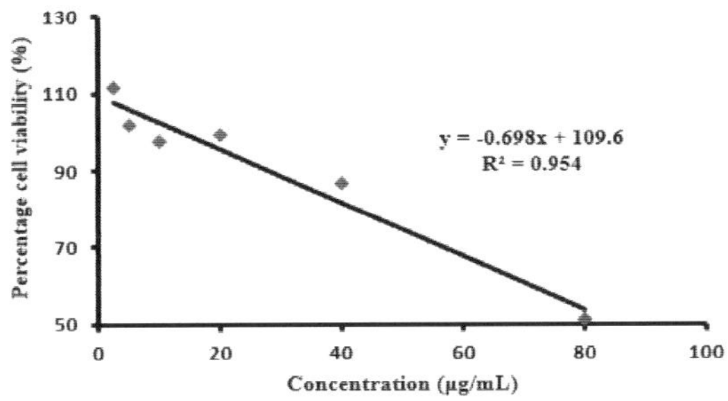

Figure 2. Percentage of cell viability after treatment with *C. longa*

3.3. *Selectivity Index (CC_{50}/IC_{50})*

After the CC_{50} was divided by IC_{50}, it was found that the selectivity index of *Curcuma longa* extract in this study was 4.8

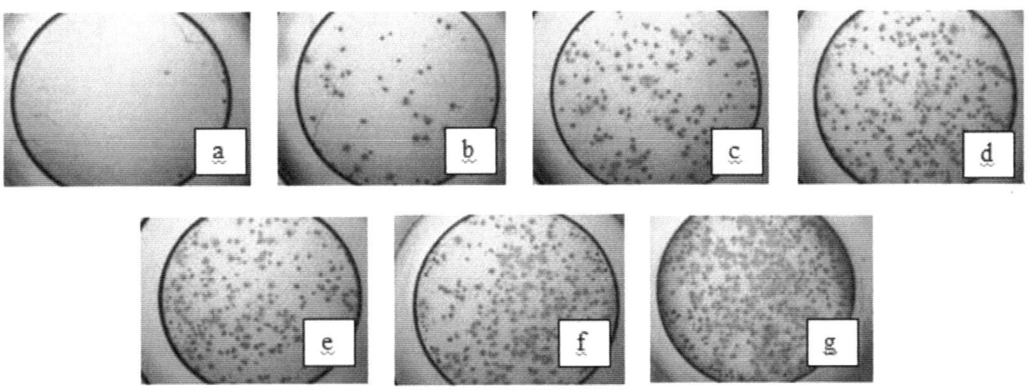

Figure 3. Focus image of DENV-2 NGC after treated with various concentration of *C. Longa* extract. (a). 80 µg/mL; (b). 40 µg/mL; (c).20 µg/mL; (d).10 µg/mL, (e).5 µg/mL;(f).2.5 µg/mL and (g).DMSO.

3.4. *Toxicity of extract to mice kidney and liver*

The results of observation of mice body weight for two weeks both male and female for each dose of *C. longa* extract showed weight gain during experiment, but statistically the administration with various doses did not affect the body weight of mice and changes in body weight was not significantly different with the group control. The macroscopic images of liver and kidney showed no change occurred compared to control. The liver and kidney were red and shiny, the surface was smooth, the consistency was springy, there was no discoloration.

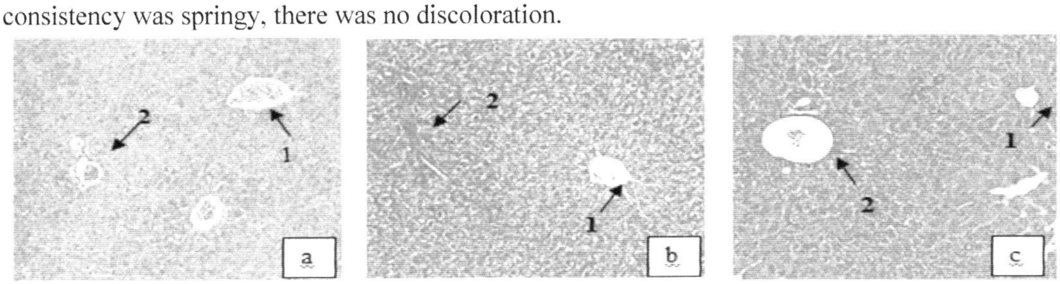

Figure 4. Histopathological staining of ddY mice liver (200x). There was not specific abnormality in the liver. (a) dose 500 mg/kgBW, (b) dose 1000 mg/kgBW orally and (c) control. The arrow shows (1) Vena centralis, (2) Tractus portalis. There was no specific abnormality in the liver.

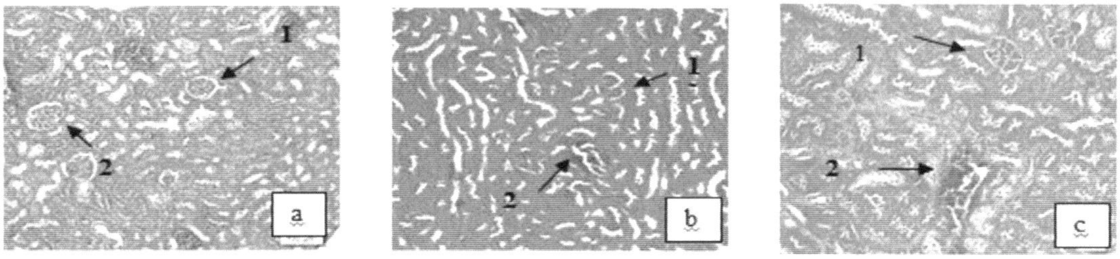

Figure 5. Histopathological staining of ddY mice kidney (200x). There was no specific abnormality in the kidney. (a) dose 500 mg/kgBW, (b) dose 1000 mg/kgBW and (c) control. The arrow shows (1) Glomerulus; (2) Tubulus proximal.

SGPT, SGOT, serum urea and creatinine levels after treatment with 500 mg/kgBW, 1000 mg/kgBW extract and CMC solvent control, both male and female, respectively, were not significantly affected by the administration of the extract. The results of further statistical calculations Tukey test showed results that did not differ significantly with the control group.

Table 4. Result of biochemial analysis in mice serum

	Male			Female		
	500 mg/kgBW	1000 kg/kgBW	Control CMC	500 mg/kgBW	1000 mg/kgbw	Control CMC
SGPT (IU/L)	74,4±22,8	84,1±9,5	72,9±11,4	64,7±00,0	88,9±3,4	83,0±1,8
SGOT (IU/L)	141,5± 38,2	141,6±39,7	153,3±59,5	152,8±36,4	79,2±20,5	102,4±16,6
Ureum (mg/dL)	57,1±12,7	62,5±5,5	59,7±6,5	57,8±9,3	64,2±9,6	63,4±21,2
Creatinine (mg/dL)	0,59±0,17	0,53±0,37	0,89±0,24	0,79±0,35	1,00±0,43	0,92±0,15

3.5. *Effect C. longa extract as DENV-2 antiviral in vivo*

In the preliminary study, the titer of DENV in ddY mice injected with DENV infected Huh7it-1 cells was detected up to 24 hours. Virus titer in mice serum was 5.8x10^3 FFU/mL at 6 hours after infection and decreased at the level of 3.6x10^3 FFU/mL at 24 hours after infection.

To determine effect of *C. longa* extract to the DENV titer in mice, we infect mice with DENV infected Huh7it-1 cells and serum was collected at 6 and 24 hours after infection. Mice serum from each group were pooled for testing because of the limitation of serum's amount.

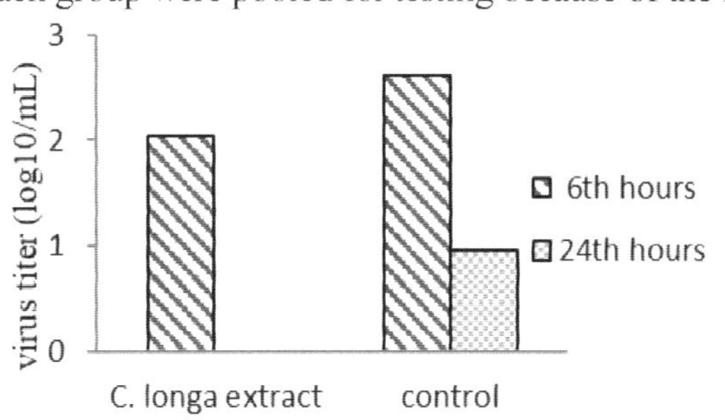

Figure 6. Level of DENV titer in DENV infected mice after treated with *C. longa* extract.

The results showed that the titer of DENV decreased in treated group compared to control (Figure 6). The titer of DENV from the group with of *C. Longa* treatment showed decreased to the level of 1.11x10^2 FFU/mLat 6 hours of infection and was not detected at 24 hours after infection. While on the control of solvent, the virus titer was detected up to 4.0x10^2 FFU/mL and up to 0.95x10^1 FFU/mL after 6 and 24 hours after infection. Treatment with *C. Longa* reduced viremia period in mice.

4. Discussion

Dengue has been a significant problem in most tropical and sub-tropical countries. There is no specific treatment yet for dengue. Treatment is only supportive based on the maintenance of the patient's body fluid since it is crucial in managing the severe form of dengue. As of today, approved antiviral agents for dengue are not available yet although studies regarding this have becoming more popular. Most of the studies directly or indirectly used natural product as sources for their antiviral study since the active substances for most therapies derived conventionally from natural sources [16]. Furthermore, the sources of around 50% of approved drugs have been using plants. However, numerous plants are cytotoxic, which means that they could kill the cell. Using a substance that is cytotoxic may not safe if it is not intended to kill the cells like the anticancer does. As a result, a substance that is going to be used for therapy needs to be effective and safe. Selectivity index is a ratio between the cytotoxic and the inhibitory capacity of a substance (CC_{50}/IC_{50}). It may help people to screen a substance that is appropriate for further development or to measure the effectiveness and safety of a product [16-17].

The purpose of this study is to find antiviral to DENV when it infect to human. Thus, Huh7it-1 cell, which is a human cell line derived from the liver, was used in this study. In addition, Huh7 could be infected by dengue virus since the characteristic of the cell is similar to hepatocyte in which basically the place of dengue virus to replicate itself [18].

It was found in this study that the CC_{50} was 85.4 μg/mL and the percentage of cell viability decreased in higher concentration. Thus, it can be said that the *C. longa* extract, in fact, killed the cells although not significantly. Half cytotoxic concentration is a concentration of a substance in which it may reduce the viability of a cell by 50%. Thus, theoretically, a good substance should have high CC_{50}. According to a review of studies about potential plants that can be used as anti-dengue, *Alternanthera philoxeroides* or also called as alligator weed had the lowest cytotoxic effect on cells (CC_{50} = 535.91) [19].

The IC_{50} of *C. longa* extract was 17.91μg/mL and there was a viral inhibition even though at lower concentration, it showed dengue virus enhancement. This enhancement may be useful for diagnosing patient, which it can be used to enhance and culture the virus of the patient. Half cytotoxic concentration is a concentration of a substance that can be used to inhibit the viral activity by 50%. Alligator weed and extracts from several plants also was shown to have high inhibitory activity towards dengue virus [19]. The IC_{50} of them was less than 5 μg/mL [19]. The data in this study was analyzed by using unpaired t-test on GraphPad Prism. It showed that the virus titer of the cells that were treated with *C. longa* extract at concentrations 80 μg/mL, 40μg/mL, and 20 μg/mL was significant (p value < 0.05) while the others were not.

One of the major active constituents of *C. longa* called curcumin. It was reported that curcumin has many pharmacological effects such as its antiviral properties. In 2013, Padilla S *et al* studied the antiviral effect of curcumin against dengue virus [20]. The selectivity index found was 2.56, which differs from this study (SI = 4.8). The method, however, were a bit different. They used Vero cell instead of Huh7it-1 cell [20]. Based on the same journal, it stated that the inhibited virus was probably due to cellular alterations from apoptosis, which induced by curcumin. Therefore, the inhibition was not directly from curcumin [20].

On the other hand, other study reported that the extract of *Curcuma longa* or turmeric is a potent inhibitor of dengue virus by targeting the NS2B/NS3 protein that coded protease enzyme DENV [21]. IMDPH (Inosine Monophosphate Dehydrogenase) is an enzyme that catalyzes the oxidation of several purine. Thus, it plays a significant role in synthesizing DNA and RNA and also signal transduction. Through inhibiting this enzyme, curcumin is suggested as a potential antiviral compound [22].

One of the obstacles in the development of DENV antiviral is the difficulty of getting experimental animals that exhibit clinical manifestations, DENV pathogenesis as in humans including viral protein identification and potential host to target treatment of DENV infection. The development of model animals is constrained because viruses cannot replicate well on models other than the natural hosts of humans and mosquitoes.

The effect of *C. longa* extract on DENV-2 in vivo was done by inducing Huh7it-1 infected DENV-

2 NGC cells in ddY mice intraperitoneally (i.p) to see the level of viremia in mice. This method is based on the development of Yamanaka S and Konishi E method in 2009 in vaccine testing [14]. Researchers used K562 cells infected with dengue and inoculated into ICR, Balb/c and ddY mice. The results explain that viremia levels last up to 34 hours after infection in mice ddY and ICR, whereas in Balb/c viremia occurs only lasts for 24 hours [14]. Laboratory of microbiology has tested the ddY mice inoculated K562 infected DENV-2 (i.p). From the results of this test, virus still detected until 24 hours in mice serum. Therefore ddY mice was selected as animal models of this study, in addition ddY mice are cheap and easy to develop.

It this study, virus titer detected up to 24 hours in mice serum that was inoculated with DENV infectedHuh7it-1 cells. The titer of DENV in mice treated with *C. longa* was decreased at 6 and 24 hours after infection. In the 24 hours after infection, mice treated with *C. longa* extract orally, no DENV-2 detected while in controls were present. In other words that treatment with *C. Longa* reduced viremia period in mice.

In the development of antivirus, the materials should be safely consumed every day orally. The literature review confirmed on safety aspects of curcumin and essential oil fractions of *C. longa* [23-26]. Since the source of plant may give different toxicity effect, therefore toxicity study of *C. Longa* from Indonesia was needed. In this study we evaluated cytotoxic effect of *C. longa* extract by determined of LD_{50} and its effect in liver and kidney of mice. Several studies have been conducted to determine LD_{50} *C. longa* extract. It was found that LD_{50} of *C. longa* extract orally was 10 g/kgBW in the mice, and in that dose not cause toxic symptoms and deaths were observed for 72 hours [27]. In addition, acute toxicity of *C. longa* extract has also been studied in mice and monkeys by Bhavani Shangkar *et al* [28]. From previous study LD_{50} of *C. longa* extract was 5 g/kgBW and no toxicity effects on histopathological observation and cytology of the liver, heart and kidneys. In this study, histopathological examination showed no specific abnormalities in liver and kidney organ after treated orally with 500 mg/kgBW and 1000 mg/kgBW of *C. longa* extract for 14 days. There was no significant increase in levels of SGPT, SGOT, urea, and creatinine

5. Conclusion

CC_{50} of *C. longa* extract was 85.4 µg/mL, IC_{50} was 17.91 µg/mL, and the SI of *C. longa* extract in this study was 4.8. *C. longa* extract was not provide toxic symptoms in mice liver and kidney. In vivo study showed that *C. longa* extract at doses of 0.147 mg/mL has antiviral effect against DENV-2 and reduce viremia period.

Acknowledgments

This study was supported by grant of Publikasi International Terindeks untuk Tugas Akhir Mahasiswa (PITTA) 2017 Universitas Indonesia, grant of Penelitian Unggulan Perguruan Tinggi (PUPT) 2015, The Development of Medical Education & Research Center and also Two University Hospital Project no IND-0137& IND-0138 and JICA-SATREP 2010-2014.

References

[1] Bhatt S, Gething P W, Brady O J, Messina J P, Farlow A W, Moyes C L *et al* 2013 The global distribution and burden of dengue *Nature* **496** (7446) 504 –7

[2] Anne N E 2013 Epidemiology of dengue: past, present and future prospects *Clinical Epidemiology* **5** 299–309

[3] WHO 2009 Dengue: guidelines for diagnosis, treatment, prevention, and control *Spec Program Res. Train. Trop. Dis.* 47

[4] Karyanti M R, Uiterwaal C S, Kusriastuti R, Hadinegoro S R, Rovers M M, Heesterbeek H *et al* 2014 The changing incidence of dengue haemorrhagic fever in Indonesia: a 45-year registry-based analysis *BMC Infect. Dis.* **14** 1 412-20

[5] CDC 2009 Dengue and dengue hemorrhagic fever: information for health care practitioners *Dengue Hemorrhagic Fever* pp 1–4

[6] WHO 2011 Prevention and Control of Dengue and Dengue Haemorrhagic Fever *Resived and expanded* pp 17-28
[7] Schul W, Liu W, Xu H Y, Flamand M, Vasudevan S G and Vasudevan S G 2007 A Dengue fever viremia model in mice shows reduction in viral replication and suppression of the inflammatory response after treatment with antiviral drugs *Keystone Symp. Cell Biol. Virus Entry J. Infect. Dis.* **195** 665–74
[8] Wilder-Smith A, Ooi E E, Vasudevan S G, Gubler D J 2010 Update on dengue: epidemiology, virus evolution, antiviral drugs, and vaccine development *Curr. Infect. Dis. Rep.* **12** 157–64
[9] Araújo C C and Leon L L 2001 Biological activities of *Curcuma longa* L. *Mem. Inst. Oswaldo Cruz.* **96** 5 723–8
[10] Benzie I F F and Wachtel-Galor S 2011 *Herbal medicine: Biomolecular and clinical aspects* 2nd ed (New York: CRC Press) pp 262-82
[11] Guerrant R L, Walker D H and Weller P F 2011 *Tropical infectious diseases: principles, pathogens, and practice* 3rd ed (London: Elsevier) pp 504-10
[12] Payne A F, Gajewska I B, Kauffman E B and Kramer L D 2006 Quantitation of flaviviruses by fluorescent focus assay *J. Virol Methods* **134** 2 183-9
[13] Badan Pengawas Obat dan Makanan Republik Indonesia 2014 *Pedoman Uji Toksisitas Nonklinik Secara In vivo* (Jakarta: Pusat Riset BPOM RI) pp 15-27
[14] Yamanaka A and Konishi E 2009 A simple method for evaluating dengue vaccine effectiveness in mice based on levels of viremia caused by intraperitoneal injection of infected culture cells *Vaccine* **27** 28 3735–43
[15] Soemadji and Lenny S 2003 Uji aktivitas anti diabetes dari ekstrak kulit jamblang pada mencit diabetes aloksan Indonesia *Research Report*
[16] Muhamad M, Kee L Y, Rahman N A, Yusof R 2010 Antiviral actions of flavanoid-derived compounds on dengue virus type-2 *Int. J. Biol. Sci.* **6** 3 294-302
[17] Idrees S and Ashfaq U A 2013 RNAi: antiviral therapy against dengue virus *Asian Pac. J. Trop. Biomed.* **3** 3 232–6
[18] Fuzo C A and Degrève L 2013 New pockets in dengue virus 2 surface identified by molecular dynamics simulation *J. Mol. Model* **19** 3 1369–77
[19] Kadir S L A, Yaakob H and Zulkifli R M 2013 Potential anti-dengue medicinal plants: a review *J. Nat. Med.* **67** 4 677–89
[20] Padilla S L, Rodríguez A, Gonzales M M, Gallego-G J C and Castaño J C 2014 Inhibitory effects of curcumin on dengue virus type 2 infected cells in vitro. *Arch. Virol.* **159** 573–9
[21] Tan S K, Pippen R, Yusof R, Rahman N A, Ibrahim H and Khalid D 2006 Screening of selected zingiberaceae extracts for dengue-2 virus protease inhibitory activities. *Sunw. Acad. J.* **3** 1-7
[22] Moghadamtousi S Z, Kadir H A, Hassandarvish P, Tajik H, Abubakar S and Zandi K 2014 A review on antibacterial, antiviral, and antifungal activity of curcumin *Biomed. Res. Int.* 1-12
[23] Chainani-Wu N 2003 Safety and anti-inflammatory activity of curcumin: a component of tumeric (*Curcuma longa*) *Journal of Alternative and Complementary Medicine* **9** 161-8
[24] Liju V B, Jeena K and Kuttan R 2012 Acute and subchronic toxicity as well as mutagenic evaluation of essential oil from turmeric (*Curcuma longa* L) *Food and Chemical Toxicology* **53** 52-61
[25] Balaji S and Chempakam B 2010 Toxicity prediction of compounds from turmeric (*Curcuma longa* L.) *Food and Chemical Toxicology* **48** 1 2951-9
[26] Joshi J, Ghaisas S and Vaidya A 2003 Early human safety study of turmeric oil (*Curcuma longa* oil) administered orally in healthy volunteers. *Journal of Association of Physicians of India* **51** 1055-60
[27] Ammon H P and Wahl M A. 1991 Pharmacology of *Curcuma longa.Planta Med.* **57** 1 1–7
[28] Bhavani S T N, Shantha N V, Ramesh H P, Indira M A S and Sreenivasa M 1980 Toxicity studies on turmeric (*Curcuma longa*): acute toxicity studies in rats, guineapigs and monkeys. *Indian J. Exp. Biol.* **18** 1 73–5

IOP Conf. Series: Earth and Environmental Science **101** (2017) 012006 doi:10.1088/1755-1315/101/1/012006

Anti-hyperglycemic effect and glucose tolerance of guajava (*Psidium guajava* L.) leaf ethanol extract in diabetic rats

Muhammad Yanis Musdja[1], Feizar Mahendra[1], Ahmad Musir[2]

[1]Department of Pharmacy, Faculty of Medicne and Health Sciences, State Islamic University, Syarif Hidayatullah, Jakarta
[2]Faculty of Pharmacy, University of Pancasila, Jakarta

E-mail: myanis88@gmail.com

Abstract. Traditionally guava (*Psidium guajava* L) leaf is used for treatment of various ailments like diarrhea, wounds, rheumatism, anti-allergy, ant-spasmodic, etc, as folk medicine. The aim of this research is to know the effect of hypoglycemia and glucose tolerance of ethanol extract of guava leaf against male white rat. The guajava leaf was obtained from Balitro Bogor. Preparation of guajava leaf extract was done by cold maceration extraction technique using ethanol 70%. Male albino rats were made into diabetics using the alloxan method. Rats were divided into 6 groups, as a comparative drug for anti-hyperglycemic used glibenclamid and as a comparative drug for glucose tolerance used acarbose. The result of blood glucometer test showed that ethanol extract 70% of guajava leaf had effect as anti-hyperglycemic and glucose tolerance with no significant difference with glibenclamid drug as anti-hyperglycemic and acarbose as glucose tolerance drug.

1. Introduction

Diabetes is a chronic health problem with devastating, yet preventable consequences. It is characterized by high blood glucose levels resulting from defects in insulin production, insulin action, or both. Globally, rates of type 2 diabetes were 15.1 million in 2003, the number of people with diabetes worldwide is projected to increase to 36.6 million by 2030. Diabetes is on the rise. No longer is a disease of predominantly rich nations, the prevalence of diabetes steadily increasing everywhere, most markedly in the world's middle-income countries [1].

Insulin therapy has been shown to benefit the prognosis in patients with type 2 diabetes, but its initiation and intensification is often delayed through concerns about hypoglycemia and weight gain. In addition, weight gain is linked to the pathophysiology of type 2 diabetes and contributes to the overall risk for adverse cardiovascular outcomes. So are other synthetic Diabetes medicines that have many adverse side effects. Therefore it is necessary to find a drug that comes from nature that has a smaller side effect [2].

Psidium guava. L. is popularly known as guava (family Myrtaceae) and has been used traditionally as a medicinal plant throughout the world for a number of ailments. All parts of this tree, including fruits, leaves, bark, and roots, have been used for treating stomachache and diarrhea in many countries. Leaves, pulp and seeds are used to treat respiratory and gastrointestinal disorders, and as an antispasmodic, anti-inflammatory, as a cough sedative, anti-diarrheic, in the management of hypertension, obesity and in the control of diabetes mellitus. It also possesses anticancer properties. The seeds are used as antimicrobial, gastrointestinal, anti-allergic and anticarcinogenic activity [3]. The important active constituents of guava are essential oils, flavonoids, carotenoids, polyphenolic compounds, pentacyclic triterpenoids, esters, and aldehydes etc. This paper explains the Evidence-based information regarding the phytochemistry and pharmacological activity of this plant [4].

2. Materials and Method

2.1. Preparation of the extract

Fresh guava leaves were made into a slurry using a blender with 70% ethanol. Then immersed in a glass beaker with a height of 70% ethanol there were 2.5 cm above the surface of the slurry. The maceration process is carried out for 3 hours while stirring. Then filtered using cotton to filter the dregs. This process is done repeatedly until no more extracts are extracted, which is marked with a clear color of solvent. The obtained filtrate was collected, filtered using filter paper and evaporated ethanol until a viscous ethanolic extract was obtained. A test solution was then prepared and administered to the test animal.

2.2. Preparation of experimental animals
An acclimatization of test animals was performed to male albino rats of 7-9 weeks age, weighing 170-190 g, were used. The animals were kept in clean and dry plastic cages, with light-dark at room temperature. The animals were fed with standard pellet diet and water was given ad libitum. This study was carried out in the animal house of State Islamic University, Syarif Hidayatullah, Jakarta and this study was approved by the Institutional Ethical Committee. Animals were divided into 6 groups and each group had 5 rats.

2.3 Administration of sample and determination of rat blood glucose level
Rats are fasted for 16 hours. Prior to treatment, rat blood was taken through the rat tail vein and measured as fasting glucose using glucometer. Then the rat was given extract with high doses using gastric sonde. 30 minutes after administration, the mice were given 50% glucose solution at a dose of 1 g / kg BW, then immediately took the blood of the mouse and its glucose level was measured as blood glucose level at minute 0. Furthermore, rat blood was taken at 30, 60, 90, 120, 150 and 180. The data obtained is the result of oral glucose tolerance test. Then the rats were again fed and drank normally, with the administration of high-dose extracts daily. On day 3, 8 and 14 after oral glucose tolerance test, rat blood was taken and measured glucose levels.

2.4. Statistical analysis
Prior to the ANOVA test, first test of normality and Levene homogeneity test was performed. If the normality test and Levene homogeneity test is worth ($p > 0.05$), then it is continued with ANOVA test. However, if the significance of the test for normality and homogeneity test Levene value of less than 0.05, ANOVA test cannot be performed, so proceed with Kruskal Wallis test.

3. Results
The results of this study give results that the highest effect of hypoglycemia extract of guava leaf ethanol (*Psidium guajava* L.) with alloxan diabetes test method is 1,300 mg / kg BW which can decrease blood glucose level up to 31.66%, indicated by no significant difference between test dose With normal and positive control, and a significant difference between the test dose and the negative control ($P < 0.05$). As shown in table 1.
The highest effect of hypoglycemia extract of guava ethanol extract (*Psidium guajava* L.) by oral glucose tolerance test method was proven to inhibit glucose absorption, indicated by no significant difference between 1300 mg / kg BW dosage with normal and positive control, and presence A significant difference between the test dose and the negative control ($P < 0.05$). As shown in table 2.

Table 1. Effect of ethanolic (70%) leaf extract of guava as anti-hyperglycemic on blood glucose (mg/dL) alloxan-induced diabetic rats

Time (Days)	NC	C(-)	C(+) Glibenclamid	Low Doses	Middle Doses	High Doses
Day 0, BI	84.75	85.5	83.75	84.75	86.25	82.75
Day 0,AI	88.5	114.25	172.25	135.25	134.5	139.5
Day 3, AI	121.5	130.25	72.5	166	151.5	143.5
Day 8, AI	104.25	158.25	77	139	163.5	179.5

Day14, AI	90.5	152	78.5	110.75	96.25	97

Note: BI = Before Induction; AI = After Induction; NC = Normal Control; C(-) = Negative Control; C(+) = Positive Control (Glibenclamid)

Table 2. Effect of ethanolic (70%) leaf extract of guava on test of glucose tolerance (mg/dL) alloxan-induced diabetic rats

Time (Minutes)	NC	C(-)	C(+) Acarbose	Low Doses	Middle Doses	High Doses
Fasting	88.5	122.75	108.25	135.25	136.5	156.25
Minute 30 AGA	95.75	161.75	114.25	294	213.5	146.75
Minute 60 AGA	92.75	280.75	103.75	252.5	161.75	154.5
Minute 90 AGA	86.5	285.25	118.5	203	155.25	134.75
Minute 120 AGA	93.25	244.5	106.5	169.5	128.75	127.5
Minute 150AGA	94.5	237	110.5	162.5	120.5	116.25
Minute 180 AGA	115	229.5	132.25	154.25	140.75	131.25

Note: AGA = After Glucose Administration; NC = Normal Control; C(-) = Negative Control; C(+) = Positive Control (Acarbose)

4. Discussion

Bhanu et al [5] conducted a study of the the antihyperglycemic and antihyperlipidemic effects of guava on alloxan-induced diabetic rats by making extracts of guava with 96% ethanol at hot temperatures, but they did not test the glucose tolerance of guava against alloxan-induced diabetic rats.The results of this study for antihyperglycemic are similar to the results of research conducted by Bhanu S, et al, [5].

In this research, preparation of guava leaf extract (*Psidium guajava* L.) using cold way extraction method by maceration and using ethanol 70% as solvent. Cold-way extraction has the advantage in the total extraction process, which minimizes the likelihood of damage to the thermolabile compounds present in the sample. Most of the compounds can be extracted by cold-way extraction, although there are some compounds that have solubility limitation to the solvent at room temperature. The maceration method is chosen because the process is easy, the equipment used is fewer and simpler, and does not require any special skills. While 70% ethanol is used because ethanol is commonly used in total extraction. The advantage of using ethanol is that most of the lipophilic and polar compounds can be extracted. Guava leaf used is a bit old leaves because it contains the most tannins with the characteristics of dark green leaves, clear and hard leaf bones, the top surface is slippery and has a little stiff.

From the results of decreased blood glucose levels showed that the greater the dose of thick guava leaf extract test the greater the ability to lower blood glucose levels. The result of percentage of glucose decrease after 14 days given test preparation is as follows: low dose 162,5 mg / kg BW equal to 18,11%, 325 mg / kg BW equal to 28,49%, and 1300 mg / kg BW equal to 31,66 %.

Blood glucose data obtained from each measurement is then processed statistically. The statistic method used is One Way Variance Analysis (ANOVA) to blood glucose level. First, normality test and homogeneity test are done. Such testing is a common practice before a statistical method is applied. The method used in the normality test is by Kolmogorov-Smirnov test, and its homogeneity uses Levene test.

Based on the normality test showed that blood glucose levels throughout the initial animal test group before treatment were normally distributed. While homogeneity tests showed blood glucose levels throughout the initial animal test group before homogenized treatment

In the alloxan diabetes test, the one-way Variance Analysis Test showed the 8th day blood glucose level after treatment between the 162.5 mg / kg BW dose group and the normal control did not differ significantly, while the blood glucose level between the 325 mg / kg BW dose group and the dose of 1300 mg / kg BW with normal control differed significantly at the 0.05 test level. Blood glucose level on day 14 after treatment showed between group 325 mg / kg BW and dose 1.300 mg / kg BW with normal control and positive control did not differ significantly while there was significant difference at test level 0.05 between group 162.5 mg / kg BW with normal control and positive control. The decrease in blood glucose level of the group given glibenclamide was caused by increased insulin stimulation but decreased blood glucose levels of the group which were given test of viscous ethanol extract of guava leaf (*Psidium guajava* L.) unknown mechanism of action.

In the oral glucose tolerance test, fasting blood glucose levels across all animal groups differed significantly. This is because blood glucose levels throughout the test group had hyperglycemia, whereas normal blood glucose control levels remained within the normal range. Blood glucose levels between the low-dose assay groups with normal control and positive control varied significantly in the 30th, 60th, 90th and 120th minutes, but at minutes 150 and 180 did not differ significantly at the 0.05 test level. Blood glucose levels between the moderate-dose test group and the normal control and positive control varied significantly in the 30th minute, but at 60, 90, 120, 150 and 180 minutes did not differ significantly at the 0.05 test level. However, the highest inhibitory ability was obtained in the high dose test group. This was demonstrated by statistical results indicating that blood glucose levels between high dose test groups with normal control and positive control in the 30th to 180th minutes did not differ significantly at the 0.05 test level.

Inhibition of group blood glucose absorption inhibited by acarbose caused by inhibition mechanism of α-glucosidase enzyme. While the inhibition of glucose absorption in the group that was given test of viscous ethanol extract of guava leaf (*Psidium guajava* L.) was allegedly caused because of one of the compounds contained in guava leaf viscous extract which can inhibit the absorption of glucose. Guava leaf extract (*Psidium guajava* L.) contains tannins. Based on previous research, tannin compound is astringent that can minimize skin surface pores [6].

5. Conclusion

Ethanol extract of 70% guava leaf (*Psidium guajava* L.) has the potential to lower blood glucose levels, either by alloxan diabetes test or oral glucose tolerance test.

References

[1] Kusuma Neela Bolla, Santhi Sri.K.V, K.N.Varalakshmi, Diabetes Mellitus & Its Prevention, *Int. J. Sci & Tech Res*, 2015, Vol. 4, Issue **08**, August
[2] WHO, *Global report on diabetes*, 2016
[3] Sandra M. Barbalho, Flávia M.V, Farinazzi-Machado, Ricardo de Alvares Goulart, , Anna Cláudia Saad Brunnati,, Alda Maria Machado, Bueno Ottoboni, and Cláudia Cristina Teixeira Nicolau, Psidium Guajava (Guava): A Plant of Multipurpose Medicinal Applications, *Med Aromat Plants*, **1:4**, 2012
[4] Payal Mittal, , Vikas Gupta, Gurpreet Kaur, , Ashish K Garg, and Amarjeet Singh, Phytochemistry and Pharmacological Activities of Psidium Guajava: A Review, *IJPSR* (2010), Vol. **1**, Issue 9 (Suppl.)
[5] Banu S M, Sujatha K, Sridharan G and Manikandan R, Antihyperglycemic and Antihyperlipidemic Potentials of Psidium Guajava In Alloxan-Induced Diabetic Rats, *Asian J. Pharm and Clin Res*, Vol **6**, Suppl 1, 2013
[6] Hagerman, Ann E. 2002. *Tannin Handbook*. Department of Chemistry and Biochemistry. Miami University. Oxford

IOP Conf. Series: Earth and Environmental Science **101** (2017) 012007 doi:10.1088/1755-1315/101/1/012007

Evaluation Lactogenic Activity of Ethyl Acetate Fraction of Torbangun (*Coleus amboinicus* L.) Leaves

R M Damanik[1], L Kustiyah[1], M Hanafi[2] and A C Iwansyah[3]

[1]Department of Community Nutrition, Faculty of Human Ecology, Bogor Agricultural University, Jl. Meranti Dramaga Bogor 16680, Indonesia
[2]Research Center for Chemistry, Indonesian Institute of Sciences, Puspiptek-Serpong, Indonesia.
[3]Center for Appropriate Technology Development, Indonesian Institute of Sciences, Jl. KS. Tubun No.5 Subang 41213, West Java, Indonesia

E-mail address: damanik@ipb.ac.id

Abstract. This study aimed to assess the lactogenic property of ethyl acetate fraction of torbangun (*Coleus amboinicus* L.) leaves and to identify the compounds that responsibility as 'milk booster' using LC- MS approach. Lactagogue activity was evaluated in terms of quantity of milk produced from the rats treated with commercial milk booster (AF), ethyl acetate fraction of torbangun leaves (EA), water extraction of torbangun (AQ) and kaempferol (KP). The feed was given orally every two days and starting from Day 2 after giving birth until Day 28. The performance of milk production was measured along the experimental period by weight-suckle-weight method. The level of prolactin serum was determined by ELISA methods. Histopathological analysis of mammary gland, liver, intestines and kidney tissues was carried out. Moreover, in order to profiling and identification of compounds of ethyl acetate fraction, ultra-performance liquid chromatography quadrupole time of flight to electrospray ionization mass spectrometry (UPLC-QTOF-ESI-MS) in the positive-ion mode was performed. The ethyl acetate fraction of torbangun leaves (EA) was induced milk production about 17%, and AF 22% and KP 51% compared to the control group. Meanwhile, the EA was not significantly stimulate the synthesis of serum prolactin at Day 14 and Day 28 ($p>0.05$). Administration of EA did not cause any signs or symptoms of toxicity. In addition, a total of ten compounds was identified by UPLC-QTOF-ESI/MS in the ethyl acetate fraction of the leaves of *C. amboinicus*, mostly phenolic compounds, flavonols and some of their glycoside derivatives, such as: digiprolatone, and kaempferol-3-7-O-di-rhamnopyranoside. The present study reveals the ethyl acetate fraction of torbangun leaves and its bioactive compounds has the potency as a remedy for stimulating and improving milk production.

1. Introduction

In recent decades, bioactive compounds derived from nature were beginning to gain attention to be developed as alternative therapies. A secondary metabolite of the plant is widely investigated for its bioactivity. According to [1] reported that about 80% of the world's population (4 billion people) used plant products to improve quality of health (nutraceuticals).

A systematic analysis of non-communicable diseases (NCD) risk factors, globally in 21 countries in 1999-2010, showed that not optimal of breastfeeding was one of the risk factors contributing to the development of NCD [2]. World Health Organization [3] recommended breastfeeding infants for 6 months. According to [4] reported that limited of milk production is the main reason for mother's not giving exclusive breastfeeding to the infant. Limitations of milk production are influenced by factors, i.e., nutritional and endocrine factors (hormones) [5]. As an alternative solution to treat these problems, used galactopoetics herbs are highly recommended [6].

Prolactin plays an important role in the development of mammary glands during early gestation and the postpartum period [7]. Interaction of prolactin and its receptor (PRLR) on the glandular epithelial of membrane cell will determine the amount of milk production. The prolactin receptor is needed for the expression of milk proteins and lactation [8]. The medicinal-herb interactions associated with galactagogues have concrete evidence, but the evidence explaining the mechanism of action of herbs as galactagogues is still very limited and scares. One of the pharmacological-used plants has been used as galactagogues, is *Coleus amboinicus* L [9].

Torbangun (*Coleus amboinicus* L.), family Lamiaceae, believed can be used as food, food additives and medicines in treating various diseases, such as diarrhoea, flatulence, constipation, cough, chronic asthma, bronchitis, kidney, liver and malaria [10]. In vitro studies reported that the extract of *C. amboinicus* has antioxidant capacity, antibacterial and cytotoxic, anti-inflammatory and hepatoprotective [11-14]. Meanwhile, in vivo studies using animals and human experiments showed that this plant had anti-diabetic activity, immunomodulator, analgesic and lactagogue [15-21]. Many studies have examined extensively the bioactivity of these plants, but studies of exploration and identification of responsible components are scarce and rare.

In our previous studies, ethanolic extract and its fractions of *C. amboinicus* showed potent antioxidant activities [22]. Ethyl acetate fraction of this plant also has potential to increase milk production by 17% comprised with control and increasing rat pup's performance [21]. Hence, the present work to evaluate the lactogenic property of ethyl acetate fraction of torbangun (*C. amboinicus* L.) leaves and to identify the compounds that responsibility as 'milk booster' using liquid chromatography – mass spectrophotometry (LC-MS) approach.

2. Material and methods

In this section, material and methods of this study are discussed.

2.1. Plant material

Coleus amboinicus (Lour.) leaves were collected from Sukabirus Garden-Bogor, Indonesia, ("S 06°39'58.3"; "E 106°52'28.7"; 558 mdpl), during the month of October 2015 – March 2016. The cutting interval of torbangun leaves were 60 days, and each cutting was cultivated with 25 cm radius from the other [23-24]. Botanical identification was performed by Dr. J. S Rahajoe, Botanist from "Herbarium Bogoriense", Research Center for Biology, Indonesian Institute of Sciences, where the voucher specimen has been deposited (No.145/IPH.1.01/II.8/II/2015).

2.2 Procedure analysis

2.2.1 Quality control of medicinal plant materials

Standardization (quality control) of medicinal plant material, i.e., simplisia, comprises various analytical and phytochemical procedures. The analytical such as moisture content, ash values were described by [25], microbial content, heavy metal analysis measured by atomic absorption spectrophotometer (AAS) and phytochemical analysis.

2.2.2 Preparation samples of C. amboinicus

Samples of *C. amboinicus* (200 g) were washed and then extracted with ethanol 96% for 24 h for three times. The supernatant was filtered through filter paper (Whatman No. l), then dried by vacuum evaporator. The ethanolic extract (20 g) was dissolved in 10 ml of methanol: water (with ratio 1:1), and filtered again, thus obtaining the crude extract. This crude extract was then partitioned with hexane (500 ml fractions repeatedly up to decoloration of the organic solvent), thus obtaining both the hexane fraction

(1.4 g) and the "defatted" crude extract. The defatted crude extract was then successively partitioned with ethyl acetate (as in the hexane partition), thus obtaining the ethyl acetate (7.5 g), and aqueous fractions (2.7 g). The fractions, then dried by vacuum evaporator.

2.2.3 Preliminary phytochemical screening

The powdered of ethanolic extract and its fraction of *C. amboinicus* (Lour.) spreng were tested for the presence of sterols or triterpenes, flavonoids, tannins, saponin and alkaloids [26-28]. The results are given in Table 2.

2.2.4 Effect of oral treatment on milk production

Twenty five lactating dams weighing 250-300 g at the beginning of lactation and suckling six pups were used for this experiment [29]. The females were divided into five groups of five animals each. Group I (KO) was given aqueous as a control with vehicle orally, Group II (AF) was orally administered commercial milk booster (composition: 114 mg katuk leaf extract, 20 µg vitamin B12, 25 mg vitamin B2, 10 mg vitamin B1) at a dose of 50 mg/kg BW. Group III (EA) and Groups IV rats were orally administered 30 mg/kg of ethyl acetate fraction of torbangun and 80 mg/kg BW of aqueous extract of torbangun, respectively [18-19, 30]. Group V (KP) animals were treated orally kaempferol compound at a dose of 60 mg/kg BW. Milk production was measured from day 3 to day 15 of lactation by using method described by [30]. The ethical clearance approved by The Commission on Supervision of Animal Welfare and Veterinary Research, Research Institute and Community Service (LPPM), IPB No. 37-2016 IPB.

2.2.5 Serum prolactin and histopathological examination

On 14th day and 28th day of lactation period, lactation rats were adjusted to blood sampling procedure through a lateral tail vein. The blood samples were centrifuged (14.000 rpm, 15 minutes) and serum was stored at -20°C. Serum prolactin was estimated by enzyme-linked immunosorbent assay (ELISA) method. Histopathological analysis, after termination of experiment, whole mammary gland and other vital organs *viz.*, intestines, liver and kidney were excised from each experimental group of animal. Fibrosis as the effect of treatments was estimated by blind scoring method [31]. Similar fibrosis levels were subjectively grouped into 4 groups of scoring criteria, i.e., 0 (none, very little), +1 (mild), +2 (moderate), +3 (severe).

2.2.6 LC-MS analyses

An aliquot of ethyl acetate fractions (10 mg) of *C. amboinicus* was dissolved in methanol (8 ml) and 2 ml of ultra pure water for LC, respectively. The extract and fractions were filtered through 0.2 µm filters, and 1 µl of each extract and fraction were automatically injected into the LC system (UPLC-QTOF-ESI-MS: Acquity; Electrospray Ionization system; column UPLC@HSS T_3 1.8µm C_{18} (2.1 mm x 100 mm)), and the absorbance changes were monitored at 280 nm. Mobile phase: 0.1% formic acid in water solution (A) and 0.1% formic acid in methanol solution (B). The gradient elution: 5% B (1 minute), 5-40% B (7 minutes), 40-100% B (3 minutes), 100% B (2 minutes), 100-5% B (2 minutes). Total running time was 15 minutes with the flow rate used was 0.3 ml/min. The MS analyses were performed with the following conditions: positive mode scanning, the electrospray voltage was 4.7 Kv, and the capillary temperature and voltage were maintained at 250°C and 15v, respectively. The molecular ions were scanned from 100 to 1500 m/z. Peak identification: the bioactive compounds were identified by comparing their retention time and their mass spectra with those reported in literature.

2.3 *Statistical Analysis*

Data were presented in mean ± standard deviation, and then tested for normality. All statistical analysis was performed using Microsoft Excel 2010. The differences between treatments were analysed using ANOVA. Significant differences between mean values were determined using Duncan's Multiple Range Test (α=5%).

3. Results

3.1 Quality control of Coleus amboinicus L.

In recent year, standardization and quality control of medicinal plants of therapeutic potential has been emphasized. Controlling of starting material is most important. The evaluation of chemical composition, microbial and heavy metal analysis should be carried out before any tests are undertaken. The qualities of simplisia of *C. amboinicus* were shown in Table 1.

Table 1. Chemical composition and microbial content of simplisia of *Coleus amboinicus* (Lour.)

Parameters	Values	Indonesia Food and Drug Regulation*
Water content	10.3 %	≤ 10 %
Total ash	14.2%	Max. 33.3%
Heavy metal:		
Arsenic	< 0.003 mg kg^{-1}	≤ 5 mg kg^{-1}
Lead	< 0.040 mg kg^{-1}	≤ 10 mg kg^{-1}
Mercury	< 0.005 mg kg^{-1}	≤ 0.5 mg kg^{-1}
Cadmium	< 0.005 mg kg^{-1}	≤ 0.3 mg kg^{-1}
Microbial content:		
Total Plate Count	1.58 x 10^2 CFU ml^{-1}	Max. 10^6 CFU ml^{-1}
E.Coli/ S. aeurus	Negative	Negative
Mold or yeast	Negative	Max. 10^4 CFU ml^{-1}

*Source: Indonesia National Agency of Drug and Food Control (INADFC) [32]

3.2 Phytochemical screening

Phytochemical screening of ethanolic extract and it's fraction of *C. amboinicus* (L.) showed that ethanolic extract is rich in phenolic compounds, this indicated by the presence of flavonoids, alkaloids, tannin, triterpenoids and saponin (Table 2).

Table 2. Phytochemical screening of extract of *Coleus ambionicus* (L.) spreng

Phytoconstituents	EtOH	FHex	FEA	Fair
Alkaloid				
Bouchardat	++	-	++	-
Mayer	+	+	+	+
Dragendorf	+	++	++	++
Triterpenoids	++	+++	+	-
Saponin	++	+++	-	+
Tannins	+	-	-	++
Flavonoids	++	-	+++	+

(+) = present; (-) = absent. EtOH: ethanolic extract; FHex: hexane fraction; FEA: ethyl acetate fraction; Fair: aqueous fraction

3.3 Milk production

The influences of lactagogue compounds with the ability to increase milk production were stimulated by different treatments are shown in Fig. 1.

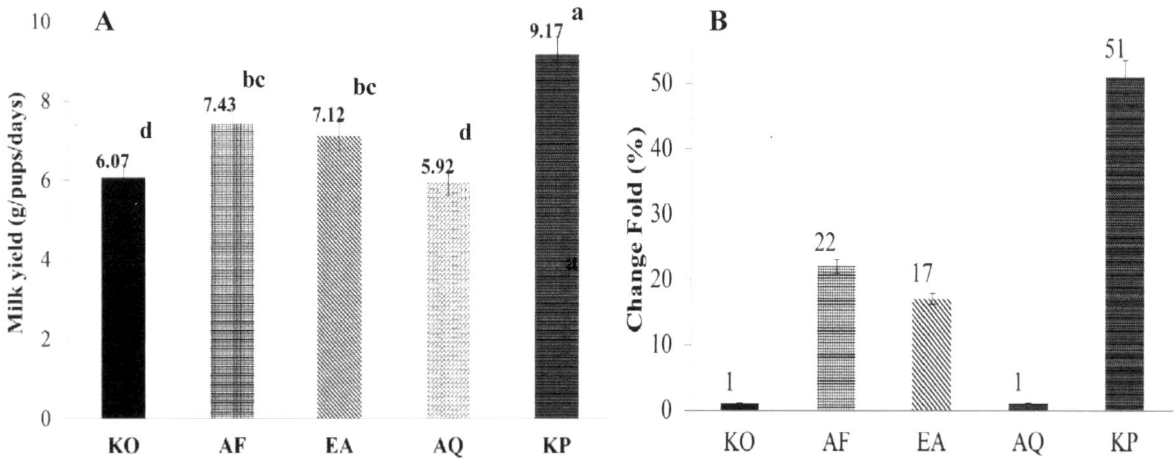

Figure 1. Average of milk yield (A) and change fold of milk production (B) during lactation period (days 2 – 14) (n=5). KO: control groups, AF: commercial milk booster groups, EA: ethyl acetate fraction of CA groups, AQ: aqueous fraction of CA groups, KP: kaempferol groups. a>b>c>d>e; the difference alphabetic showed significantly ($p<0.05$).

3.4 The levels of serum prolactin

The effect of lactagogue compounds on the level of serum prolactin is shown in Fig. 2.

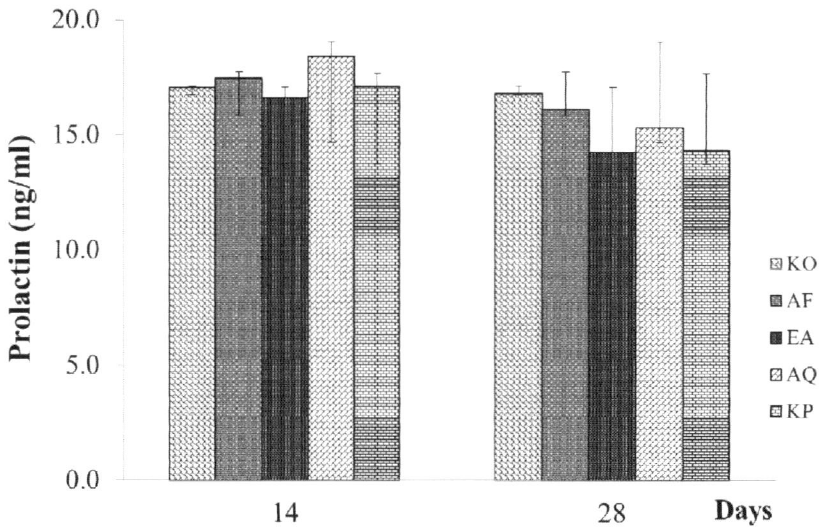

Figure 2. The levels of serum prolactin of lactating rats during days 14 and days 28. KO: control group, AF: commercial milk booster group, EA: ethyl acetate fraction of CA group, AQ: aqueous fraction of CA group, KP: kaempferol group. a>b>c>d>e; difference alphabetic showed significantly ($p<0.05$).

3.5 Toxicity profile and histological

Table 3. Scoring level of organ damage (fibrosis) at day 28 of lactation period

Samples	Mammary gland	Liver	Intestine	Kidney
KO	0	0	0	0
AF	0	0	0	+2
EA	0	0	0	0
AQ	0	0	0	0
KP	0	0	0	0

(+2): more than one (mineralized) multifocal focus (2-5 focus of mineralization) in the cortex or renal medulla. KO: control group, AF: commercial milk booster group, EA: ethyl acetate fraction of CA group, AQ: aqueous fraction of CA group, KP: kaempferol group.

3.6 LC-MS analyses of ethyl acetate fraction of C. amboinicus

The ethyl acetate fraction of *C.amboinicus* leaves was injected into LC-MS to identify bioactive compounds. Ten major peaks were detected in the chromatogram from the ethyl acetate fraction of *C.amboinicus* leaves (Fig.3).

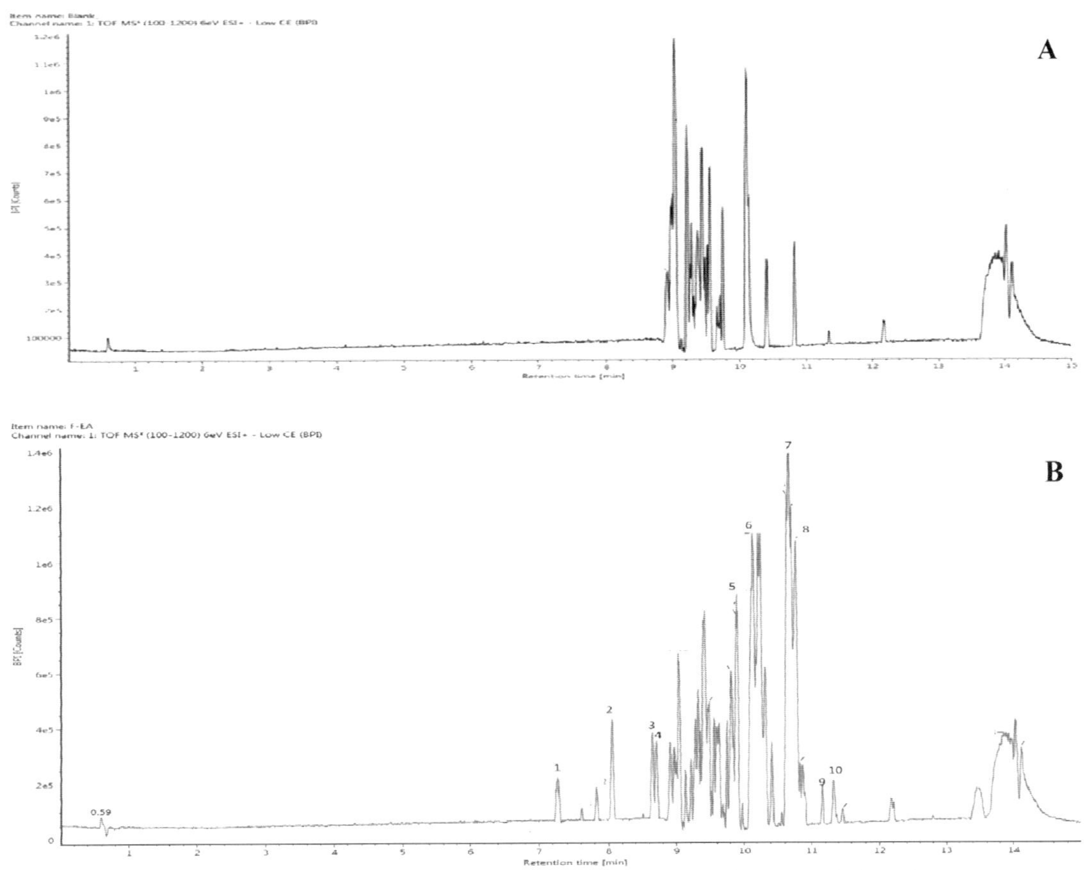

Figure 3. LC-MS chromatogram of blank (A) and ethyl acetate fraction of *C. amboinicus* L. leaves (B) with retention time (R_t) at λ= 280 nm.

Table 4. Predicted compounds from ethyl acetate fraction of *C. amboinicus* L. leaves

No.	RT	Mass Spectra Ion	*m/z*	Compounds	References
1	7.266	$[M+H]^+$	225.109	9-((2-Aminoethoxy)methyl)guanine	[41]
2	8.042	$[M+H]^+$	197.118	Digiprolactone	[41]
3	8.641	$[M+H]^+$	225.109	9-((2-Aminoethoxy)methyl)guanine	
4	8.705	$[M+H]^+$	197.118		
		$[M+Na]^+$	219.098	Digiprolactone	[41]
5	9.876	$[M+H]^+$	609.278	Reserpine	[42]
6	10.126	$[M+H]^+$	623.296	Pectolinarin	[43]
7	10.644	$[M+H]^+$	608.304	Methyl phaeophorbide a	[44]
8	10.746	$[M+H]^+$	607.299	unknown	-
9	11.142	$[M+H]^+$	775.535	N9-Allylazithromycin	[41]
10	11.307	$[M+H]^+$	579.259	Kaempferol-3, 7-O-di-rhamnopyranoside (kaempferitin)	[45]

4. Discussion

Phytochemical screening of ethanolic extract and its fraction of *C. amboinicus* (L.) showed that ethanolic extract is rich in phenolic compounds, this indicated by the presence of flavonoids, alkaloids, tannin, triterpenoids and saponin (Table 2). Phenolic compounds are accumulated at a fraction with high polarity, i.e. ethyl acetate and aqueous fractions. These results were consistent with the finding of [33] which reported *C. amboinicus* contained flavonoids, tannin and saponin. Table 2 also shows that the ethyl acetate fraction of *C.amboinicus* were negatively on saponin and tannin compounds. Moreover, the preliminary phytochemical screening results agreement with previous works that we have done, which showed the highest total phenolics content and antioxidant activity were present in the ethanolic extract followed by aqueous fraction and ethyl acetate fractions [22].

Fig. 1a showed that the milk yield of the ethyl acetate fraction of *C.amboinicus* (EA), kaempferol (KP) and commercial 'milk booster' (AF) groups was significant difference compared to the control group ($P<0.05$). The kaempferol group contained the highest on average of yields produced (9.17±2.00 gram/pup/day) by the rats during lactation period followed by EA and AF>KO and AQ. The figure also showed that the total milk production for 14 days of breastfeeding observation, kaempferol group (KP) had the highest total milk production (55 gram/14 days), followed by AF (45 gram/14 days)> EA (43 gram/14 days)> KO (36 grams/14 days) and AQ (35 grams/14 days) ($P<0.05$). Ethyl acetate fraction of torbangun leaves (EA) was induced milk production about 17%, AF 22% and KP 51% compared to the control group (Fig. 1b). These results agreement with [30] and [34], which reported administration extract and soup of *C.amboinicus* was given positive respond to increase milk production in lactating rats. It can be assumed that in the ethyl acetate fraction of *C.amboinicus* contained bioactive compounds as "galactagogue" which can modulate hormone in lactogenesis and lactation period.

According to [35] reported in galactopoetics herb plants, phytochemical compounds *viz.*, polyphenols, tannins, and alkaloids have an effect to increase milk production, protein concentration and ovulation rate, bioavailability of protein digestion and secretion milk. Moreover, [36] suggested phytoestrogenic action and some molecules of phytoestrogen in herbal galactagogues may have effects similar to 17β-estradiol (E2) that promotes the proliferation of mammary epithelial cells (MEC) [37]. Some phytoestrogen molecules are mentioned, *viz.*, diosgenin, estragole, silybin, shatavarine, kaempferol, and quercetin, could induce the expression of prolactin receptor (PRLR), upregulate casein production, and lactose synthetase activity in MEC [38]. In these results, kaempferol positively induced milk production about 51% compared to the control group (see Fig. 1b). In addition, present work shows that the ethyl acetate fraction of *C.amboinicus* contained phenolic compounds i.e., digiprolactone and kaempferol derivatives.

Fig. 2 showed that concentration of serum prolactin not affected by all treatment in day 14 and day 28 (P>0.05). In contrast with these results, [39] reported that fenugreek increased milk production and concentration of serum prolactin was found higher in the fenugreek fed goats compared to control group. This difference results are probably due to the ethyl acetate fraction of *C.amboinicus*, kaempferol, or phytoestrogen molecules could not affect the prolactin (PRL) expression in the anterior pituitary, but they induce the expression of PRL receptor (PRLR), upregulate casein production, and lactose synthetase activity in MEC to increase milk production. According to [8] prolactin receptor (PRLR) is needed for the expression of milk proteins and lactation. Interaction of prolactin and its receptor (PRLR) on the glandular epithelial of membrane cell will determine the amount of milk production.

In this study, all treatment did not cause any behavioural alterations or deaths in rat when oral administration. In histological profile, generally no significant changes were observed in ethyl acetate fraction of *C. amboinicus* and all treated animals and controls (Table 3). It concludes that ethyl acetate fraction *C. amboinicus* and all treatment is safe and non toxic.

The ethyl acetate fraction of *C. amboinicus* leaves was injected into LC-MS to identify bioactive compounds. Ten major peaks were detected in the chromatogram from the ethyl acetate fraction of *C. amboinicus* leaves (Fig.3). The eluted peak at retention time of 10.64 minutes is the most numerous peak based on peak size. In the structure of components, a limitation of the database on LC-MS has forced the mass spectra analysed by manually [40]. Predicted compounds of the ethyl acetate fraction of *C. amboinicus* presented in Table 4.

According to [46] that digiprolactone had oestrogen activity, which can selectively activate the oestrogen receptor (ER). This compound identified may be used to increase proliferation of any cell that is capable of proliferation. In addition, kaempferol-3,7-O-di-rhamnopyranoside (kaempferitin) had phytoestrogenic action and have effects similar to 17β-estradiol (E2) that promotes the proliferation in MEC [38]. Meanwhile, as an antioxidant, pectolinarin also possess anti-inflammatory activity and might inhibit eicosanoid formation in inflammatory lesions [43]. The presence of digiprolactrone and kaempferol derivative following the synergic action between the two components in the ethyl acetate fraction is a factor that can be predicted to determine the ability to stimulate and increase milk production as shown by the ethyl acetate fraction of *C. amboinicus.*

5. Conclusion

The ethanolic extract is rich in phenolic compounds, this indicated by the presence of flavonoids, tannin and saponin. Kaempferol group (KP) with dose 60 mg/kg body weight had the highest total milk production followed by AF>EA> KO and AQ. Ethyl acetate fraction of torbangun leaves (EA) with dose 30 mg/kg body weight was induced milk production about 17% compared to the control group. Meanwhile, there was not significantly affected to stimulate the synthesis of serum prolactin at Day 14 and Day 28. Generally, administration of EA and other treatment did not cause any signs or symptoms of toxicity. The presence of digiprolactrone and kaempferol derivative following the synergic action between the two components in the ethyl acetate fraction is a factor that can be predicted to determine the ability to stimulate and increase milk production as shown by the ethyl acetate fraction of *C. amboinicus.*

Acknowledgement

The authors are grateful to the Ministry of Research and Technology and Higher Education, Republic of Indonesia (KEMENRISTEKDIKTI-RI) for financial assistance and to Bogor Agricultural University (IPB) and Indonesian Institute of Sciences (LIPI) for providing access and technical support.

References

[1] Ekor M. 2013. The growing use of herbal medicines: issues relating to adverse reactions and challenges in monitoring safety. *Front Pharmacol*, 4(177):1-10.

[2] Lim S. et al. 2012. A comparative risk assessment of burden of diseases and injury attribute to 67 risk factors and risk factors clusters in 21 regions, 1999-2010: systematic analysis for the global burden of disease study 2010. *Lancet* 380: 2224-60.

[3] [WHO] World Health Organization. 2001. Global Strategy for Infant and Young Child Feeding: The Optimal Duration of Exclusive Breast Feeding. 54th World Health Assembly. Document A54/INF.DOC./4. World Health Organization; Geneva.

[4] Mannion C, Mansell D. 2012. Breastfeeding self-efficacy and the use of prescription medication : a pilot study. *Obstet Gynecol Int*. 2: 562704.

[5] Zuppa AA, Sindico P, Orchi C, et al. 2010. Safety and efficacy of galactogogues: substances that induce, maintain and increase breast milk production. *J Pharm Pharmaceut Sci* 13(2): 162 – 74.

[6] Walker M. 2011. Breastfeeding Management for the Clinician: Using the Evidence. 2nd ed. Sudbury, MA: Jones & Bartlett Publishers.

[7] Eyal O, Jomain JB, Kessler C, Goffin V, Handwerger S. 2007. Autocrine prolactin inhibits human uterine decidualization: a novel role for prolactin. *Biol Reprod*, 76:777-83

[8] Saunier E, Dif F, Kelly PA, Edery M. 2003. Targeted expression of the dominant negative prolactin receptor in the mammary gland of transgenic mice results in impaired lactation. *Endocrinology*,144:2669e7

[9] [Kemenkes-RI] Kementerian Kesehatan Republik Indonesia. 2016. Peraturan menteri kesehatan republik indonesia no. 6 tahun 2016. Formularium obat herbal asli indonesia. Kementerian Kesehatan Republik Indonesia, Jakarta:1-225.

[10] Warrier PK, Nambier VP. 1996. Indian Medicinal Plants: A compendium of 500 species. Chennai: Orient Longman Limited: 315.

[11] Kumaran RJ, Kuranakaran. 2006. Antioxidant and free radical scavenging activity of an aqueous extract of *Coleus aromaticus*. *Food Chemistry*, 97:109-14.

[12] Khan-Khattak MM, Taher M, Damanik R., Abdulrahman S, Bakar IA, Yahaya A. 2013. Torbangun (*Coleus amboinicus* Lour) extracts affect microbial and fungus activities. *Journal of Nutrition Theraphatic* 2(4): 194-200.

[13] Khare RS, Banerjee S, Kundu K. 2011. *Coleus aromaticus* Benth-A nutritive medicinal plant of potential therapeutic value. *International Journal of Pharma and Bio sciences* 2(3): 488-500.

[14] Patel R, Mahobia N, Waseem N, Upwar N, Singh S. 2010. Phyto-physicochemical investigation of leaves of *Plectranthus amboinicus* (Lour) Spreng. *Phcog J.* 2(13): 536-542.

[15] Suryowati T, Rimbawan, Damanik R, Bintang M, Handharyani E. 2015. Identifikasi komponen kimia dan aktivitas antioksidan dalam tanaman torbangun (*Coleus amboinicus* Lour). *J Gizi Pangan*, 10(3):217-24.

[16] Sari LP. 2008. Kajian histopatologi efek irnunomodulator dari daun Torbangun (*Coleus amboinicus* Lour) pada organ sistem limforetikular mencit (*Mus musculus*) pada masa laktasi. Disertasi Sekolah Pascasarjana IPB. Bogor.

[17] Devi M, Syarief H, Damanik R, Sulaeman A, Setiawan B, Dewi R. 2010. Suplementasi daun Torbangun (*Coleus amboinicus* Lour.) untuk menurunkan keluhan sindrom pramenstruasi pada remaja putri. *Penelitian Gizi dan Makanan* 33(2): 180-194.

[18] Santosa CM. 2001. Khasiat konsumsi daun bangun-bangun (*Coleus amboinicus* L) sebagai pelancar sekresi air susu ibu menyusui dan pemacu pertumbuhan bayi. Tesis Sekolah Pascasarjana IPB. Bogor.

[19] Damanik R, Wahlqvist ML, Wattanapenpaiboon N. 2006. Lactagogue effects of torbangun, a bataknese traditional cuisine. *Asia Pacific Journal Clinical Nutrition*, 15(2): 267-74.

[20] Doloksaribu TH, Syarief H, Damanik MRM, Marliyanti SA. 2015. The development of torbangun flour-based functional supplementary food for breastfeeding mother. *Int J Sci: Basic Appl Res*, 1:348-55.

[21] Iwansyah AC, Damanik RM, Kustiyah L, Hanafi M. 2017. Potensi fraksi etil asetat daun torbangun (*Coleus amboinicus* L.) dalam meningkatkan produksi susu, bobot badan induk dan anak tikus. *Jurnal Gizi Pangan* 11(3):1-8.

[22] Iwansyah AC, Damanik RM, Kustiyah L, Hanafi M. 2016. Relationship between antioxidant properties and nutritional composition of some galactopoietics herbs used in indonesia: a comparative study. *Int J Pharm Pharm Sci*, 8(12): 236-43.

[23] Sajimin ND, Purwantari E, Sutedi, Oyo. 2011. Effect of cutting interval to productivity and quality of bangun-bangun (*Coleus amboinicus* L) as aforage promising commodity. *JITV* 16(4): 288-93.

[24] Andarwulan N, Yuliana ND, Hasna E, Aziz SA, Davis TD. 2014. Comparative analysis of three torbangun clones (*Plectranthus amboinicus* (Lour.) spreng) based on phenotypic characteristics and phenolic content. *American Journal of Plant Sciences* 5: 3673-83.

[25] [AOAC] Association of Official Analytical Chemistry. 1990. Official Methods of Analysis. Association of Official Analytical Chemist. Arlington, Virginia (US): AOAC.

[26] Harborne JB . 1973. Phytochemical methods. Chapman and Hall, Ltd London.: 49-188.

[27] Sofowara A. 1993. Medicinal plants and traditional medicine in Africa. Spectrum Books Ltd, Ibadan, Nigeria. p. 289.

[28] Trease GE, Evans WC. 1989. Pharmacognsy. 11th edn. Brailliar Tiridel Can Macmillian publishers.

[29] Frederer WT. 1967. *Experimental design: theory and application.* Calcutta(IN):Oxford & IBH Publ. Co.: 591

[30] Silitonga M. 1993. Efek laktogogum daun jinten (*Coleus amboinicus* L) pada Tikus Laktasi. Tesis Sekolah Pascasarjana IPB. Bogor.

[31] Gibson-Corley KN, Oliver AK, Meyerholz DK. 2013. Principles for valid histopathologic scoring in research. *Veterinary Pathology*, 50(6): 1007-15.

[32] [INADFC] Badan Pengawas Obat dan Makanan - Republik Indonesia. 2014. Peraturan kepala badan pengawas makanan nomor 12 tahun 2014 tentang persyaratan mutu obat tradisional. Badan Pengawas Obat dan Makanan - Republik Indonesia, Jakarta.

[33] Yulianto W, Andarwulan N, Giriwono P, Pamungkas J. 2016. HPLC-based metabolomics to identify cytotoxic compounds from *Plectranthus amboinicus* (Lour.) spreng against human breast cancer MCF-7 cells. *Journal of Chromatography B*, S1570-0232(16): 30551-7. http://dx.doi.org/10.1016/j.jchromb.2016.10.024.

[34] Permana D. 2008. Studi histopatologi pengaruh pemberian daun torbangun (*Coleus amboinicus* Lour) terhadap produksi susu kelenjar susu mencit (*Mus musculus*). Skripsi Fakultas Kedokteran Hewan, IPB.

[35] Mohanty I, Senapati MR, Jena D, Behera PC. 2014. Ethnoveterinary importance of herbal galactogogues - a review. *Veterinary World* 7(5): 325-30.

[36] Turkyilmaz C, Onal E, Hirfanaglu M, et al. 2011. The effect of galactagogue herbal tea on breast milk production and short-term catch-up of birth weight in the first week of life. *The Journal of Alternative and Complementary Medicine*, 17(2): 139-42.

[37] Foidart M, Colin C, Denoo X, et al. 1998. Estradiol and progesterone regulate the proliferation of human breast epithelial cells. *Fertility and Sterility*, 69(5): 963-70.

[38] Tabares FP, Jaramillo JVB, Ruiz-Cortes ZT. 2014. Pharmalogical overview of galactagogues. *Veterinary Medicine International*, ID602894:1-20.

[39] Janabi AK. 2012. Feeding effect of fenugreek seeds (*Tringonella foenum-graceum*) on lactation performance, some serum constituents and prolactin hormone level in damascus crossbred goats. *Diyala Agricultural Sciences Journal*, 4(1): 1-8.

[40] Verpoorte R, Choi YH, Kim HK. 2005. Ethnopharmacology and system biology: a perfect holistic match. *J. Ethnophamacol*. Doi:10.1016/j.jep2005.05.033.

[41] National Center for Biothecnology Information. Compound. http:// pubchem.ncbi.nlm.nih.gov.

[42] Pandey DK, Radha, Dey A. 2016. A validated and densitometric HPTLC method for the simultaneous quantification of reserpine and ajmalicine in *Rauvolfia serpentina* and *Rauvolfia tetraphylla*. *Revista Brasileira de Farmacognosia* 26: 553–557.

[43] Lim H, Son KH, Chang HW, Bae K, Kang SS, Kim HP. 2008. Anti-inflammatory activity of pectolinarigenin and pectolinarin isolated from *Cirsium chanroenicum*. *Biol Pharm Bull*. 31(11): 2063-7.

[44] Tan PJ, Appleton DR, Mustafa MR, Lee HB. 2012. Rapid identification of cyclic tetrapyrrolicphotosensitisers for photodynamic therapy using on-line hyphenated LC-PDA-MS coupled with photo-cytotoxicity assay. *Phytochem. Anal*. 23: 52–59

[45] Matsuda F, Suzuki M, Sawada Y. 2009. Kaempferol-3-glucoside-2-rhamnoside,-7-rhamnoside. Plant Science Center, RIKEN. http://spectra.psc.riken.jp/menta.cgi/respect/datail/datail?accession=PT209173

[46] Huang, Wang, Kuo *et al.* 2011. Estrogenic-active compounds and application thereof. US Patent: No.US 20100125102 A1.

Effects of Cultivating Conditions on the Water Soluble Polysaccharides Content of *Ganoderma lucidum* Mycelium in Submerged Flask Culture

V T Rosyida, S N Hayati, W Apriyana, C Darsih, Hernawan, and C D Poeloengasih

Research Unit for Natural Product Technology of Indonesian Institute of Sciences, (BPTBA-LIPI). Jl. Jogja-Wonosari Km 31.5, Gading, Playen, Gunung Kidul, 55861 D.I Yogyakarta, Indonesia

E-mail: hayanurrizal@gmail.com

Abstract. The carcinostatic substance in *Ganoderma lucidum* (Fr.) Karst (Polyporaceae) is a water soluble polysaccharides (WSP) which might be useful in immunotherapy. Attempt to produce effective substances from cultured mycelia is important to carry out since solid cultivation is a time consuming and quality fluctuating. The effects of cultivating conditions on the water soluble polysaccharides content of *G. Lucidum* mycelium were investigated in submerged flask cultures. Culture from fruiting bodies was maintained on potato dextrose-agar slope. Slopes were inoculated and incubated at 30°C for 7 days, and stored at 4°C. The flask experiments were performed in 100 ml erlenmeyer flasks containing 20 ml of the sterilized media. Actively growing mycelia (1 piece, 5 mm X 5 mm) from a newly prepared slant culture (about 7 days incubation at 30°C) were inoculated into the flask. The pH was measured and adjusted to the desired value by addition of either 4 M HCl or 2.5 M NaOH. Incubation temperature were 20, 25, and 30°C. At the end of inoculation period (14 days) mycelium consisting of individual pellets was harvested and wash for the analysis. WSP content was analysed using phenol-sulfuric acid method. The optimal initial pH for metabolite production would depend on the culture medium. Generally, high values of pH, such as 9, negatively affect both cell growth and WSP production. The optimum temperature range for the high *G. lucidum* mycelium and WSP production were found to be 25 – 30 °C at pH values 5 – 7 in both of media.

1. Introduction

Ganoderma lucidum (Fr.) Karst (Polyporaceae) is a medicinal mushroom which contains water-soluble polysaccharides as a bioactive substance. Water-soluble polysaccharides (WSP) in the *G. lucidum* fruiting bodies has been discovered to be medically active as immunomodulatory, antioxidant, anti-inflammatory, and antitumor. *G. lucidum* polysaccharides have represented a variety of immune modulating effects. Polysaccharides from *G. lucidum* was a promising biological response modifier and immune potentiator [1]. It was found that the polysaccharide extracts from the mycelium of *G. lucidum* had anti-tumor effects against fibro sarcoma in male and female mice and inhibited the metastasis of a tumor to the lung [2]. The fruit bodies of *G. lucidum* grow slowly in nature, their growth is restricted to a specific area. It is necessary to develop an efficient method for the mass production of water-soluble polysaccharides and the mycelial biomass. Some attempts have been made to obtain useful cellular materials or to produce effective by a submerged mycelial culture since conventional cultivation is a time consuming and quality fluctuating. Submerged culture by fermentation is an alternative approach for an efficient production of higher fungus metabolites. There are several advantages of submerged culture over solid culture for WSP production: high productivity, low costs, availability of convenient control and easy downstream processing [3]. The fundamental information obtained in this work is complementary to those of previous investigations on the submerged culture of *G. lucidum* for the production of bioactive metabolites.

2. Material and methods

2.1 Materials

The fresh mature fruiting bodies of *G. lucidum* were supplied by Gubug Jamur Sitiram from Sleman, Yogyakarta and have been taxonomically verified at Pharmaceutical Biology Division, Faculty of Pharmacy, Universitas Gadjah Mada Yogyakarta. The chemicals and microbiological materials included distilled water, ethanol, sulfuric acid pa, glucose standard, hydrochloric acid, organic potatoes, dextrose bacteriological, agar, and Potato Dextrose Agar, etc.

2.2 Inoculum and media

Inoculum culture from fruiting bodies was maintained on potato dextrose-agar slope. Slopes were inoculated and incubated at 30°C for 7 days, and stored at 4°C. Microbiological media included commercial Potatoes Dextrose Agar and formulated media made from organic potatoes extract, dextrose bacteriological, agar.

2.3 Cultivation of Microorganism

The flask experiments were performed in 100 ml erlenmeyer flasks containing 20 ml of the sterilized media. Actively growing mycelia (1 piece, 5 mm X 5 mm) from a newly prepared slant culture (about 7 days incubation at 30°C) were inoculated into the flask. The pH was measured and adjusted to the desired value by addition of either 4 M HCl or 2.5 M NaOH (pH 5, 7, 9). Incubation temperature are 20, 25, and 30°C. At the end of inoculation period (14 days) mycelium was harvested and wash for the analysis.

2.4 Sample preparation

Mycelia collected from flasks were separated from media, the mycelia was washed twice with distilled water and then weight of dry mycelia was obtained by filtering culture samples through a preweighed filter paper and dried at 80°C to constant weight.

2.5 Water-soluble polysaccharides content

The total water-soluble polysaccharides were determined based on the color reaction of polysaccharides and their derivatives with phenol and concentrated sulfuric acid (Phenol – sulfuric acid method) [4-6]. The total water-soluble polysaccharides were determined as glucose with hydrolyze polysaccharides into glucose monomer. Dried *G. lucidum* mycelia were ground into a powder. The powder of mycelia (0.50 g) was subjected to hot water extraction at 95°C and hydrochloric acid (HCl) 2 M to accelerate the extraction rate. The extract solution was pipetted into a 10-mL centrifuge tube and 1 mL of 5% phenol was added. The mixture was shaken for 2 min and the 5 mL of concentrated sulfuric acid (H_2SO_4) (98% v/v) was added to the solution and shaken for another 5 min. The concentration of water-soluble polysaccharides in the extract solution was determined quantitatively by measuring the absorbance at 490 nm. The standard curve for quantitative analysis of total water-soluble polysaccharides content in the dried *G. lucidum* mycelium was plotted using standard glucose (Sigma, Milwaukee, WI, USA) as a standard solution and distilled water was used as blank solution.

3. Result and discussion

Mycelium weight of *G. lucidum* cultivated in both formulated and commercial media at same pH and temperature had no significant difference. But, WSP production from commercial media was higher than formulated media. It showed that commercial media provide better nutrition than formulated media for WSP production.

Temperature is a very important ecological factor for mycelium growth of fungi. The optimum temperature for the highest yield of *G. lucidum* mycelia in both formulated and commercial medium at incubation pH 5, 7, and 9 was 25 °C. The optimum temperature range for the high *G. lucidum* WSP production were 25 – 30 °C at pH values 5 – 7 in both of media.

Table 1. Dry weight (gram) of mycelial *G. lucidum* in formulated medium at various pH and temperature

Initial pH	Temperature		
	20°C	25°C	30°C
pH 5	2.77 ± 0.06	3.15 ± 0.12	3.11 ± 0.13
pH 7	3.04 ± 0.07	3.31 ± 0.08	2.85 ± 0.09
pH 9	2.89 ± 0.11	2.75 ± 0.06	2.70 ± 0.11

Table 2. Dry weight (gram) of mycelial *G. lucidum* in commercial medium at various pH and temperature

Initial pH	Temperature		
	20°C	25°C	30°C
pH 5	2.92 ± 0.07	3.16 ± 0.09	2.97 ± 0.09
pH 7	3.09 ± 0.15	3.24 ± 0.10	2.89 ± 0.05
pH 9	3.05 ± 0.08	3.11 ± 0.05	2.40 ± 0.06

Table 3. Water-soluble polysaccharides content (%) of mycelial *G. lucidum* in formulated medium at various pH and temperature

Initial pH	Temperature		
	20°C	25°C	30°C
pH 5	38.48 ± 1.15	24.02 ± 1.21	40.44 ± 1.24
pH 7	39.77 ± 1.17	23.03 ± 0.29	36.28 ± 1.09
pH 9	32.62 ± 1.25	25.23 ± 1.11	34.59 ± 1.14

Table 4. Water-soluble polysaccharides content (%) of mycelial *G. lucidum* in commercial medium at various pH and temperature

Initial pH	Temperature		
	20°C	25°C	30°C
pH 5	39.25 ± 1.12	42.46 ± 1.13	45.66 ± 1.20
pH 7	45.46 ± 1.11	53.99 ± 1.14	54.66 ± 1.17
pH 9	40.92 ± 0.97	39.64 ± 1.04	36.03 ± 0.85

The mycelia of various mushrooms species will grow over a wide range of pH values. Nevertheless, for most organisms, the most preferred pH range is from 5 to 7. Lower pH values would be useful to inhibit the growth of bacterial contaminants. pH is also one of the critical factors for controlling mycelial growth and polysaccharide production [8]. Each mushroom has its optimal pH range for development, and it is variable; for example, pH between 4.0 and 7.0 for the mycelium and 3.5 to 5.0 for formation of basidiocarp. The optimum pH for mycelial growth and subsequent fruiting body development is obtained at between 6.5 and 7.0 [9].

The optimum pH for the highest yield of *G. lucidum* mycelia in both formulated and commercial medium at incubation temperature 20 and 25°C was 7. But, the optimum pH for *G. lucidum* growing at incubation temperature 30 °C in both formulated and commercial medium was found to be 5. The optimum pH for the highest *G. lucidum* WSP production in formulated medium was different at each incubation temperature. At incubation temperature 20, 25, and 30°C, the optimum pH was 7, 9, and 5, respectively. However, it is interesting to note that the optimum pH for *G. lucidum* WSP production in commercial medium at various incubation temperature (20, 25, and 30°C) was found to be 7. It demonstrated that optimal initial pH for metabolite production would depend on the culture medium. Generally, high values of pH, such as 9, have negatively affect both cell growth and WSP production.

4. Conclusions

The optimal initial pH for metabolite production would depend on the culture medium. Generally, high values of pH, such as 9, negatively affect both cell growth and WSP production. The optimum temperature range for the high *G. lucidum* mycelia and WSP production were found to be 25 – 30 °C at pH values 5 – 7 in both of media. Commercial media provide better nutrition than formulated media for water soluble polysaccharides production. However, most of commercial media utilized in WSP production are not yet economically available at low cost. In order to meet the requirement of large scale production, further study about media formulation in order to optimize the growth of mycelia of *G. lucidum* and water soluble polysaccharides production in fermenter cultures would be necessary.

Acknowledgements

The authors acknowledged Sanggar Tani Media Agro Merapi and Gubug Jamur Sitiram for supply of fresh *G. lucidum*. This research was supported financially by the Indonesian Institute of Sciences through the program of Penguatan Kompetensi Inti Satuan Kerja.

References

[1] Paterson, R.R.M 2006 *Phytochemistry* **67**, 1985–2001, Review Ganoderma– A therapeutic fungal biofactory

[2] Zhu, X., Lin, Z. 2006 Modulation of cytokines production, granzyme B and perforin in murine CIK cells by *Ganoderma lucidum* polysaccharides. *Carbohyd. Polym.* **63**, 188–197

[3] Tang YJ, Zhong JJ. Fed-batch fermentation of Ganoderma lucidum for hyperproduction of polysaccharide and ganoderic acid. *Enzyme Microb Technol* 2002;**31**:20–8.

[4] Ian W Dawes 1990 *Biotechnology of microbial exopolysaccharides*, Cambridge University press, New York, , pp. 1-11, 20-23, 38-42, 117-151

[5] Graeme WM 1992 *Yeast Physiology and piotechnology*, John Wiley and Sons, New York , pp. 122-131

[6] Poedjiadi Anna 1994 *Dasar-dasar biokimia*, UI Press, Jakarta , pp. 8-50, 81-124

[7] David HJ, Peck Hazel 1993 *Analytical biochemistry*, 2nd Ed., John Wiley and Sons Inc., New York, pp. 338-431, 408-411

[8] Catiey, B. J.: Role of pH and nitrogen limitation in the elaboration of the extracellular polysaccharide pullulan by Pullularia pullulans. *Appl. Microbial.*, **22**, 650-654 (1971)

[9] E. Kalmis, N. Azbar, H. Yıldız, F. Kalyoncu 2008 Feasibility of using olive mill effluent (OME) as a wetting agent during the cultivation of oyster mushroom, Pleurotus ostreatus, on wheat straw, *Bioresour. Technol.*, **99**, pp. 164-169

Lactic Acid Bacteria Producing Inhibitor of Alpha Glucosidase Isolated from Ganyong (*Canna Edulis*) and Kimpul (*Xanthosoma sagittifolium*)

Rifa Nurhayati[1], Miftakhussolikhah[1], Andri Frediansyah[1], Desy Lailatul Rachmah[2]

[1]Research Unit for Natural Product Technology, Indonesian Institute of Sciences Jl. Yogya-Wonosari km 31.5 Gading, Playen, Gunungkidul, Yogyakarta, Indonesia.
[2]Faculty of Biology, Sepuluh November Institute of Technology, Surabaya, Indonesia.he author addresses here

E-mail: rifa.lipi@gmail.com

Abstract. Type 2 diabetes is a disease that caused by the failure of insulin secretion by the beta cells of the pancreas and insulin resistance in peripheral levels. One therapy for diabetics is by inhibiting the activity of α-glucosidase. Lactic acid bacteria have the ability to inhibit of α-glucosidase activity. The aims of this research was to isolation and screening of lactic acid bacteria from ganyong tuber (*Canna Edulis*) and kimpul tuber (*Xanthosoma sagittifolium*), which has the ability to inhibit the activity of α-glucosidase. Eightteen isolates were identified as lactic acid bacteria and all of them could inhibit the activity of α-glukosidase. The GN 8 isolate was perform the highest inhibition acivity.
Keywords. Lactic acid bacteria, inhibitor α-glucosidase, *Canna Edulis*, *Xanthosoma sagittifolium*

1. Introdustion

Diabetes mellitus (DM) is a disease characterized by the occurrence of hyperglycemia and disfunction metabolism of carbohydrates, fats, and proteins. DM is associated with an absolute deficiency or relative deficiency of insulin or insulin secretion [1,2]. Indonesia is the 4th country with the highest prevalence of diabetes in the world after India, China and the United States. Even the number of people with diabetes continues to increase from year to year, especially for type 2 diabetes is caused by the failure of insulin secretion by the beta cells of the pancreas and insulin resistance in peripheral levels [3]. Based on the latest estimate of IDF (International Diabetes Federation) [4], in 2013 diabetics in the world reached 382 million people. By 2035, diabetics are expected to increase to 592 million people. In Indonesia, diabetes mellitus reaches 12 million people counted in 2013 [5].

Various therapeutic strategies used to treat type 2 diabetes, including the stimulation of insulin secretion, increase insulin action in target tissues, the use of oral hypoglycemic agents such as sulfonylureas and biguanids and inhibition of starch degradation by the enzyme α-glucosidase. Alpha glucosidase is an enzyme that plays a role in the hydrolysis of food carbohydrates into glucose and other monosaccharides [6,7]. The excess production of monosaccharides and the lack of sugar removal by the body can cause diabetes [8]. Therefore, one way to inhibit postpradial hyperglycemia is to inhibit the activity of the α-glucosidase enzyme [9]. Some synthetic drugs have been used to inhibit the activity of α- glucosidase, for example acarbosa as a treatment for type 2 diabetes mellitus. These synthetic compounds have limited, non-specific and cause side effects such as indigestion, bloating, nausea, abdominal discomfort and diarrhea [10,11,12,13].

Some recent researches say that lactic acid bacteria has the ability to inhibit the action of the enzyme α- glucosidase [14,15,16,17,18]. Therefore, this study aims to isolation and sreening of lactic acid bacteria from ganyong tuber and kimpul tuber, which has the ability to inhibit the activity of α-glucosidase enzyme. Such lactic acid bacteria can be developed as probiotics with antidiabet ability.

2. Material and Methods

2.1. Materials

The samples used in this study were Ganyong (*Canna Edulis*) and kimpul (*Xanthosomasagittifolium*). The sample was obtained from farmers in Mertelu village, Gedangsari district Gunungkidul Regency.

2.2. Isolation of Acid Producing Bacteria

The tuber sample was peeled to remove the skin, then sterilized the surface with 70% alcohol soaked for 1 minute. The sample was smoothed, then taken as much as 20 grams and fermented with MRS Broth medium 80 ml at 37°C for 48 hours. Isolation was done by plating using Pour Plate method on MRSA + $CaCO_3$ medium. Colonies that form clear zones are thought to be acid producing bacteria. Clear zones are formed because the acid produced by the bacteria will react with $CaCO_3$, so that initially cloudy will be clear. Several colonies of different shapes and sizes were selected and purified on the MRSA medium by streak plate method. Incubation at 37°C for 48 hours. The experiment is done 2-3 times until the pure or single isolate was obtained.

2.3. Screening of Lactic Acid Bacteria

The screening of lactic acid bacteria refers to Bergey"s Manual Determination of Bacteriology by biochemical test. There are 4 key characteristis of lactic acid bacteria: gram positive, non-motile, negative catalase, and do not form spores.

2.4. Preparation of Cell Free Supernatant

The isolates of Lactic acid bacteria were inoculated in 5 ml MRSB, then incubated for 24 h at 37°C. Supernatant was obtained by centrifugation at 4°C, 5000 rpm for 15 min. Then separated between supernatant and cell biomass. The supernatant was used for measurement of α-glucosidase inhibitor.

2.5. Measurement of Alpha Glucosidase Inhibitor

This method refers to Zheng *et.al.* [19]. The test was performed by adding 25 μl 10 mM p-nitrophenyl-α-D-glucopyranoside (PNPG), 25 μl PBS (pH 6.8) and 50 μl sample or 50 μl acarbose 1.5% (15mg/ml), then incubated for 10 min at 37°C. Thereafter, 50 μl α-glucosidase from Saccaromyces cerevisiae 1 U/ml was added and incubated for 30 min at 37°C. The reaction was terminated by adding 100 μl 0.1M Na2CO3, and the absorbance of sample was measured at 405 nm with microplate Reader. Each test sample was analyzed in duplo, and the absorbance value was corrected against the blank sample where α-glucosidase was replaced with PBS (0.1 M, pH 6.8). Positive control (α-glucosidase activity without inhibitor) and negative control (no α-glucosidase activity) was performed using PBS (0.1 M, pH 6.8), instead of samples and α-glucosidase solution. The results of α-glucosidase inhibition were calculated by :

$$\% \text{ inhibition} = \left(1 - \frac{\text{sample} - \text{blank}}{\text{positive control} - \text{negative control}}\right) \text{x } 100\%$$

3. Result and Discussion

3.1. Isolation and Screening of Lactic Acid Bacteria

There were 9 isolates of acid producing bacteria from ganyong (*Canna edulis*) and 10 isolates of acid producing bacteria from kimpul (*Xanthosoma sagittifolium*). These acid producing bacteria were obtained by formation of clear zone in MRS agar + CaCO3 medium. However, one isolate (GN 6) from ganyong was known not include to lactic acid bacteria because its positive result on catalase test and gram negative. Screening for lactic acid bacteria based on Bergey"s Manual Determination of Bacteriology: gram staining, catalase test, endospore staining and motility.

Among 19 isolate, there were 18 isolates identified which perform gram positive, catalase negative, non motil and negative endospore. The screening result are presented on Table 1.

Table 1. Screening of LAB isolated from Ganyong (*Canna edulis)* and Kimpul (*Xanthosoma sagittifolium*)

Isolate	Gram staining	Catalase test	Motility	Endospore staining	identity
GN 1	+	-	-	-	LAB
GN 2	+	-	-	-	LAB
GN 3	+	-	-	-	LAB
GN 4	+	-	-	-	LAB
GN 5	+	-	-	-	LAB
GN 6	-	+	-	-	Not LAB
GN 7	+	-	-	-	LAB
GN 8	+	-	-	-	LAB
GN 9	+	-	-	-	LAB
KM 3.7	+	-	-	-	LAB
KM 6.7	+	-	-	-	LAB
KM 9.7	+	-	-	-	LAB
KM 4.7	+	-	-	-	LAB
KM 1.7	+	-	-	-	LAB
KM 5.7	+	-	-	-	LAB
KM 10.7	+	-	-	-	LAB
KM 2.7	+	-	-	-	LAB
KM 7.7	+	-	-	-	LAB
KM 8.7	+	-	-	-	LAB

The characteristics of lactic acid bacteria are the cells reacting positively to gram staining, reacting negatively to catalase, not forming spore and non-motile [20]. Positive catalase is characterized by forming air bubbles when dropping hydrogen peroxide to lactic acid bacteria. This indicates that lactic acid bacteria have catalase enzymes. The catalase enzyme serves to neutralize the bactericidal effect of hydrogen peroxide. Catalase accelerates the breakdown of hydrogen peroxide (H_2O_2) into water and oxygen ($2H_2O_2$ + catalase $\rightarrow$ $2H_2O + O_2$). This reaction is evidenced by the rapid formation of bubbles. In gram staining, lactic acid bacteria include gram-positive bacteria. Gram positive bacteria will retain the violet crystalline dye and will therefore look purple under the microscope. In the motility test,the lactic acid bacteria only growing around the puncture. This was showed negative test results [21]. For endospora staining, all isolates are red. Endospores will look green due to malachite green coloration. Lactic acid bacteria do not form endospores.

3.2. Measurement of alpha glucosidase inhibitor

Cell free supernatant of each lactic acid bacteria isolates was used for measurement of α-glukosidase inhibitor. The result showed that 8 isolates from ganyong and 10 isolates from kimpul were able to inhibit the activity of α-glukosidase at varying degrees (Table 2).

Table 2. Inhibition of Alpha glukosidase by LAB from Ganyong (*Canna edulis)* and Kimpul (*Xanthosoma sagittifolium*)

Isolate	% Inhibition	Isolate	% Inhibition
GN 2	$80{,}97 \pm 0{,}95^{def}$	KM 9.7	$56{,}26 \pm 4{,}18^{g}$
GN 3	$78{,}60 \pm 0{,}06^{ef}$	KM 4.7	$7{,}48 \pm 1{,}87^{i}$
GN 9	$81{,}48 \pm 4{,}45^{def}$	KM 1.7	$84{,}88 \pm 0{,}23^{cde}$
GN 5	$34{,}95 \pm 0{,}21^{h}$	KM 5.7	$96{,}19 \pm 0{,}13^{ab}$
GN 7	$32{,}61 \pm 1{,}50^{h}$	KM 10.7	$90{,}22 \pm 6{,}34^{bc}$
GN 1	$63{,}10 \pm 0{,}10^{g}$	KM 2.7	$80{,}60 \pm 5{,}05^{def}$
GN 4	$93{,}82 \pm 2{,}04^{b}$	KM 7.7	$90{,}68 \pm 8{,}32^{bc}$
GN 8	$103{,}18 \pm 3{,}96^{a}$	KM 8.7	$93{,}75 \pm 2{,}70^{b}$
KM 3.7	$75{,}62 \pm 5{,}99^{f}$	Acarbosa	$87{,}60 \pm 0{,}01^{bcd}$
KM 6.7	$84{,}15 \pm 0{,}16^{cde}$	15mg/ml	

Each value represent mean ± SD from 2 independent treatments

The α-glukosidase inhibitor of each isolates is different. The highest inhibition activity is found in GN 8 isolate, while the lowest inhibition activity is KM 4.7. Even the inhibition activity of GN 8 is higher than inhibition activity of acarbosa 15 mg/ml. This may be due to the difference in concentration between acarbosa and the cell free supernatant of each lactic acid bacteria. The cell free supernatant content extracelluer metabolite which associated with α-glukosidase inhibitor.

Based on the research of Bajpai et.al [8], that lactic acid bacteria capable of inhibiting alpha glucosidase. Inhibition of α-glucosidase is derived from exopolisakarida and inulin produced by lactic acid bacteria [15]. The ability of lactic acid bacteria isolated from ganyong (*Canna edulis*) in inhibiting α-glucosidase may be caused by ganyong including essential tuber crops in herbal medicines as anti-diabetic [22,23]. Ganyong is a plant that has a high fiber content, high fiber serves to prevent diabetes. While the ability of lactic acid bacteria isolated from kimpul in inhibiting α-glucosidase can be caused by bioactive compounds from kimpul that positively affect the absorption of blood glucose such as water-soluble polysaccharide (PLA), dietary fiber and diosgenin [24]. Kimpul has low-carbohydrate, low fat, and low glucose content so it can be used for diet of diabetic people [25]. Endophytic microorganisms from a plant have the same metabolites as their host which potentially for exploitation in medicine, agriculture and industry [26].

The percentage of inhibitor activity in this study showed higher results when compared with studies conducted by Zeng et al. [19]. The study reported that the percentage of activity of α-glucosidase inhibitors from lactic acid bacteria, especially the Lactobacillus strains, ranged from 25%-34%. Other studies have also reported that lactic acis bacteria strains have a 9% -26% of α-glucosidase inhibitor activity [17].

4. Conclusion

The current study showed that there were 8 isolates that isolated from Ganyong and 10 isolates that isolated from kimpul. Eightteen isolates were identified as lactic acid bacteria and all of them could inhibit the activity of α-glukosidase. The GN 8 isolate was perform the highest inhibition acivity (103,18 %). These isolates were potentially to be developed as antidiabet probiotic.

References

[1] Ieyama, T., Maria, D. P. T., Gunawan, P. & Kawabata, J. 2011. ELSEVIER Journal Food Chemistry 128: 308–311.
[2] Fatimah, R. Noor. 2015. *J Majority* 4(5):93-101
[3] Rahmawati, G., Rachmawati, F. N., Winarsi, H. 2014. *Scripta Biologica*. Vol.1 No. 3 pp : 19-23.
[4] IDF. 2013. IDF Diabetes Atlas Sixth Edition, International Diabetes Federation 2013
[5] Kementrian Kesehatan RI. 2014. Pusat Data dan Informasi Kementrian Kesehatan RI. Jakarta
[6] Sudha, P., Zinjarde, S. S., Bhargava, S. Y., & Kumar, A. R. 2011. *BMC. Complementary and Alternative Medicine* 11:5,2
[7] Mataputun, San Paris., J. A. Rorong., J. Pontoh. 2013. *Jurnal MIPA Unstrat Online* 2(2): 119-123.
[8] Bajpai, K. Vivek., J. H. Han., G. J. Nam., R. Majumder., C. Park., J. Lim., W. K. Paek., I. A. Rather., Y. H. Park. 2016. *Journal of Pharmaceutical Sciences* 24(8):1-12.
[9] Mugiyanto
[10] Wehmeier, U. F. & Piepersberg, W. 2004. *J Appl Biotechnol* 63:613-625
[11] Chiasson, J.L., Josse, R.G., Gomis, R., Hanefeld, M., Karasik, A., and Laakso, M. 2002. *Lancet*, 359, 2072-2077.
[12] Febrinda, A. Early., M. Astawan., T. Wresdiyati., N. D. Yuliana. 2013. *Jurnal Teknologi dan Industri Pangan*. 24(2): 161-167
[13] Andayani, Tri Murti., M. I. M. Ibrahim., A. H Asdie. 2009. *Majalah Farmasi Indonesia* 20(4): 224-230.
[14] Chen, P., Zhang, Q.X., Dang, H., Liu, X.M., Tian, F.W., Zhao, J.X., Chen, Y.Q., Zhang, H. and Chen, W. 2014. *Food Control*. 35,65-72.
[15] Ramchandran, L, & Shah, N. P. 2008. *Jurnal of Food Science* 73(2): M75-M81.
[16] Panwar, H., D. Calderwood, I. R. Grant, S. Grover, and B. D. Green. 2014. *European Journal of Nutrition*. 53: 1465–1474.
[17] Muganga, L., Liu, X., Tian, F., Zhao, J., Zhang, H. and Chen, W. 2015. *Journal of Functional Foods*. 16 : 125-136.
[18] Nurhayati, Rifa., A. Frediansyah., F.Rahmawati., E. Retnaningrum., L. Sembiring. 2015. *Proceeding of Internationalon Appropriate Technology Development*. Vol (1).
[19] Zheng, Z., Luo, J., Zuo, F., Zhang, Y., Ma, H., and Chen, S. 2016. *Journal of Functional Foods*. 20 : 486-495.
[20] Mitsuoka, T. 1990. Japan : Yakult Honsha Co. Ltp
[21] Lay, B. W. 1994. Jakarta: PT Raja Grafindo Persada.
[22] Mishra, Tanmayee., A. K. Goyal., S. K. Midha., A. Sen. 2011. *Indian Journal of Natural Products and Resources* 2(3): 315-321
[23] Noriko, Nita., A. Pambudi. 2014. *Jurnal Al-Azhar Indonesia Seri Sains dan Teknologi* 2(4): 248-252.
[24] Yofananda, Olivia., T. Estiasih. 2016. *Jurnal Pangan dan Agroindustri* 4(1): 410-416.
[25] Puspitaningrum, Ika., L. Kusmita., Mutmainah. 2014. *Prosiding Seminar Nasional*.
[26] Strobel, G. and Daisy, B. 2003. *Microbiology and Molecular Reviews* Vol. 67, No. 4

Preliminary study on biosynthesis and characterization of bacteria cellulose films from coconut water

A W Indrianingsih, V T Rosyida, T H Jatmiko, D J Prasetyo, C D Poeloengasih, W Apriyana, K Nisa, S Nurhayati, Hernawan, C Darsih, D Pratiwi, A Suwanto, D Ratih

Research Unit for Natural Product Technology, Gading, Playen, Gunungkidul, Yogyakarta 55861, Indonesia

E-mail: anastasia_wheni_i@yahoo.com / anas004@lipi.go.id

Abstract. Bacterial cellulose produced by *Acetobacter xylinum* is a unique type of bacterial cellulose. It contains more than 90% of water. A preliminary study had shown that bacterial cellulose films has remarkable mechanical properties. The aim of this study was to investigate the optimum condition such as percentage of carbon source, time of cultivation, and pH to produce bacterial cellulose films from local coconut water, and its characterization on morphology, swelling ability and tensile strength of dried bacterial cellulose. *A. xylinum* was grown on coconut water culture medium with addition of 3%, 5%, and 7% of sugar, while the cultivation time was vary from 3 days, 5 days and 7 days. pH condition was conducted in pH 3, pH 5 and pH 7. Bacterial cellulose samples were dried using oven with temperature of 100°C until the moisture content reached 4-5%. This study showed that several parameters for optimum condition to produce bacterial cellulose films from local waste of coconut water had been obtained (5% of carbon source; pH 5; and 7 day of incubation period). The electron microscopy also showed that dried bacterial cellulose films had pores covered by fibrils on the surface. Therefore, the present work proposes the optimum formula and condition that can be used based on properties of end product needed.

1. Introduction

Cellulose is an insoluble water compound normally found in plant cell walls, especially in branches, stems, stems, other woody parts. It plays an important role in the integrity of plant cell walls [1]. Cellulose fibers from plants are known as microfibrils that are 2-20 nm in diameter and form a strong network within the cell wall of the plant. Groups of polysaccharides are arranged in parallel to form cellulose microfibrils that cluster into macrofibrils. Cellulose molecules can be arranged up to thousands of microfibrils chain units of D-glucose linked by hydrogen bonding [2].

Cellulose can be produced by a fermentation process. Bacteria that can be used to produce cellulose are Acetobacter strains, Gluconacetobacter, Agrobacter, Sarcina, among others which includes non-pathogenic bacteria and is known as cellulose bacteria [3, 4]. *Acetobacter xylinum* has been known to produce pure cellulose for more than 100 years. *A. xylinum* has been isolated from rotting sugary fruits, vegetables and fermenting coconut water [5]. Bacterial cellulose is cellulose produced using the help of cellulose bacteria. Actually cellulose from plants and bacteria has the same chemical structure, but the cellulose produced by the bacteria has better fiber composition. Bacterial cellulose is a polysaccharide consisting of several hundred to thousands units of D-glucose associated with β bond (1→4). The fibers of the bacterial cellulose usually have a diameter of more than 50 nm. The cellulose fiber units are interconnected to form a network structure, consisting of a three dimensional network of microfibrils and nanofibrils [6, 7]. Bacterial cellulose is pure and exhibits a higher degree of crystallinity than the cellulose obtained from plant in which the cellulose fibrils are embedded with lignin and hemicelluloses [3]. As already known, the lignin and hemicellulose that make up the cell wall components of plants is difficult to remove in the purification process. Thus, bacterial cellulose could be processes further without hazardous by-products and using low energy on the purification process.

Bacterial celluloses have transparent properties found by Klemm et al 2001 [8]. These properties can be identified during the purification and drying process of cellulose microbial on the surface of culture. After the filtering and drying process, a thin layer of bacterial cellulose can be observed transparently by placing colored paper. Bacterial cellulose has a greater water absorption capacity than plant cellulose. This is because the hydrogen bond in the bacterial cellulose is stronger and the bacterial cellulose polymer chain is also longer [9]. Applications of bacterial cellulose are extensive in many industries, such as in biomedical industries, paper production, food industries, and cosmetics. In the field of biomedical industries, bacterial cellulose can be used as biosensor, biofill, transdermal drug delivery, and wound cover [9, 10, 11]. In paper making, bacterial cellulose is used in the production of synthetic paper, insulation materials, and surface coatings. In the process of food manufacturing, bacterial cellulose is used as a stabilizer agent, making Nata de Coco, and baking [12], whereas in the cosmetic field, bacterial cellulose has been used as a health tonic drink and artificial nails [13]. In this study, the synthesis of bacterial cellulose by utilizing local waste of coconut water had been conducted. This study may be used as preliminary study in the evaluation of bacterial cellulose film as a promising candidate for application in cosmetic and biomedical industries.

2. Material and Methods

2.1. Materials and General Instrumentations

Acetobacter xylinum (strain FNC-0001), was purchased from Gadjah Mada University (UGM). Coconut water was obtained fromYogyakarta Province, Indonesia.Yeast extract, sucrose , acetic acid, distilled water, urea, NaOH, universal pH paper, plate count agar were purchased from Merck. The equipments used in this study were glassware, scales, autoclave, Scanning Electron Microscopy (SEM), oven, desiccator, hot plate, a pH meter, Universal Testing Instrument and calipers.

2.2. Media culture and bacterial cellulose production

The procedure used is a modification of Almeida et al, 2014 [14] with several modifications. The culture medium used for the fermentation of *A. xylinum* to produce bacterial cellulose consist of coconut water (1 litre), sucrose (50 g), urea (10 g), starter of *A. xylinum* (10%) and the pH is set to be pH 5 with acetic acid. The media was then sterilized by autoclave at 121°C for 15 minutes. Bacterial cellulose was grown at room temperature and static conditions for 7 days in a Petri dish. After incubation, the resulting bacterial cellulose was harvested and boiled in distilled water for 2 hours, and rinsed in running water until neutral. The drying of bacterial cellulose was conducted using oven for 5 hours with temperature of 100°C.

2.3. Characterization of bacterial cellulose morphology and mechanical performance

Scanning Electron Microscopy (SEM) of the surface of dried bacterial cellulose was performed using a Hitachi SU-70 instrument operating at 4 kV. Tensile assays were performed on a Universal Testing Mechine (Zwick 0.5, Lloyd's Universal Testing Instrument), operating at a deformation rate of 10 mm/min, under ambient conditions. The films samples were cut into strip-shaped specimens 1 cm width and 10 cm long. At least 5 specimens were tested for each composite.

2.4. Swelling Measurement
The swelling percentage was determined using method by Maneerung et al, 2008 [15] with little modifications. The samples were cut into a size of 1 cm width x 2 cm length. Then the samples were soaked in distilled water for 72 hours.

3. Results and Discussion

3.1. Physical appearances
The physical appearances of wet and dried samples of bacterial cellulose films were demonstrated in Figure 1. From the observation, physical appearances of the samples before and after drying were different.

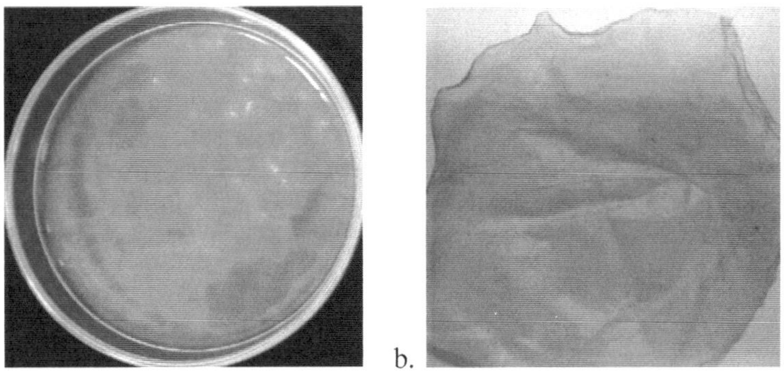
a. b.

Figure 1. The physical appearances of wet (a) and dried (b) samples of bacterial cellulose films.

At the end of drying process, the bacterial cellulose film exhibited brownish color. It probably caused by the heat applied during the drying process. The surface was also shrunk that indicating the removal of water content in drying process. In this study the water content of the obtained bacterial cellulose film is about 90-95%.

3.2. Morphology structure of bacterial cellulose film
The morphology of the surface structure of the bacterial cellulose film was analyzed using SEM (SU-3500 from Hitachi, Tokyo, Japan) with 10,000 magnification. Figure 2 shows the SEM image of the surface of the bacterial cellulose film.

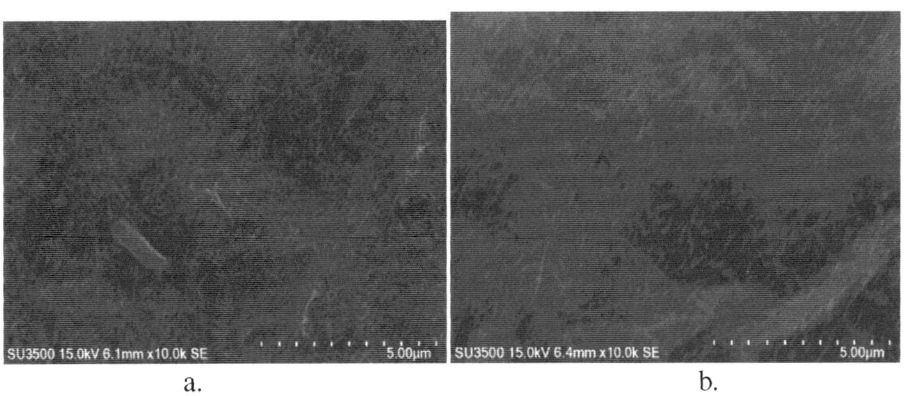

a. b.

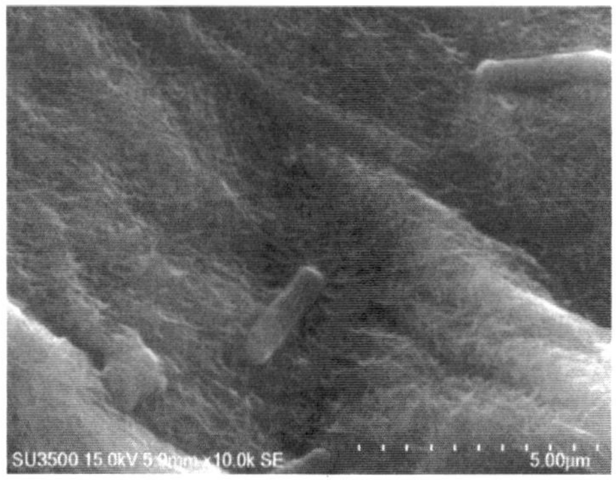

c.

Figure 2. SEM images of the surface morphological structure of dried bacterial cellulose film with 5-day incubation time and percentage of carbon source of 3% (a), 5% (b), and 7% (c).

The surfaces of bacterial cellulose were composed of many intertwined strings which produced an aggregated structure. Figure 2 indicates that the network of thin layers of bacterial cellulose structures is composed of very neat fibrils. As the carbon source concentration increases, it appears that the fibril density formed is also higher (from Figure 2.a to 2.c.). The width of the formed fibrils is less than 100 nm in size. This result is in agreement with the previous study [16] that reported of fine fibrils which overlapping each other as a layer of cellulose ribbons.

3.3. Yield

Yield is the result of the percentage of the division between the weight of bacterial cellulose produced and the weight of the starting material. The more sugar concentrations added to the medium, the yield of the resulting bacterial cellulose also increases to a certain concentration. pH is also an important factor for the growth and formation of bacterial cellulose film.

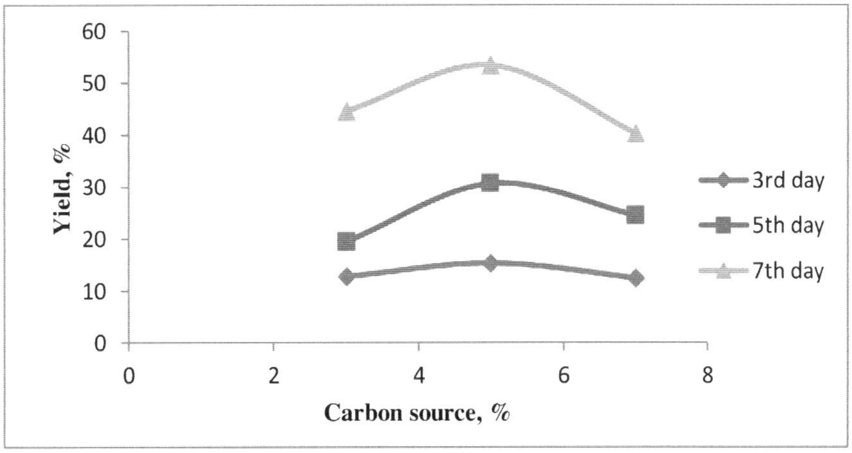

Figure 3. The relationship between the incubation period and the concentration of carbon source on the yield.

Sugar as a carbon source was used in the bacterial cellulose fermentation process. The relationship between culture time and sugar concentration to bacterial cellulose yield can be seen in Figure 1. Figure 1 shows that the addition of 5% of sugar give the maximum yield of bacterial cellulose. The yield of 5% sugar on the 3rd, 5th or 7th day were 15.4%; 30.7%; and 53.4%, respectively. The yield were decreased at 7% of sugar i.e. 12.5%; 24.5%; 40.3%; on the 3rd, 5th or 7th day respectively, while the addition of 3% sugar resulting in low yield. This indicates that the addition of 5% sugar concentration is the optimum condition for bacteria *Acetobacter xylinum* growth. The addition of more than 5% of sugar cause the plasmolysis in these bacterial cells, thus decreasing the formation of cellulose and also resulting in low yield.

3.4. Thickness

Table 1 indicates that the optimum pH in this study was pH 5, in accordance with the literature that the optimum pH for *A. xylinum* growth was pH 5.5 [17].

Table 1. The relationship of pH and bacterial cellulose thickness in addition of 5% sugar concentration for 7 days.

pH	Thickness, cm
3	0.031± 0.001
5	0.072± 0.003
7	0.058± 0.002

The thickness of the bacterial cellulose film obtained at pH 5 is 0.072 cm (Table 1). The thickness of the bacterial cellulose layer at pH 3 was only 0.031 cm and increased up to pH 5. The thickness of the bacterial cellulose layer decreased at pH 7 of 0.058 cm. This means that pH 5 is the optimum pH of the formation of bacterial cellulose film.

3.5. Tensile Strength

The characteristic of the thin layer of bacterial cellulose is highly dependent on the tensile strength and elongation of the thin film. Tensile strength is one of the mechanical properties to measure the strength of the thin film. Tensile strength is the maximum pull force that can be retained by a thin layer during the measurement takes place until the thin layer is disconnected.

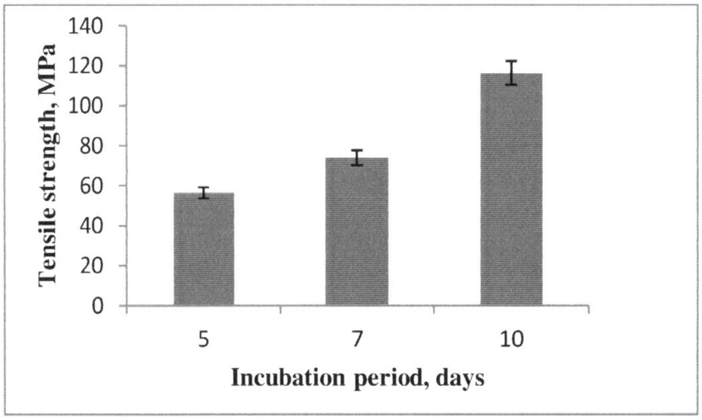

Figure 4. Tensile strength of bacterial cellulose film with culture time of 5, 7 and 10 days.

Figure 4 shows the tensile strength of the bacterial cellulose film. It can be seen that the incubation time for 10 days gives the highest tensile strength of 116.3 MPa, followed by incubation time for 7 days and 5 days respectively of 73.9 MPa and 56.4 MPa. The high and low of tensile strength of bacterial cellulose films can be linked to its morphology structure. A few pores that covered by fibrils string at the bacterial cellulose surface increased the surface area where larger surface area may lead to lower strength, this result is also in agreement with the literature [18].

3.6. Elongation
The elongation at break of the bacterial cellulose films was shown in Fig. 5.

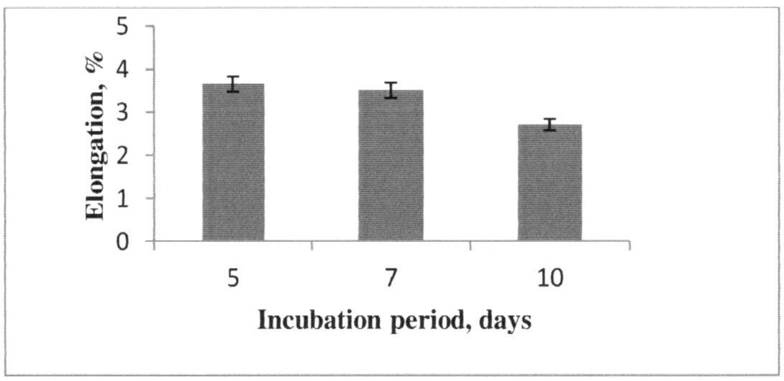

Figure 5. Elongation of bacterial cellulose film with culture time of 5, 7 and 10 days.

Figure 5 shows the elongation of the bacterial cellulose film. It can be seen that 5 days incubation time gives the highest elongation of 3.65%, followed by incubation time for 7 days and 10 days respectively 3.5% and 2.7%. The elongation at break of bacterial cellulose films show the inverted pattern of tensile strength, this result is also in agreement with the literature [18, 19].

3.7. Swelling ability
Swelling is the ability of water to penetrate into the structure of the thin layer of bacterial cellulose. In general, amorphous structures have greater hydrophilic properties than materials with high crystallinity.

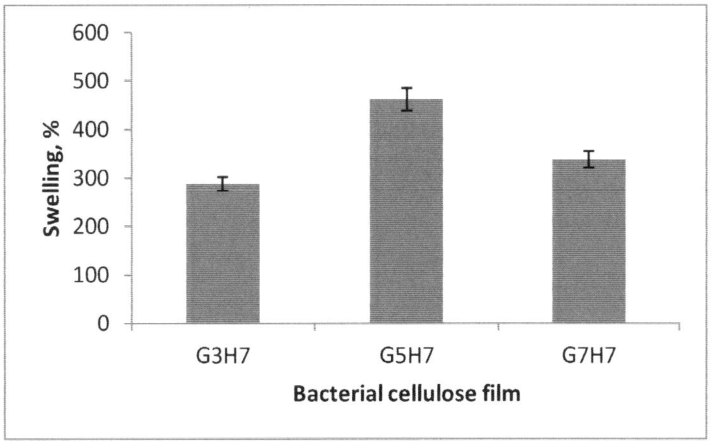

Figure 6. Percentage of swelling from bacterial cellulose film with 7 days incubation time and sugar variation of 3%, 5% and 7% (G3H7, G5H7 and G7H7).

Percentage of swelling ability was determined by Maneerung et al, 2008 method [15] with slight modification, i.e. dried sample with size of 1 x 2 cm was soaked in aquades for 72 hours. Figure 6 shows that the highest percentage of swelling is owned by a thin layer of bacterial cellulose with 5% sugar percentage and 7 days incubation time. The swelling of G3H7 (3% of sugar) and G7H7 (7% of sugar) were lower than G5H7 sample (5% of sugar). This might due to the lower structure of rigidity that samples could not retain more water. This result also in agreement with the yield that the optimum of carbon source is around 5%. As fibrous films, bacterial cellulose can hold water by physically entrapping it in fine capillaries and internal pores [18].

4. Conclusion

This study showed that several parameters for optimum condition to produce bacterial cellulose films from local waste of coconut water had been obtained (i.e. 5% of carbon source; pH 5; 7 day of incubation period). The electron microscopy also showed that dried bacterial cellulose films had pores covered by fibrils on the surface. This result on preliminary study of bacterial cellulose film (surface morphology, mechanical and swelling properties) may be used as a consideration promising candidate for application of bacterial cellulose film in cosmetic and biomedical industries.

Acknowledgement

The authors are grateful for the financial support by "Riset Unggulan LIPI" Project (Indonesian Institute of Sciences) 2017.

References

[1] Delmer D P and Amor Y 1995 Cellulose biosynthesis *Plant Cell* **7** 987–1000
[2] Nishiyama Y, Langan P and Chanzy H 2002 Crystal Structure and Hydrogen-Bonding System in Cellulose Iβ from Synchrotron X-ray and Neutron Fiber Diffraction *J. Am. Chem. Soc.* **124** 9074–82
[3] Ross P, Mayer R and Benziman M, 1991 Cellulose biosynthesis and function in bacteria *Microbiol. Rev.* **55** 35–58
[4] Klemm D, Heublein B, Fink H P and Bohn 2005 A Cellulose: fascinating biopolymer and sustainable raw material *Angew. Chem. Int. Ed.* **44** 3358–3393
[5] Jesus E G, Andres R M and Magno E T 1971 A Study on the isolation and screening of microorganism for production of diverse textured nata *Philip. J. Sci.* **100** 41–53
[6] Zhang S and Luo J 2011 Preparation and properties of bacterial cellulose/alginate blend bio-

fibers *J. Eng Fiber. Fabr.* **6** 69–72

[7] Czaja W, Romanovicz D and Brown R M 2004 Structural investigations of microbial cellulose produced in stationary and agitated culture *Cellulose* **11** 403–411

[8] Klemm D, Schumann D, Udhart U and Marsch S 2001 Bacterial synthesis cellulose- artificial blood , vessel for microsurgery *Prog. Poly. Sci.* **26** 156–1603

[9] Watanabe K, Tabuchi M, Morinaga Y and Yoshinaga F 1998 Structural features and properties of bacterial cellulose produced in agitated culture *Cellulose* **5** 187–200

[10] Czaja W, Krystynowicz A, Bielecki S and Brown Jr R M 2006 Microbial cellulose – the natural power to heal wounds *Biomaterials* **27** 145–151

[11] Trovatti E, Silva N H, Duarte I F, Rosado C F, Almeida I F, Costa P, Freire C S , Silvestre A J, and Neto C P 2011 Bacterial cellulose membranes as supports for dermal release of lidocaine, *Biomacromolecules* **12** 4162–4168

[12] Prashant R C, Ishwar B J, Shrikant A S and Rekha S S 2008 Microbial cellulose: fermentative productions and applications *Food Technol. Biotechnology* **47** 107–124

[13] Vandamme E J, Baets S D, Vanbaelen V, Joris K and Wulf P D 1998 Improved production of bacterial cellulose and its application potential *Polym. Degrad. Stab.* **59** 93–99

[14] Almeida I F, Pereira T, Silva N H C S, Gomes F P, Silvestre A J D, Freire C S R, Sousa Lobo J M, and Costa P C 2014 Bacterial cellulose membranes as drug delivery systems: An in vivo skin compatibility study *Eur. J. Pharm. Biopharm.* **86** 332–336

[15] Maneerung T, Tokura S and Rujiravanit R 2008 Impregnation of silver nanoparticles into bacterial cellulose for antimicrobial wound dressing *Carbohydr. Polym.* **72**: 43–51

[16] Nadia H, Amin M C I M, and Ishak A 2012 Physicochemical properties and characterization of *nata decoco* from local food industries as a source of cellulose *Sains Malaysiana* **41** 205–211

[17] Hwang J W, Yang Y K, Hwang J K, Pyun Y R, and Kim Y S 1999 Effects of pH and dissolved oxygen on cellulose production by Acetobacter xylinum BRC5 in agitated culture *J. Biosci. Bioeng.* **88** 183–188

[18] Pa'e N, Hamid N I A, Khairuddin N, Zahan K A, Seng K F, Siddique B M and Muhamad I I 2014 Effect of different drying methods on the morphology, crystallinity, swelling ability and tensile properties of *Nata De Coco Sains Malaysiana* **43**767–773

[19] Phisalaphong M and Jatupaiboon N 2008 Biosynthesis and characterization of bacteria cellulose-chitosan film *Carbohydr. Polym.***74** 482–488

IOP Conf. Series: Earth and Environmental Science **101** (2017) 012011 doi:10.1088/1755-1315/101/1/012011

Thin Layer Drying Model of Bacterial Cellulose Film

Tri Hadi Jatmiko, Vita Taufika Rosyida, Anastasia Wheni Indrianingsih, and Wuri Apriyana

Research Unit for Natural Product, Indonesia Institute of Sciences, Gunungkidul, DI Yogyakarta, Indonesia

E-mail: trih011@lipi.go.id

Abstract. The bacterial cellulose film produced by *Acetobacter xylinum* using coconut water as a carbon source was dried at a temperature of 60 to 100 C. The drying process of bacterial cellulose film occur at falling rate drying period. Increasing drying temperature will shorten the drying time. The drying data fitted with thin layer drying models that widely used, Newton, Page and Henderson and Pabis models. All thin layer drying models describe the experimental data well, but Page model is better than the other models on all various temperature with coefficients of determination (R^2) range from 0.9908 to 0.9979, chi square range from 0.000212 to 0.000851 and RMSE range from 0.014307 to 0.0289458.

1. Introduction

The interest in bacterial cellulose began to grow since the discovery of the unique physical properties of bacterial cellulose. Applications of bacterial cellulose use were extensive, from the fields of food, medicine, electronics and cosmetics [1–7]. Bacterial cellulose produced by the bacterial species *Acetobacter xylinum* has different properties than cellulose commonly found in plants.Film or sheet from bacterial cellulose has a mechanical strength that is better than film from a synthetic polymer.[8,9]. Bacterial cellulose film can be used as a skincare ingredient, facial mask, food wrapping, drug delivery, wound dressing and as the membrane material [2,6,7,9].

Previous studies have reported that the nata de coco film produced by *Acetobacter xylinum* contains major components of water and cellulose, so for further utilization, the bacterial cellulose needs to be dried. For practical uses, bacterial cellulose in dried form is more portable and convenient than the wet one [7].

The drying of material is a complicated process that needs much energy and time. Drying process at industrial scale consumes up to 15% total energy requirements for entire processes [10-12]. Drying of materials can increase shelf-life and reduces the bulk and mass of the material, thus simplifying storage, packaging, and transport. Sometimes, drying procedure can reduce the product volume, leading to a decrease space requirements for storage. Therefore knowledge of the drying characteristics of the material is necessary. So this research was done to study the drying nature of bacterial cellulose film and found the appropriate model to describe the characteristic drying of bacterial cellulose film.

2. Materials and methods

2.1. Materials
Acetobacter xylinum (strain FNC-0001) obtained from Gadjah Mada University (UGM). Coconut water was from local market at Yogyakarta, Indonesia. Yeast extract, acetic acid, sucrose, distilled water, NaOH, urea, universal pH paper, plate count agar were from Merck.

2.2. Bacterial cellulose film production
The production of bacterial cellulose adopted from Radiman and Yuliani (2008) with slight modifications [13]. The culture medium used to produce bacterial cellulose consist of 1 liter of coconut water, 50 g of sucrose, 10 g of urea, starter of *A. xylinum* (10%) and the pH is set to 5 with acetic acid. The media was then sterilized by autoclave at 121 °C for 15 minutes. Bacterial cellulose was grown at static conditions for 10 days at room temperature. After incubation, the bacterial cellulose was boiled in distilled water for 2 hours, and washed in running water until neutral.

2.3. Drying process
The bacterial cellulose film with a diameter of 3.25 cm was dried in moisture analyzer A&D MX-50 at a temperature of 60, 70, 80 and 100 °C. The drying data was collected automatically using PC with A&D WinCT-Moisture software installed.

2.4. Mathematical Modelling
The moisture ratio (MR) of the bacterial cellulose was calculated using the following equation:

$$MR = \frac{M_t - M_e}{M_0 - M_e} \quad (1)$$

where M_t is the moisture content at any time of drying, M_0 is initial moisture content and M_e is equilibrium moisture content. The experimental data fitted with widely used thin layer drying model i.e. Newton, Page and Henderson and Pabis models [12,14–19].

Newton model is also known as Exponential or Single exponential model or the Lewis model. The Newton model is the simplest thin layer model because it has only one constant.

$$MR = \exp(-kt) \quad (2)$$

where MR is moisture ratio, k constant and t is time.

The Page model is a modification of the Newton model to overcome the deficiencies of the model by adding the dimensionless empirical constants (n) at the time parameter, so this model gives better moisture loss prediction results.

$$MR = \exp(-kt^n) \quad (3)$$

where MR is moisture ratio, k, n constant and t is time.

Henderson and Pabismodel is based on Fick's second law of diffusion and widely used to predict the drying nature of biological materials.

$$MR = a\exp(-kt) \quad (4)$$

where MR is moisture ratio, a, k constant and t is time.

A widely used statistical evaluation to assess goodness of fit of the mathematical equations to the experimental data are correlation coefficient (R^2), root mean square error (RMSE) and chi-square (χ^2) [12,16,19–22]. The correlation coefficient (R^2) is widely used for selecting the best equation to describe the equation of the drying curve and the highest r value is required. Another criterion for selecting the best model is chi-square, and root mean square error (RMSE). The chi-square value is the mean square of the deviations between the experimental and predicted values for the models and was used to determine the goodness of the fit. The lower the values of the chi-square were, the better the goodness of the fit. The RMSE value shows the deviation between the experimental and predicted values.

$$\chi^2 = \frac{\sum_{i=1}^{N}(MR_{exp,i}-MR_{pre,i})^2}{N-z} \quad (5)$$

$$RMSE = \left[\frac{1}{N}\sum_{i=1}^{N}(MR_{pre,i} - MR_{exp,i})^2\right]^{1/2} \quad (6)$$

whereMR_{exp} is moisture ratio from experimental data, MR_{pre} is moisture ratio prediction, N is number of data observed, z is number of constant parameter and t is time.

3. Results and Discussion

The experimental results of the cellulose drying process show that the moisture content and the drying rate decrease with the drying time. The moisture loss is high at the initial period and becomes slow in the last period of drying process. The rate of drying is also high during the initial period but becomes very slow in the last stage of the drying process.In the early period of drying, free moisture content located on the surface of the material is still high. The free water content at the surface of the material is very easily released into the environment resulting in a large decreased of moisture content from material at the beginning of drying period.At a later stage, the moisture from within the material move to the surface by diffusion resulting in a mass transfer of water from the material to the environment becomes slower [23].

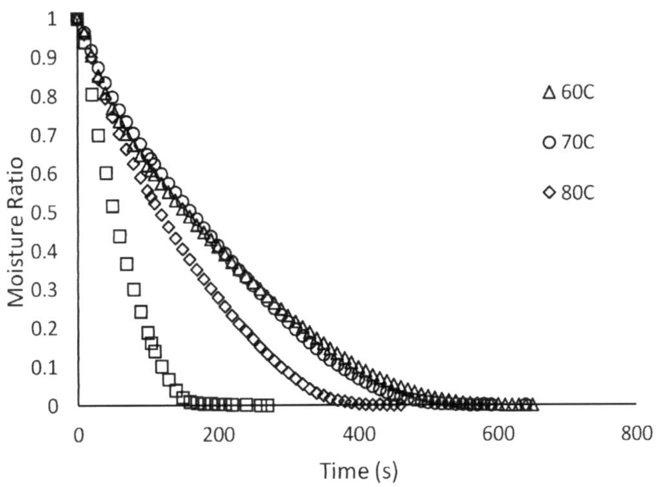

Figure 1. Moisture ratio over time in different drying temperatures

Figure 1 shows the drying kinetic of bacterial cellulose film at different temperatures. Figure 1 indicates that the drying process becomes shorter as the drying temperature increases. Previous studies on drying of biological materials show the same result [17,20,22]. Increasing the drying temperature would increase the drying rate, so at higher drying temperature, drying process becomes faster. Increasing the drying temperature from 60 to 100 °C shortened the drying time from 664 minutes to 291 minutes.

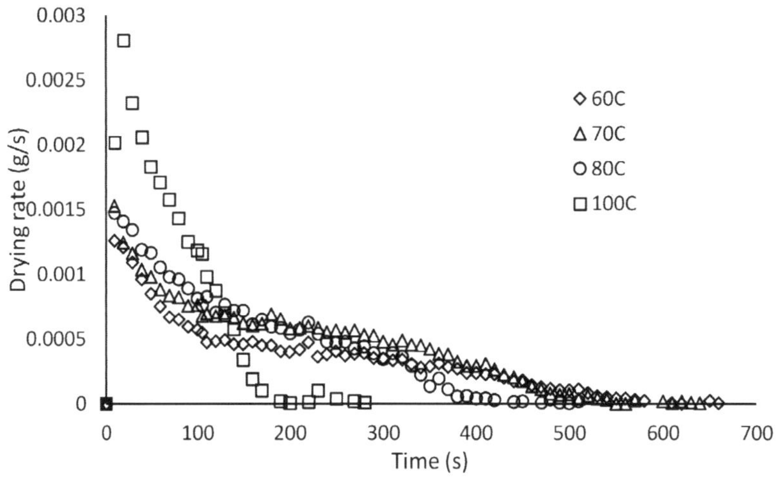

Figure 2. Drying rate of bacterial cellulose film

Figure 2 shows the graph of the drying rate relationship over time, where the drying rate decreases over the drying time. From Fig. 2, it can be seen that the drying process of bacterial cellulose occurs in the falling rate drying period. According to previous studies, the biological materials commonly dried at falling rate drying period [11,22,24]. The drying process at a falling rate period is more noticeable at 100 °C than at lower temperature. A high drying temperature causes the process of moving moisture from the surface of the material to the environment to be faster than the moisture transfer in the material to the surface of the material. Fan et. al. (2011) found this phenomenon generally occur in material with high moisture content [17].

Table 1. Parameter of thin layer model

Model		60°C	70°C	80°C	100°C
Newton	k	0.005073	0.005128	0.006829	0.015831
	R^2	0.9892	0.9864	0.9909	0.9901
	χ^2	0.001422	0.002557	0.001962	0.002436
	RMSE	0.037564	0.050374	0.044085	0.048931
Page	k	0.002097	0.001026	0.001757	0.002807
	n	1.160085	1.292747	1.26007	1.396144
	R^2	0.9908	0.9918	0.9941	0.9979
	χ^2	0.00085	0.000851	0.000607	0.000212
	RMSE	0.028927	0.0289458	0.024411	0.014307
Henderson &Pabis	a	1.03548	1.075744	1.075726	1.121695
	k	0.005244	0.005486	0.007304	0.017522

R^2	0.987	0.9815	0.986 7	0.987
χ^2	0.001 33	0.0021 01	0.001 533	0.001 498
RM SE	0.036 195	0.0454 795	0.038 776	0.038 04

The thin layer drying models used show good prediction for the bacterial cellulose drying process, with R^2 range from 0.9815 to 0.9979, χ^2 range from 0.000212 to 0.002557 and RMSE range from 0.014307 to 0.050374 as shown in Table 1. The Newton, Page and Henderson and Pabis models could use to fit with experimental data very well. The Newton model has R^2 value 0.9864-0.9909, χ^2 0.001422-0.002557 and RMSE 0.037564-0.050374. The Page model has R^2 value 0.9908-0.9979, χ^2 0.000212-0.000851 and RMSE 0.014307-0.0289458. The Henderson & Pabis model has R^2 value 0.9815-0.987, χ^2 0.00133-0.002101 and RMSE 0.036195-0.0454795. From the fitting assessment value between experimental data and models as shown in Table 1, it's clear that Page model was better than Newton and Henderson &Pabis models. These results confirm the results of previous studies that Page equations are well suited to predict the drying of biological materials [11].Furthermore American Standard has adopted this model as their standard for thin layer drying model of agricultural and biological products (ANSI/ASAE 2014)[11].

4. Conclusion

The drying of bacterial cellulose film takes place in the falling drying period and is strongly influenced by temperature. The thin layer drying equation can predict the kinetic drying of bacterial cellulose very well. The thin layer drying model proposed by Page give best prediction of drying bacterial cellulose film.

Acknowledgment

This study was financially supported by Indonesian Ministry of Research, Technology, and Higher Education through INSINAS Grant 2017 No: 04/INS-2/PPK/E/E4/2017.

References

[1] Esa F, Tasirin S M and Rahman N A 2014 Overview of bacterial cellulose production and application *Agric. Agric. Sci. Procedia* **2** 113–9

[2] Keshk S M 2014 Bacterial cellulose production and its industrial applications *J. Bioprocess. Biotech.* **4**

[3] Jozala A F, de Lencastre-Novaes L C, Lopes A M, de Carvalho Santos-Ebinuma V, Mazzola P G, Pessoa-Jr A, Grotto D, Gerenutti M and Chaud M V 2016 Bacterial nanocellulose production and application: a 10-year overview *Appl. Microbiol. Biotechnol.*

[4] Leitão A, Silva J, Dourado F and Gama M 2013 Production and characterization of a new bacterial cellulose/poly (vinyl alcohol) nanocomposite *Materials (Basel).* **6** 1956–66

[5] Almeida I F, Pereira T, Silva N H C S, Gomes F P, Silvestre A J D, Freire C S R, Sousa Lobo J M and Costa P C 2014 Bacterial cellulose membranes as drug delivery systems: An in vivo skin compatibility study *Eur. J. Pharm. Biopharm.* **86** 332–6

[6] Ullah H, Santos H A and Khan T 2016 Applications of bacterial cellulose in food, cosmetics and drug delivery *Cellulose* **23** 2291–314

[7] Wei B, Yang G and Hong F 2010 Preparation and evaluation of a kind of bacterial cellulose dry films with antibacterial properties *Carbohydr. Polym.* **84** 533–8

[8] Yamanaka S, Watanabe K, Kitamura N, Iguchi M, Mitsuhashi S, Nishi Y and Uryu M 1989 The structure and mechanical properties of sheets prepared from bacterial cellulose *J. Mater. Sci.* **24** 3141–5

[9] Iguchi M, Yamanaka S and Budhiono a. 2000 Bacterial cellulose — a masterpiece of nature ' s arts *J. Mater. Sci.* **35** 261–70
[10] Barbosa de Lima A G, da Silva J V., Pereira E M A, dos Santos I B and de Lima W M P B 2016 Drying of Bioproducts: Quality and Energy Aspects pp 1–18
[11] Onwude D I, Hashim N, Janius R B, Nawi N M and Abdan K 2016 Modeling the thin-layer drying of fruits and vegetables: A review *Compr. Rev. Food Sci. Food Saf.* **15** 599–618
[12] Erbay Z and Icier F 2010 A review of thin layer drying of foods: theory, modeling, and Eexperimental results *Crit. Rev. Food Sci. Nutr.* **50** 441–64
[13] Radiman C and Yuliani G 2008 Coconut water as a potential resource for cellulose acetate membrane preparation *Polym. Int.* **57** 502–8
[14] Kucuk H, Midilli A, Kilic A and Dincer I 2014 A Review on thin-layer drying-curve equations *Dry. Technol.* **32** 757–73
[15] Ertekin C and Firat M Z 2017 A comprehensive review of thin-layer drying models used in agricultural products *Crit. Rev. Food Sci. Nutr.* **57** 701–17
[16] Fernando J A K M and Amarasinghe A D U S 2016 Drying kinetics and mathematical modeling of hot air drying of coconut coir pith. *Springerplus* **5** 807
[17] Fan H, Wu Y, Hu X, Wu J and Liao X 2011 Characteristics of thin-layer drying and rehydration of nata de coco *Int. J. Food Sci. Technol.* **46** 1438–44
[18] Onwude D I, Hashim N, Janius R B, Nawi N M and Abdan K 2016 Modeling the Thin-Layer drying of fruits and vegetables: A Review *Compr. Rev. Food Sci. Food Saf.* **15** 599–618
[19] Jan K, Cs R and Dc S 2014 Food Processing & amp; technology mathematical modelling of thin layer drying kinetics of biodegradable pellets *J Food Process Technol* **5**
[20] Ertekin C and Yaldiz O 2004 Drying of eggplant and selection of a suitable thin layer drying model *J. Food Eng.* **63** 349–59
[21] Hii C L, Law C L and Cloke M 2008 Modelling of thin layer drying kinetics of cocoa beans during artificial and natural drying *J. Eng. Sci. Technol.* **3** 1–10
[22] Doymaz İ 2011 Thin-layer drying characteristics of sweet potato slices and mathematical modelling *Heat Mass Transf.* **47** 277–85
[23] Fithriani D, Assadad L and Arifin Z 2017 Karakteristikdan Model Matematika Kurva Pengeringan Rumput Laut *Eucheuma cottonii J. Pascapanen dan Bioteknol. Kelaut. Dan Perikan.* **11** 159
[24] Ertekin C and Firat M Z 2017 A comprehensive review of thin-layer drying models used in agricultural products *Crit. Rev. Food Sci. Nutr.* **57** 701–17

Drying Characteristics and Water-soluble Polysaccharides Evaluation of Kidney Shape *Ganoderma lucidum* Drying in Air Circulation System

D J Prasetyo[1], T H Jatmiko[1], C D Poeloengasih[1], M Kismurtono[1]

[1]Research Unit for Natural Product Technology, Indonesian Institute of Sciences, Yogyakarta, Indonesia 55861

E-mail: prasetyo_dwijoko@yahoo.co.id

Abstract. In this project, drying kinetic of kidney shape *Ganoderma lucidum* fruiting body in air circulation system was studied. The drying experiments were conducted at 40, 50 and 60°C with air flow rate of 1.3 ms^{-1}. Samples were weighted periodically until no change in sample weight was recorded, and then the samples were analyzed for its moisture content. Four different thin-layer mathematical models (Newton, Page, Two-term, Midilli) were used and compare to evaluate the drying curves of kidney shape *G. lucidum*. The water-soluble polysaccharides were evaluated in order to find the best drying temperature condition. The results indicates that Midilli model was the fittest model to describe the characteristic of kidney shape *G. lucidum* in the air circulation drying system and temperature of 50°C was the best drying condition to get highest value of water-soluble polysaccharides.

1. Introduction

Ganoderma lucidum (Fr.) Karst (Polyporaceae), is a traditional Chinese mushroom which commonly known as Ling-Zhi in Chinese and Reishi in Japanese language. *G. lucidum* contains some bioactive ingredients, such as water soluble polysaccharides, triterpenoids, adenosine, and lectin. Previous studies have reported that water soluble polysaccharides in Ganoderma fruiting body have pharmacological effects as anti-tumor (through immunomodulation and anti-angiogenesis) and antioxidant [1].

As an agricultural commodity with high catabolism process, preservation of G. lucidum is a must. Drying is a common method to preserve *G. lucidum*. Several drying methods have been conducted to preserve *G. lucidum*, *i.e.*, drying under direct sunlight with and without black fabric coverage [2], drying in the shades [2], drying using oven without circulation [2-4], and drying using oven with air circulation [2-6]. To find an appropriate drying method, some drying aspects such as drying time, energy consumption and minimum quality reduction need to be considered. Drying under direct sunlight is the easiest and cheapest drying method. However, this method has some drawbacks, such as unhygienic aspect and poor quality product [7]. Meanwhile, the drying process of *G. lucidum* under the shades or in direct sunlight with black fabric coverage needs more time to reach its equilibrium moisture content [2]. On the other hand, the drying process using oven with and without air circulation would reduce the drying time in which drying with air circulation will require a shorter time [2-4]. According to previous studies, it was found that oven drying with air circulation was a better method than others to preserve *G. lucidum* [2]. It was also found that drying *G. lucidum* in kidney shaped resulted a product with better quality than in sliced shape [2].

Optimization of drying methods including drying conditions, machine design, and product quality is important to minimize losses in the drying process. The evaluation of drying characteristics and determination of drying time are necessary for drying optimization [8]. Since the drying process of *G. lucidum* in oven with air circulation was found to be an appropriate method [2], it is important to evaluate its drying characteristics and the effect of temperature on dried *G. lucidum* quality (in water-soluble polysaccharides content). Therefore, the aim of this study was to investigate the drying characteristics and water-soluble polysaccharides contents of *G. lucidum* dried by oven with air circulation.

2. Materials and methods

2.1. Materials

Kidney shape *G. lucidum* mature fruiting body was obtained from cultivation area in Gunungkidul, Yogyakarta, Indonesia. The fruiting body was swiped to remove dirt soon after being harvested, then packed in plastic bag in order to prevent evaporation of its volatile content. The fruiting body was taken from plastic bag soon before drying process.

2.2. Drying Procedure

Drying process was performed using air circulation drying technique at hot air temperatures of 40, 50, and 60°C with air velocity of 1.30 ms^{-1} in a laboratory scale hot air circulation oven (Memmert, Germany). The oven was conditioned into selected temperature for about 30 minutes before the samples were placed in. The weight of samples were measured at initial condition, at 1 hour interval (at 1st until 12th hour), at 2 hour interval (after 12th until 24th hour), and 3 hour interval (after 24th hour). The drying process was stopped after the weight of samples reach constant. The data of sample weight was then converted into moisture content. Moisture content was stated on dry basis in unit of gram of water per gram of bone dry sample.

2.3. Mathematical Modelling

Drying curves of this experiment were fitted using non-linear least square regression solved by Microsoft Excel into four models of moisture ratio thin–layer drying. The chosen models were Newton, Page, Two-term, and Midilli (Table 1). The fitted curves were evaluated from the root mean square error (RMSE), coefficient of correlation (r) and reduced chi square (χ^2). The better goodness of fitted curves were indicated with higher value of r and lower value of RMSE and χ^2.

Table 1. Thin-layer Drying Models Applied to Drying Curves

Model name	Model equation
Newton[3]	MR = exp(-kt)
Page[3]	MR = $exp(-kt^n)$
Two-term[3]	MR = a exp(-kt) + (a_1) exp($-k_1t$)
Midilli[3,9]	MR = a $exp(-kt^n)$ + a_1t

2.4. Water-soluble Polysaccharides Content Evaluation

The evaluation of total water-soluble polysaccharides were based on the color of reaction of polysaccharides and their derivatives with phenol and concentrated sulfuric acid [10-12]. Total water-soluble polysaccharides were determined as total glucose as a results of polysaccharides hydrolysis. The sample of *G. lucidum* were ground into powder using a mechanical grinder. The polysaccharides were extracted in hot water at temperature of 95°C with the addition of hydrochloric acid 2 M to accelerate the extraction rate. The 1 mL of extract solution were pipetted and 0.5 mL of 5% phenol solution was added. The mixture was shaken for 2 min. The 2.5 mL of 98% v/v sulfuric acid was added into mixture and were shaken for another 5 min. The determination of total water-soluble polysaccharides concentration were conducted by measuring the absorbance of solution at wavelength 490 nm. Standard glucose (Sigma, Milwaukee, WI, USA) was used as standard solution and distilled water was used as blank solution for plotting the standard curve.

3. **Results and discussions**

3.1. Drying Rate

The average drying rate of kidney shape *Ganoderma lucidum* at air velocity 1.30 ms^{-1} were ranged from 0.0373-0.0536 g_{water} $g_{drymatter}^{-1}$ h^{-1} at temperature 40, 50 and 60°C. The drying rate as function of time was shown in Figure 1. The higher drying rate was obtained at initial phase of drying and then decrease as drying time reach end phase. The results showed that there was no constant drying rate period. Falling rate drying period dominated this process. This result was in line with previous study in oyster mushroom drying conducted by Tulek [13].

The effect of moisture content and drying temperature on drying rate was shown in Figure 2. The highest drying rate was obtained at drying temperature of 60°C. Drying time was also decrease as drying temperature increased. As drying temperature increased, the temperature difference between *G. lucidum* body and air temperature was increasing. This phenomenon stimulate the increasing of evaporation driving force which further enhanced the drying rate [3]. Drying rate would reach its maximum as drying temperature increase [14]. Drying process reached minimum time (48 hours) at temperature of 50 and 60°C. Despite of same drying time, drying temperature of 60°C slightly had lower final moisture content (0.1247 g_{water} $g_{drymatter}^{-1}$) than at 50°C (0.1261 g_{water} $g_{drymatter}^{-1}$). Further increasing of drying temperature would decrease drying time. However further increasing of drying temperature was not recommended as the harmful effect on product quality [3,13]. Therefore in this study, drying process at temperature of 50°C obtained better result than 60°C.

Moisture content of *G. lucidum* also affected its drying rate. Drying rate decreased as moisture content of *G. lucidum* decreased. Similar results were obtained at each drying temperature, this phenomenon was in line with previous study in eggplant drying conducted by Ertekin and Yaldiz [14]. Drying rate reached its highest value at the start of drying process which moisture content was still high, and then decrease until its lowest value at the end of drying process which moisture content of samples reached equilibrium moisture content. At the start of drying process, the kidney shape *G. lucidum* still had large amount of moisture at its surface which accelerated drying rate. As time went on, moisture content in surface area decrease and then it triggered moisture in the interior part of the body to travel into surface area. At this stage, moisture required longer distance which it increased time required by moisture before being evaporated at the surface.

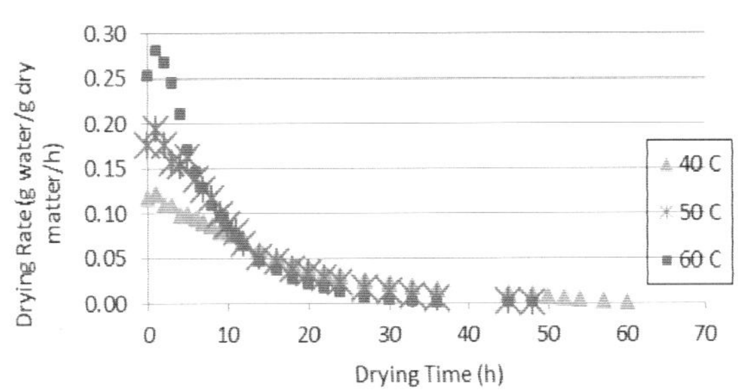

Figure 1. Drying rate as function of time

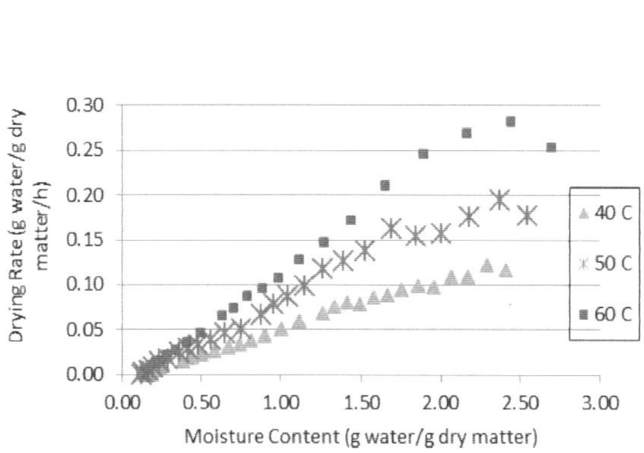

Figure 2. Effect of drying temperature and moisture content on drying rate

3.2. Evaluation of thin layer mathematical models

The correlation between moisture ratio (MR) and drying time (t) of kidney shape *G. lucidum* was shown in Figure 3. These correlations then were fitted into four thin layer drying models (Newton, Page, Two-term, Midilli). The evaluation of each model were shown in Table 2. The values of r for all models were ranged from 9.9967 x 10^{-1} to 9.9996 x 10^{-1}. Meanwhile RMSE were varied from 1.9161 x 10^{-4} to 1.1597 x 10^{-2} and χ^2 from 2.1959 x 10^{-9} to 2.0008 x 10^{-6}. The results showed that Midilli model was found the most fitted model. It has highest r value and lowest value of RMSE and χ^2. This results were in line with drying of *Ganoderma tsugae* conducted by Chin et al. [3]. Similar results was also found in previous results at drying of some agricultural products [13-15].

Table 2. Evaluation of Thin-layer Drying Models

Drying Condition	**Model**	**Model Parameters**		**R**	**RMSE**	χ^2
40°C	Newton	k = 0.0611		9.9967E-01	1.3886E-02	2.0008E-06
	Page	k = 0.0487	n = 1.0829	9.9973E-01	1.5197E-02	1.4247E-07
	Two-term	a = 4.1761	k = 0.0850	9.9978E-01	9.8347E-03	9.4962E-08
		a_1 = -3.1781	k_1 = 0.0951			
	Midilli	a = 1.0006	k = 0.0528	9.9996E-01	1.9161E-04	2.1959E-09
		n = 1.0424	a_1 = -0.0004			
50°C	Newton	k = 0.0938		9.9972E-01	2.2044E-03	6.6791E-07
	Page	k = 0.0791	n = 1.0721	9.9983E-01	4.5753E-04	3.0199E-08
	Two-term	a = 1.9716	k = 0.1170	9.9979E-01	1.7989E-03	4.8842E-08
		a_1 = -0.9620	k_1 = 0.1483			
	Midilli	a = 1.0062	k = 0.1216	9.9984E-01	1.7143E-03	2.1714E-08
		n = 1.0539	a_1 = -0.0001			
60°C	Newton	k = 0.1322		9.9978E-01	1.9642E-03	2.4667E-07
	Page	k = 0.1168	n = 1.0592	9.9986E-01	1.9807E-03	1.9175E-08
	Two-term	a = 1.0452	k = 0.1388	9.9995E-01	4.1431E-03	3.3257E-09
		a_1 = -0.0452	k_1 = 2.0649			
	Midilli	a = 1.0089	k = 0.1216	9.9987E-01	2.1362E-03	1.0750E-08
		n = 1.0440	a_1 = 0.0000			

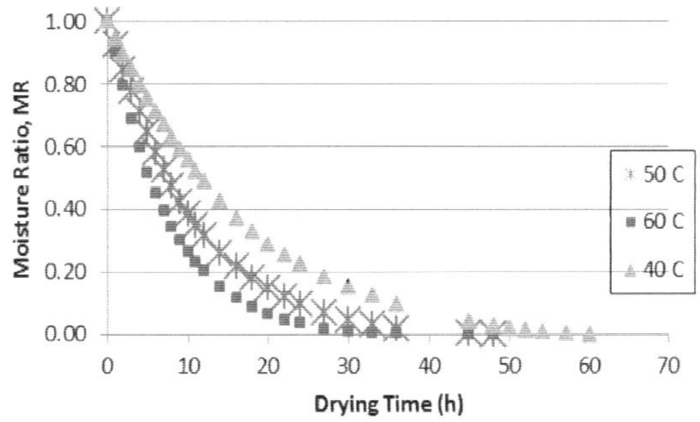

Figure 3. Correlation between moisture content (MR) and drying time (t)

3.3. Evaluation of Water-soluble Polysaccharides

The results of water-soluble polysaccharides content at different drying temperature were shown in Table 3. The worst results was found at drying temperature of 40°C which had lowest water-soluble polysaccharides retention. This result was possibly affected by the long drying time which caused *G. lucidum* exposed with hot air longer than other condition and reduced more polysaccharides content. The highest water-soluble polysaccharides retention was found at drying temperature of 50°C and shortest drying time were found at temperature 50 and 60°C. Therefore the drying temperature of 50°C was the best drying condition of kidney shape *G. lucidum*.

Table 3. The water-soluble polysaccharides of *G. lucidum* at different drying temperature

Drying temperature	Total drying time (h)	Water-soluble polysaccharides (% w/w DW)	Retention (%)
Fresh sample	–	130.79 ± 0.52	100
40°C	60	95.97 ± 0.71	73.38
50°C	48	124.01 ± 0.59	94.82
60°C	48	107.06 ± 0.72	81.86

Each value is expressed as mean ± standard deviation (n= 9)

4. Conclusion

Drying process of kidney shape *Ganoderma lucidum* in air circulation drying system has been conducted. The drying rate was affected by drying temperature. The drying rate increased as drying temperature increase. Kinetic data of this drying process were collected and then evaluated in four well-known thin layer drying model in order to determine the best fitted model. The Midilli model was found as the best model to interprete drying characteristic of kidney shape Ganoderma lucidum. The best drying condition in this study was found at temperature of 50°C which had highest water-soluble polysaccharides content.

Acknowledgments

This work was funded by Lembaga Ilmu Pengetahuan Indonesia through Penguatan Kompetensi 2017 program. The authors are grateful to Deputy of Engineering Sciences – Indonesian Institute of Sciences for support to this program.

References

[1] Boh B, Berovic M, Zhang J and Zhi-bin L 2007 *Biotechnol Annu Rev* **13** 265-301

[2] Hayati S N, Rosyida V T, Apriyana W, Darsih C, Poeloengasih C D and Hernawan 2016 *AIP Conf. Proc.* **1755** 130004
[3] Chin S K, Law C L, Supramaniam C V and Cheng P G 2009 *Dry Technol* **27** 975-84
[4] Chin S K, Law C L and Cheng P G 2011 *Int J Pharm Pharm Sci* **3** 38-43
[5] Fan L, Li J, Deng K and Ai L 2012 *Carbohyd Polym* **87** 1849-54
[6] Chin S K and Law C L 2012 *Int J Sci Res Publ* **2** 1-11
[7] Gothandapani L, Parvathi K and Kennedy Z J 1997 *Dry Technol* **15** 1995-2004
[8] Onwude D I, Hashim N, Janius R B, Nawi N M and Abdan K 2016 *Compr Rev Food Sci Food Safety* **15** 599-618
[9] Midilli A, Kucuk H and Yapar Z 2002 *Dry Technol* **20** 1503-13
[10] Walker G M 1992 *Yeast Physiology and Biotechnology* (New York: John Wiley and Sons) pp 122-31
[11] Poedjiadi A 1994 *Dasar-dasar Biokimia* (Jakarta: UI Press) pp 8-50, 81-124
[12] Holme D J and Hazel P 1993 *Analytical Biochemistry 2nd Ed* (New York: John Wiley and Sons Inc.) pp 338-431, 408-11
[13] Tulek Y 2011 *J Agr Sci Tech* **13** 655-64
[14] Ertekin C and Yaldiz O 2004 *J Food Eng* **63** 349-59
[15] Darvishi H, Asl A R, Asghari A, Azadbakht M, Najafi G and Khodaei J 2014 *J. Saudi Soc Agric Sci* **13** 130-8

The extraction of essential oil from patchouli leaves (*Pogostemon cablin* Benth) using microwave hydrodistillation and solvent-free microwave extraction methods

D K Y Putri, H S Kusuma, M E Syahputra, D Parasandi and M Mahfud

Department of Chemical Engineering, Institut Teknologi Sepuluh Nopember, Surabaya, Indonesia, 60111

E-mail address: dittakharisma21@gmail.com/mahfud@chem-eng.its.ac.id

Abstract. Patchouli plant (*Pogostemon cablin* Benth) is one of the important essential oil-producing plant, contributes more than 50% of total exports of Indonesia's essential oil. However, the extraction of patchouli oil that has been done in Indonesia is generally still used conventional methods that require enormous amount of energy, high solvent usage, and long time of extraction. Therefore, in this study, patchouli oil extraction was carried out by using microwave hydrodistillation and solvent-free microwave extraction methods. Based on this research, it is known that the extraction of patchouli oil using microwave hydrodistillation method with longer extraction time (240 min) only produced patchouli oil's yield 1.2 times greater than solvent-free microwave extraction method which require faster extraction time (120 min). Otherwise the analysis of electric consumption and the environmental impact, the solvent-free microwave extraction method showed a smaller amount when compared with microwave hydrodistillation method. It is conclude that the use of solvent-free microwave extraction method for patchouli oil extraction is suitably method as a new green technique.

1. Introduction

Essential oil is one of the potential agroindustry export commodities that can be income for Indonesia . World export -import statistic shows that the average growth of essential oil consumption and its derivatives of 5-10% per year. The increase was mainly driven by the growing need for food flavoring, cosmetic, and fragrance industries [1].

Pogostemon cablin Benth., known as patchouli, from southeast Asia has been extensively cultivated in Indonesia, Philippines, Malaysia, China, and Brazil. Essential oil produced by patchouli are one of the most important natural ingredients used in the perfume industry [2]. Patchouli plant (*Pogostemon cablin* Benth) is one of the important essential oil-producing plant, contributes more than 50% of total exports of Indonesia's essential oil. Almost all patchouli cultivation in Indonesia is a community crop involving approximately 36,461 farmers. In Indonesia, patchouli production centers are located in Bengkulu, West Sumatra, North Sumatra and Nanggroe Aceh Darussalam, and then spread to Lampung, West Java, Central Java, East Java and other provinces. The area of patchouli cultivation in 2002 was about 21,602 ha, but the oil productivity was still low with an average of 97.53 kg / ha / year. Nevertheless, Indonesia is the largest supplier of patchouli oil in the world market with a contribution of about 90%. Besides Indonesia's competitors as a supplier of patchouli oil in the world market is Malaysia, Filiphina, India, and China [3].

In the case of distillation technology it can be seen that the extraction of patchouli oil done by patchouli farmer group in Kuningan, West Java, used steam distillation system with capacity of 100 kg per boiler. The result of this distillation is 2.2 kg to 2.8 kg of patchouli oil for one patchouli oil extraction for eight hours. For once patchouli oil extraction by the steam distillation method it consumes 40 liters of kerosene [4].

Therefore, new green techniques should be considered in the extraction of essential oils with the minimum use of energy, solvent, and time. Several new methods have been developed to extract essential oils, one of these methods is using microwave (microwave-assisted extraction).

Previous research has shown that extraction with microwave is an alternative that can be developed more than conventional methods, due to high levels of product purity, minimal solvent usage, and short processing times [5]. Some extraction methods with microwave have been developed were microwave hydrodistillation method which combine hydrodistillation with microwave heating [6] and microwave steam distillation method which combine steam distillation with microwave heating [7]. Based on previous research, the extraction of essential oil from patchouli leaves has also been done by using microwave hydrodistillation method [8].

Further development of microwave hydrodistillation method is solvent-free microwave extraction method. This method has advantages over conventional methods such as have a faster extraction rate, yield, and also higher purity of extract because it does not require a solvent so it does not contact with chemicals. Based on the GC/MS (Gas Chromatography / Mass Spectrometry) analysis, the solvent-free microwave extraction method does not change the chemical components present in essential oils, and this method can be categorized as green technology because it reduces the energy requirement per mL from essential oil extraction [9].

So based on that, patchouli oil will be carried out by using microwave hydrodistillation and solvent-free microwave extraction method, which will then be compared to yield obtained, the necessary costs and the environmental impact of the extraction process from both methods.

2. Material and methods

2.1 Materials and chemicals

The materials used in this research were patchouli leaves (*Pogostemon cablin* Benth) obtained from Tulungagung, East Java, Indonesia. All samples were air dried with moisture content of 16.4%, intact size and stored at room temperature until required. The distilled water and anhydrous sodium sulphate used in the experimental work were all of analytical grade.

2.2 Solvent-free microwave extraction method

In employing solvent-free microwave extraction, we used a domestic microwave oven (EMM-2308X, Electrolux, maximum delivered power of 800 W) with wave frequency of 2450 MHz. The dimensions of the PTFE-coated cavity of the microwave oven were 48.5 cm × 37.0 cm × 29.25 cm. The microwave oven was modified by drilling a hole at the top. A round bottom flask with a capacity of 1000 mL was placed inside the oven and was connected to the three-way adapter and liebig condenser through the hole. Then, the hole was closed with PTFE to prevent any loss of the heat inside. The schematic tool can be seen on figure 1.

In the patchouli oil extraction using solvent-free microwave extraction, patchouli leaves were soaked in water for 30 min, and then removed the excess water. The wetted leaves (moisture content = 87.4%) were put in a series of round bottom flasks (1000 mL) with ratio of feed to distiller was 0.06 $g.mL^{-1}$. The extraction was carried out for 120 min under microwave power 450 W. The extraction time shall be computed from the first drop out of the condenser. The essential oils were separated using a separating funnel. To remove water, the extracted essential oils were then dried over anhydrous sodium sulphate, weighed and stored in amber vials at 4°C until they were used for analysis. The yield of patchouli oil was found by taking into moisture content factors by the following equation [10]:

$$\text{Extraction yield } (\%, w/w) = \frac{\text{Mass of extracted essential oil}}{\text{Mass of material (1-moisture content factors)}} \times 100\ \% \quad (1)$$

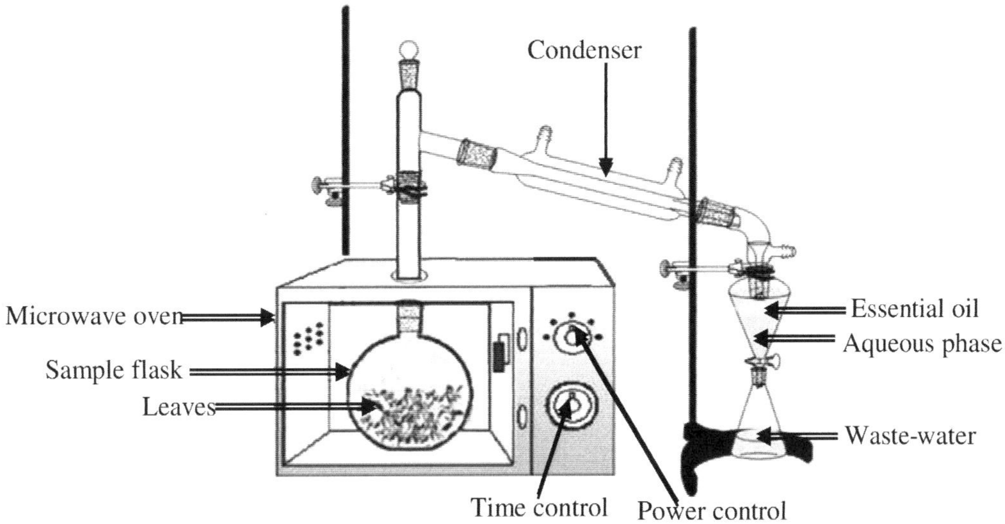

Figure 1. Schematic tool for extraction of patchouli oil using solvent-free microwave extraction method

2.3 Microwave hydrodistillation method

Patchouli leaves (60 g) were placed in a 1 L flask containing deionized water (200 mL). The flask was setup within the microwave oven cavity and a condenser was used on the top (outside the oven) to collect the extracted essential oils (figure 2). The microwave oven was operated at power 450 W for a period of 240 min. This period was sufficient to extract all the essential oils from the sample. The essential oils were separated using a separating funnel. To remove water, the extracted essential oils were then dried over anhydrous sodium sulphate, weighed and stored in amber vials at 4 °C until they were used for analysis.

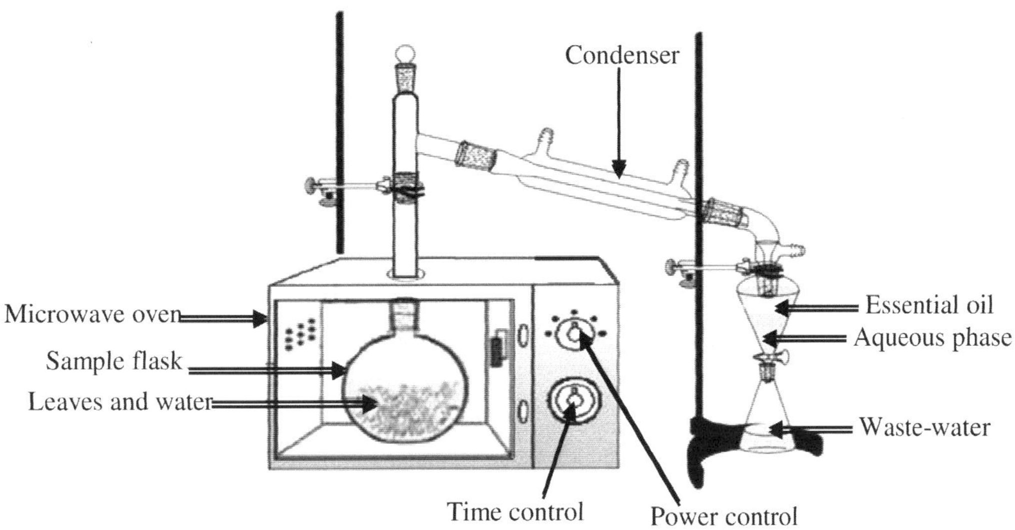

Figure 2. Schematic tool for extraction of patchouli oil using microwave hydrodistillation method

2.4 Electric consumption

The electric consumption of the different extraction methods was calculated based on the infuence of power consumption and extraction time. The general equation for electric consumption is described by equation (2):

$$E_c = \frac{Pt}{3\,600\,000} \quad (2)$$

where E_c is electric consumption (kW h), P is power consumption (W) and t is time (s).

Additionally, the relative electric consumption of the different extraction methods could be expressed by equation (3):

$$E_c^* = \frac{E_c}{m} \quad (3)$$

where E_c^* is relative electric consumption (kW h g^{-1}) and m is the mass of obtained essential oil (g) [11].

2.5 CO_2 emission

The measurements of CO_2 emitted were carried out based on the procedures mentioned in previous studies: to obtain 1 kWh of energy from coal or fossil fuels, 800 g of CO_2 will be released into the atmosphere during combustion [5]. Thus CO_2 emission is described by equation (4):

$$E_{CO_2} = \frac{E_c 800}{1000} \quad (4)$$

where E_{CO_2} is CO_2 emission (g) and E_c is electric consumption (kW h).

The relative CO_2 emission of the different extraction methods was calculated according to equation (5):

$$E_{CO_2}^* = \frac{E_{CO_2}}{m} \quad (5)$$

where $E_{CO_2}^*$ is the relative CO_2 emission (kg g^{-1}) and m is the mass of obtained essential oil (g) [11].

3. Results and discussion

3.1 Comparison of yield in patchouli oil extraction using microwave hydrodistillation and solvent-free microwave extraction methods

In general, with the longer extraction time, the yield obtained also will be greater. However, with the longer extraction time, increasing of the yield become smaller [12]. Based on figure 3, the extraction of patchouli oil by microwave hydrodistillation method has the higher yield than solvent- free microwave extraction method. This is due to the amount of water in the distiller in the microwave hydrodistillation method is less when compared with the solvent-free microwave extraction method using a soaked dry material. Where the amount of water in the distiller in microwave hydrodistillation method is about 210 mL, whereas in the the solvent -free microwave extraction method is about 260 mL. With the small amount of water in the distiller it will accelerate the temperature rising and then it will accelerate the opening of oil glands and also cause the rate of increase in yield will be faster. So the yield produced on the patchouli oil extraction by solvent-free microwave extraction method can be smaller when compared with microwave hydrodistillation method.

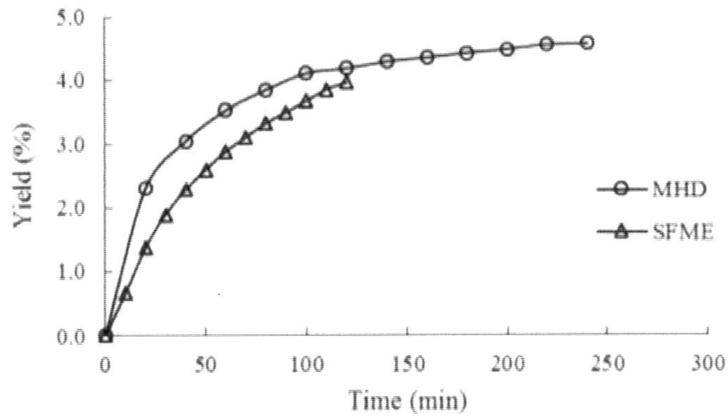

Figure 3. Comparison between the yields of patchouli oil obtained by microwave hydrodistillation (ratio of feed to solvent=0.3 gmL^{-1}, time=240 min) and solvent-free microwave extraction (ratio of feed to distiller=0.06 gmL^{-1}, time=120 min) at the microwave power was 450 W.

In the patchouli oil extraction with microwave hydrodistillation method obtained a yield of 4.6%, while the solvent-free microwave extraction method obtained a yield of 4.0%. From these data it can be seen that the difference in yield obtained between microwave hydrodistillation and solvent-free microwave extraction method is less significant, whereas in microwave hydrodistillation method with longer extraction time (240 min) only produced yield 1.2 times greater when compared with solvent-free microwave extraction method that requires faster extraction time (120 min).

3.2 Analysis of electric consumption and environmental impact of patchouli oil extraction using microwave hydrodistillation and solvent-free microwave extraction methods

In the patchouli oil extraction by the solvent-free microwave extraction method takes relatively faster than the microwave hydrodistillation method. This extraction time is related to the cost and energy required. So to know it, we conducted an analysis of electric consumption and the environmental impact of patchouli oil extraction using solvent-free microwave extraction method and microwave hydrodistillation method. An analysis of electric consumption and the environmental impact of patchouli oil extraction with both methods can be seen in figure 4.

In this study, the electric consumption requirements for patchouli oil extraction by microwave hydrodistillation method was 1.8 kWh, while for solvent-free microwave extraction method was 0.9 kWh. So it can be said that using microwave hydrodistillation method requires 2 times more electricity consumption than the solvent-free microwave extraction method. The electric consumption to obtain1 g of patchouli oil from extraction using microwave hydrodistillation method was equal to 0.8 kWh, while for solvent-free microwave extraction method was 0.5 kWh. So it can also be said that with the microwave hydrodistillation method requires an electric consumption that is 1.7 times higher when compared with the solvent-free microwave extraction method. Thus it can be assumed that the extraction of patchouli oil using solvent-free microwave extraction method requires lower operating cost when compared with microwave hydrodistillation method [8].

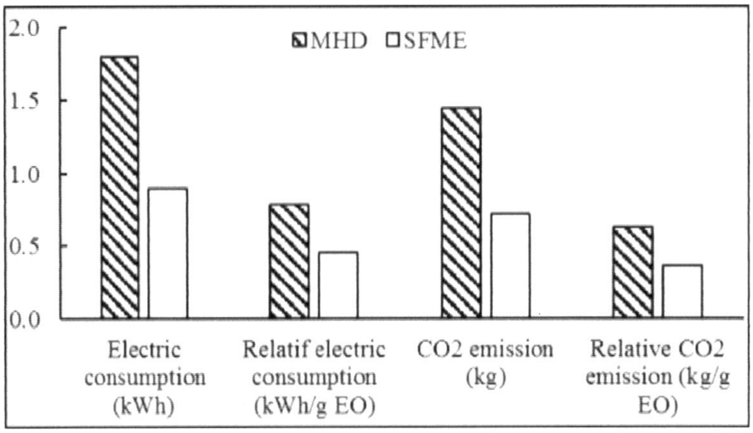

Figure 4. The electric consumption and environmental impact (CO_2 emissions) for the extraction of patchouli oil by the microwave hydrodistillation (ratio of feed to solvent=0.3 gmL^{-1}, time=240 min) and solvent-free microwave extraction method (ratio of feed to distiller=0.06 gmL^{-1}, time=120 min) at the microwave power was 450 W.

The environmental impact of the essential oil extraction can be known from the amount of CO_2 emissions produced. If 1 kWh of energy from coal or fossil fuels, 800 g of CO_2 will be released into the atmosphere during combustion [5]. The CO_2 emissions produced from patchouli oil extraction with microwave hydrodistillation method of 1.4 kg, while the solvent-free microwave extraction method was equal to 0.7 kg. It can be said that the extraction of patchouli oil by microwave hydrodistillation method produces a greater amount of CO_2 emissions than the solvent-free microwave extraction method. And to obtain 1 g of patchouli oil, CO_2 emissions produced by microwave hydrodistillation and solvent-free microwave extraction methods were 0.6 and 0.4 kg. So it can be concluded that patchouli oil extraction using microwave hydrodistillation method produces greater CO_2 emission when compared with by using solvent-free microwave extraction method. Therefore, the use of solvent-free microwave extraction method for patchouli oil extraction is suitably method as a new green technique.

4. Conclusion

Patchouli oil extraction using the solvent-free microwave extraction method is a new green technique which is better than the microwave hydrodistillation method. Where to obtain the yield, the microwave hydrodistillation method with longer extraction time (240 min) only produces yield 1.2 times greater when compared with the solvent-free microwave extraction method that requires faster extraction time (120 min). Whereas based on the analysis of electric consumption and the environmental impact, the extraction of patchouli oil with solvent-free microwave extraction method requires lower electric consumption and CO_2 emissions production is also lower when compared with microwave hydrodistillation method. So, it can be said that the use of solvent-free microwave extraction method for patchouli oil extraction is suitably method as a new green technique.

References

[1] Dewan Atsiri Indonesia and IPB 2010 *Indonesian essential oil* (in Bahasa Indonesia) (Molide R, Meika S R and Ariato M)
[2] Hu X 2014 *Rev. Bras. Farmacogn.* **24** 6 626
[3] Ministry of Trade of the Republic of Indonesia 2011 *Indonesian Essential Oil: The Scents of Natural Life* (Jakarta: Trade Policy Analysis and Development Agency, Ministry of Trade, Republic of Indonesia)

[4] Plantus 2008 *Distillation process of patchouli* (in Bahasa Indonesia) available from: http://anekaplanta.wordpress.com/2008/02/28/proses-penyulingan-nilam/ on 17 October 2016
[5] Ferhat M A, Meklati B Y, Smadja J and Chemat F 2006 *J. Chromatogr. A* **1112** 1–2 121
[6] Stashenko E E, Jaramillo B E and Martínez J R 2004 *J. Chromatogr. A* **1025** 1 93
[7] Sahraoui N, Abert M, Bornard I, Boutekedjiret C and Chemat F 2008 *J. Chromatogr. A* **1210** 229
[8] Kusuma H S and Mahfud M 2016 *RSC Adv.* **7** 1336
[9] Golmakani M T and Moayyedi M 2015 *Food Sci. Nutr.* **3** 6 506
[10] Chen F, Zu Y and Yang L 2015 *Sep. Purif. Technol.* **154** 271
[11] Kusuma H S and Mahfud M 2017 *J. Appl. Res.Med. Aromat. Plants* **4** 55
[12] Wang Y, You J, Yu Y, Qu C, Zhang H and Ding L 2008 *Food Chem.* **110** 1 161

IOP Conf. Series: Earth and Environmental Science **101** (2017) 012014 doi:10.1088/1755-1315/101/1/012014

Comparison of microwave hydrodistillation and solvent-free microwave extraction of essential oil from *Melaleuca leucadendra* Linn

A W Ismanto, H S Kusuma and M Mahfud

Department of Chemical Engineering, Faculty of Industrial Engineering, Institut Teknologi Sepuluh Nopember, Surabaya, Indonesia, 60111

E-mail: mahfud@chem-eng.its.ac.id

Abstract. The comparison of solvent-free microwave extraction (SFME) and microwave hydrodistillation (MHD) in the extraction of essential oil from *Melaleuca leucadendra* Linn. was examined. Dry cajuput leaves were used in this study. The purpose of this study is also to determine optimal condition (microwave power). The relative electric consumption of SFME and MHD methods are both showing 0,1627 kWh/g and 0,3279 kWh/g. The results showed that solvent-free microwave extraction methods able to reduce energy consumption and can be regarded as a green technique for extraction of cajuput oil.

1. Introduction

Essential oil, one of commodity export potential, could become a mainstay for Indonesia to earn foreign exchange. Essential oils are widely distilled in Indonesia, among others, patchouli, clove, nutmeg, lemongrass, vetiver, cajuput, and many others. Essential oils are the ingredients that are volatile (volatile), has a bitter taste, and smell like plant origin derived from plant parts such as leaves, fruit, seeds, flowers, roots, rhizomes, bark, even the whole plant. Furthermore, the development of essential oils are needed in the food flavoring industry, the cosmetics industry, and perfumery [1].

Melaleuca leucadendra Linn, commonly known as *cajuput*, is a plant growing at a level of 0 – 400 meter above sea level such as lowlands, along rivers, coastal areas or swamps. Cajuput can grow in barren, heat-resistant soil, and even sprout again after a fire. The spread of cajuput crops in Indonesia is found in several areas, namely Java, Irian Jaya, East Nusa Tenggara, Maluku, South Sumatra and Southeast Sulawesi. *Melaleuca leucadendra* Linn was first planted in Ponorogo, spread to Gunung Kidul in Yogyakarta, and other areas such as Gundih and Surakarta in Central Java, Mojokerto and Sukun in East Java, Cikampek, Majalengka, and Indramayu in West Java. Cajuput oil's main compound is cineole (50-60%) with some other compound like alpha-terpineol, valeraldehyde, and benzaldehyde. This oil is used in herbal remedies such as analgetic, diaphoretic, antirheumatics, and spasmolytic [2]. The 1,8-cineole compound is the main compound of eucalyptus oil that acts as antimicrobial, antioxidant, immune, analgesic, and spasmolytic [3]. In addition, 1,8-cineole compounds also have the potential to be anti-inflammatory [4]. The methyl eugenol compounds and their derivatives, methyl isoeugenol are used in the perfume and fragrance industries [5].

The prospect of essential oil is relatively good in Indonesia, should have been followed by the mastery of refining technology and the cultivation of essential oil producing plants, so that businesses of essential oil is able to thrive in terms of quality and quantity to meet the world's needs and enhances competitiveness with suppliers from countries. The process of taking cajuput oil (*Melaleuca leucadendra* Linn) is generally done by steam distillation and hydrodistillation. According to the previous research conducted by [2] using the steam-water hydrodistilation method on cajuput leaves (*Melaleuca leucadendra* Linn.) delivered yields from 0.61 % to 0.85% of 5 hours of extraction time.

The previous research conducted by [6], namely the extraction of essential oil from orange peel takes 30 min using solvent-free microwave extraction while 180 min were required by hydrodistillation method. The optimum yield of essential oil obtained was 0.42 ± 0.02% by solvent-free microwave extraction and 0.39 ± 0.02% by hydrodistillation. It has shown the microwave extraction as an alternative method that can be developed more than conventional methods, due to high levels of product purity, minimal solvent usage, and short processing times. To optimize the process of essential oil extraction, this research will use microwave hydrodistillation and solvent-free microwave extraction method. The microwave hydrodistillation method is a combination of hydrodistillation and microwave use as heating [7]. The SFME method is a method of extracting essential oils without using a microwave-heated solvent which has advantages such as having a faster extraction rate, yield, and also higher purity of extract because it does not require solvent so it does not come into contact with chemicals [8].

2. Materials and methods

2.1. Materials

The *Melaleuca leucadendra* Linn leaf used in this study was collected from a forest owned by the Department of Forestry of Mojokerto (Fig. 1). All leafs were gathered from the same plantation age at approximately 0-2 years of plantation.

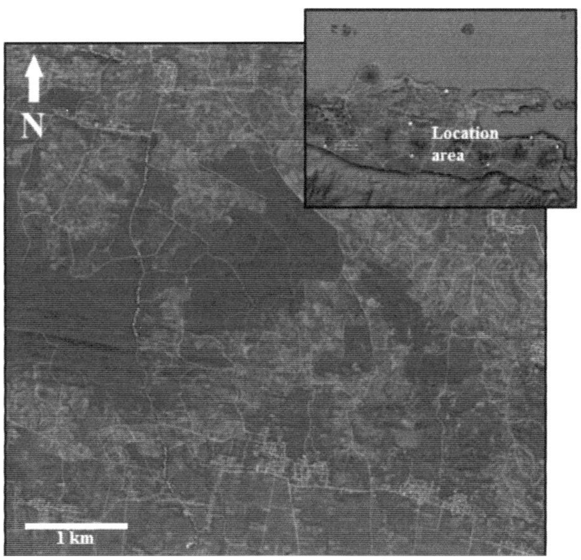

Figure 1. The location area where the *Melaleuca leucadendra* leaf was collected. The forest of the Department of Forestry is situated along Kemlagi-Mantup road and is easily accessed by any transportation vehicle.

2.2. Solvent-Free Microwave Extraction
The microwave oven used for solvent-free microwave extraction was Electrolux model EMM-2007X operating at 2450 MHz with a maximum delivered power of 600 W. The microwave oven was modified by drilling a hole at the top. Flat bottom flask having a capacity of 1000 mL was placed in the oven and connected to Liebig apparatus through the hole. After the Liebig condensor was placed into the oven, the hole around the neck of the flask was covered with polytetrafloraethylene (teflon) to prevent leakage of microwave.

For the extraction, sixty grams of dry *Melaleuca leucadendra* Linn. leaves were soaked in 700 mL distilled water at room temperature (25 °C) for an hour to hydrate the external layers of the plant material so that the excess water was drained off. The moistened plant material was placed in flat-bottom flask combined to a Liebig condensor. During the process, the vapor passed through the condenser outside the microwave cavity where it was condensed. The solvent-free microwave extraction process was performed for different mass and was continued until no essential oil obtained. The essential oil was collected, dried under anhydrous sodium sulphate and stored at 4 °C before used.

2.3. Electric Consumption
The electrical consumption of different extraction methods is calculated based on the effect of power consumption and time of extraction. The general equation for electricity consumption is explained by (Eq. (1))

$$Ec = \frac{p.t}{3600000} \quad (1)$$

Where Ec is electric consumption (kWh), P is power consumption (W) and t is time (s). In addition, the relative electric consumption of the different methods of extraction can be expressed in equation 2.

$$Ec * = \frac{Ec}{m*} \quad (2)$$

Where Ec* is the relative electric consumption (kWh / g) and m* is the period of *Melaleuca leucadendra* Linn. (g).

2.4. CO_2 emission
The CO_2 calculation has been developed, according to the procedure mentioned in the previous study [6]. However, to get 1 kWh of energy from coal or fossil fuel, 800 g of CO_2 will be released into the atmosphere during their combustion [6]. Then, the CO_2 emissions are explained by (Eq. (3))

$$E_{CO_2} = \frac{Ec.800}{1000} \quad (3)$$

Where E_{CO2} represents CO_2 emission (kg).
For relative CO_2 emissions different extraction methods are calculated according to (Eq. (4))

$$E^{*}co_2 = \frac{E_{CO2}}{M*} \quad (4)$$

Where E * co_2 is the relative CO_2 emission (kg / g)

3. Results and discussion

3.1. Extraction yield of cajuput oil

The yield of cajuput oil was measured throughout the extraction process. Fig. 2 shows the variation of essential oil yield for solvent-free microwave extraction and microwave hydrodistillation processes at same power level. The figure shows both solvent-free microwave extraction and microwave hydrodistillation yields increasing as the mass of raw materials increase. The obtained results explain that solvent-free microwave extraction has more capability to generate cajuput oil rather than microwave hydrodistillation method and even the yields value is higher at the same extraction time.

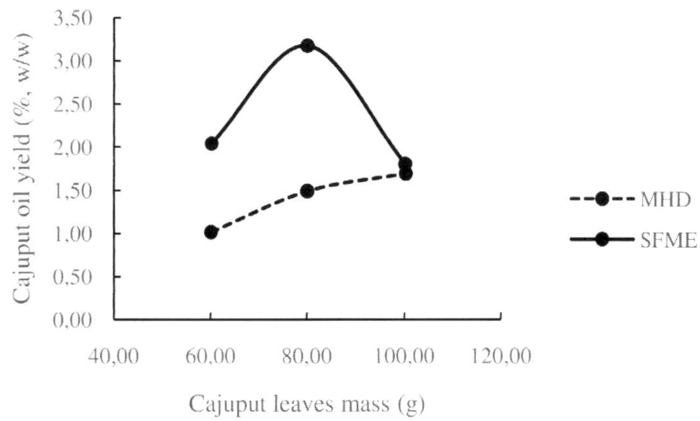

Figure 2. Extraction yield in cajuput oil extraction by solvent-free microwave extraction (SFME) and microwave hydrodistillation (MHD) methods

It is known that the power extraction of microwave will control the amount of energy received by plant material, hence to be converted into heat energy. This heat energy helps the process of removal of essential oils from plant material or sample. The microwave power used in the extraction process by microwave hydrodistillation method is closely related to the process temperature, where the greater the power used, the system temperature at the extraction process will rapidly reach the boiling point of the used solvent. The power effect on the yield for the solvent-free microwave extraction method can be interpreted as the greater the microwave power used, the greater the energy received by the plant material to be converted into heat which eventually results in lower yields of cajuput oil. This case occurs because of thermal degradation of the material.

As depicted in Fig. 3, the more mass used, the weight of cajuput oil obtained decreases. However, the large amount of raw material and the amount of cajuput oil obtained, not always positively correlated with the increase in yield of cajuput oil obtained. This is because the yield of cajuput oil is influenced by the ratio factor of cajuput oil obtained and the mass of the initial raw material. In Fig. 3, there was a decrease in yield at a mass of 100 g, a decrease in the yield of cajuput oil by solvent-free microwave extraction method because the mass of the raw material (cajuput) used was excess (solid) and almost filled the pumpkin distiller. Where this results in a difficult vapor penetrating in the material to bring the molecular oil of the volatile oil diffuses out the material. The density of the material is closely related to the large space between materials. Excessively high and uneven density of materials can lead to the formation of steam lines that can reduce yields and quality of essential oils [9].

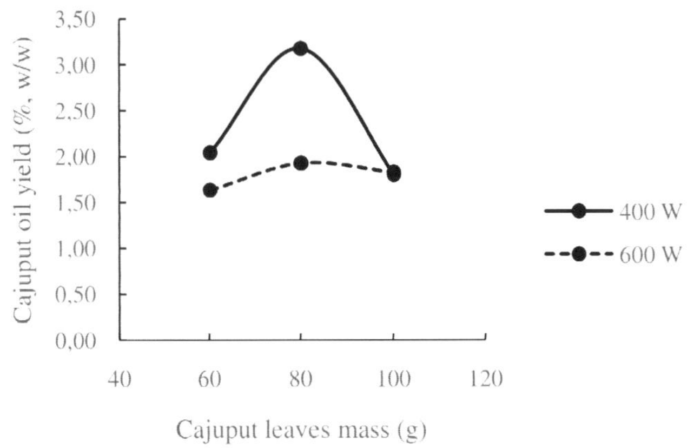

Figure 3. Extraction yield in cajuput oil extraction by solvent-free microwave extraction (SFME)

3.2. Electricity consumption and environmental considerations

The comparison of two different methods to extract cajuput oil with the same of both microwave power consumption and extraction time has been completed. The consumed energy to extract the cajuput oil with respect to the power consumption of the microwave is showed in Fig. 4. The relative electric consumption for generating 1 gram of cajuput oil using microwave hydrodistillation method is doubled the value of the solvent-free microwave extraction. The microwave hydrodistillation provide a value of 0.33 kWh/gram EO compared with 0.16 kWh/gram EO for solvent-free microwave extraction. This phenomenon suggests that the solvent-free microwave extraction is more efficient than the microwave hydrodistillation regarding the electric consumption.

The relative CO_2 emission produced of 1-gram cajuput oil is higher in the microwave hydrodistillation (0.13 kg CO_2 /gram EO) than in the solvent-free microwave extraction (0.26 kg CO_2 / g EO) for an hour extraction time (Fig. 4).

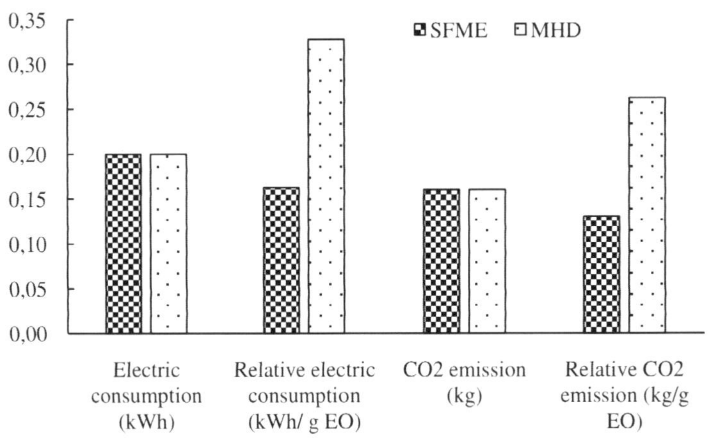

Figure 4. Electricity consumption and environmental impact in cajuput oil extraction by solvent-free microwave extraction and microwave-assisted hydrodistillation methods

4. Conclusion

The present results demonstrate that the effect of the materials used the method of microwave hydrodistillation (MHD) and solvent-free microwave extraction (SFME) is the greater the material used, the greater the yield of cajuput oil produced. The optimum yield of essential oil was obtained 3.17 % by SFME and 1.93 % by MHD. In the extraction of cajuput oil, the solvent-free microwave extraction is more effective than the microwave hydrodistillation. Moreover, the solvent-free microwave extraction consumed less energy and less CO_2 emission than the microwave hydrodistillation did. Based on the variable of microwave power, the yield of cajuput oil produced less yield generated by solvent-free microwave extraction method when microwave power increased.

References

[1] Armando R 2009 *Memproduksi 15 Jenis Minyak Atsiri Berkualitas* (Jakarta: Penebar Swadaya).
[2] Pujiarti R, Ohtani Y and Ichiura H 2011 *J. Wood Science* **57(5)** 446-451.
[3] Angela E S and Davis W L 2010 *Alternative Medicine Review* **15(1)** 33-47.
[4] Juergens U R, Dethlefsen U, Steinkamp G, Gillissen A, Repges R and Vetter H 2003 *Respiratory Medicine* **97** 250-256.
[5] Brophy J J and Lassak E V 1988 *Flavour and Fragrance Journal* **3** 43-46.
[6] Ferhat M A, Mekalti B Y, Smadja J and Chemat F 2006 *J. Chromatography* A **1112(1)** 121-126.
[7] Golmakani M T and Rezaei K 2008 *Food Chemistry* **109(4)** 925-930.
[8] Bayramoglu B, Sahin S and Sumnu G 2008 *J. Food Engineering* **88(4)** 535-540.
[9] Guenther E 1987 *The Essential Oils Volume One: History-Origin in Plants Production-Analysis*. Lancaster Press, USA.

Alternative to conventional extraction of vetiver oil: Microwave hydrodistillation of essential oil from vetiver roots (*Vetiveria zizanioides*)

H S Kusuma, A Altway, M Mahfud

Department of Chemical Engineering, Faculty of Industrial Engineering, Institut Teknologi Sepuluh Nopember, Surabaya, Indonesia 60111

E-mail: heriseptyakusuma@gmail.com; mahfud@chem-eng.its.ac.id

Abstract. In this study the extraction of essential oil from vetiver roots (*Vetiveria zizanioides*) has been carried out by using microwave hydrodistillation. In the extraction of vetiver oil using microwave hydrodistillation method is studied the effect of microwave power, feed to solvent (F/S) ratio and extraction time on the yield of vetiver oil. Besides, in this study can be seen that microwave hydrodistillation method offers important advantages over hydrodistillation, such as shorter extraction time (3 h vs. 24 h for hydrodistillation); better yields (0.49% vs. 0.46% for hydrodistillation); and environmental impact (energy cost is appreciably higher for performing hydrodistillation than that required for extraction using microwave hydrodistillation). Based on the analysis using GC-MS can be seen 19 components on vetiver oil that has been extracted using microwave hydrodistillation. In addition, GC-MS analysis showed that the main components of vetiver oil that has been extracted using microwave hydrodistillation method were β-Gurjunene (30.12%), α-Vetivone (20.12%), 4-(1-cyclohexenyl)-2-trimethylsilylmethyl-1-buten-3-yne (13.52%) and δ-Selinene (7.27%).

1. Introduction

Recently there has been an increase in demand for essential oils extracted from plant. This can be attributed to the fact that essential oils have multifunctional properties and hence are playing an increasing role in the food, fragrance, agricultural and pharmaceutical industries [1].

Vetiver grass also known by the Latin name *Vetiveria zizanioides* or its traditional name Khus, is a perennial grass which is part of the Poaceae family [2]. The root is the most valuable part of the grass as it forms an intertwined network that stops erosion and it contains the majority of the essential oils which has valuable aromatic and biological properties [3]. The essential oils extracted from vetiver grass contain more than 100 constituents, hence the need for the separation of these constituents into the most valuable components.

In developing countries, agricultural grasses are very popular types of crops for rural outreach programmes due to the high value of the essential oils extracted from the grass. Vetiver grass has many uses; it is very easy to grow as it can withstand harsh environmental conditions and does not require large amounts of fertiliser or pruning and therefore it is easily maintained by the user at low cost. It is capable of growing in extreme soil types; this includes sands, shale, gravels, mine tailings and even more toxic soils. It is also very easy to control the spread of vetiver grass as it propagates by root division or slips and is easily removed when no longer required [4].

Currently the extraction of vetiver oil is mostly done using conventional methods such as hydrodistillation, steam distillation and solvent extraction [5]. However, this conventional method requires a long time of extraction, a lot of solvents and requires a lot of energy [6]. However, this method has been controversial for subsequent determination of the oil chemical composition because of the possible transformation of aroma-active compounds by heat, steam, and pH [7]. In addition losses and degradation of some volatile compounds due to long extraction times, degradation of unsaturated or ester compounds through thermal or hydrolytic effects are the principal disadvantages of this extraction method [8]. Thus, developing an alternative rapid, sensitive, safe, and energy conserving extraction method is highly desirable. There is an increased demand for new extraction methods with shortened extraction times and reduced organic solvent consumption, preventing pollution and reducing sample preparation costs. Recently, much attention has been given to the application of microwave dielectric heating in many various processes in the chemical and food industry in order to replace the conventional extraction methods. Extraction processes performed under the action of microwave selective heating [9]. Driven by reducing energy and waste water or solvent, advances in microwave extraction have resulted in some methods such as microwave hydrodistillation and microwave steam distillation [6,9].

Therefore, in this study the extraction of vetiver oil is done using microwave hydrodistillation method. There is no information about studied parameters that affect the extraction of vetiver oil from vetiver roots using microwave hydrodistillation method has been reported in the previous study. In addition, this study will also compare chemical composition, energy consumption and environmental considerations using hydrodistillation method.

2. Materials and methods

2.1. Materials and chemicals

The dried vetiver roots used in this study were obtained from Garut, West Java, Indonesia and has size of about 35-60 mesh. The distilled water and anhydrous sodium sulphate used in the study were of analytical grade.

2.2. Microwave hydrodistillation method

In employing microwave hydrodistillation, we used a domestic microwave oven (EMM2308X, Electrolux, maximum delivered power of 800 W) with wave frequency of 2450 MHz. The dimensions of the PTFE-coated cavity of the microwave oven were 48.5 cm x 37.0 cm x 29.25 cm. The microwave oven was modified by drilling a hole at the top. A round bottom flask with a capacity of 1000 mL was placed inside the oven and was connected to the Clevenger apparatus through the hole. Then, the hole was closed with PTFE to prevent any loss of the heat inside.

Some feed to solvent ratio (0.30; 0.40; 0.50 g/mL) were placed in the reaction flask and heated by microwave irradiation with various power (300; 450; 600 W) and various extraction time (60; 120; 180 min). The different densities and their immiscibility required that the water and vetiver oil be separated from each other by separating funnel and the excess water be refluxed to the extraction vessel in order to provide uniform conditions of feed to solvent ratios for extraction. The vetiver oil was collected in amber vials, dried under anhydrous sodium sulphate and stored at 4°C. The extraction yield of vetiver oil was calculated according to the equation given:

$$\text{Extraction yield } (\%, w/w) = \frac{\text{Mass of extracted vetiver oil}}{\text{Mass of dried vetiver roots x } (1-\text{water content})} \text{ x } 100 \quad (1)$$

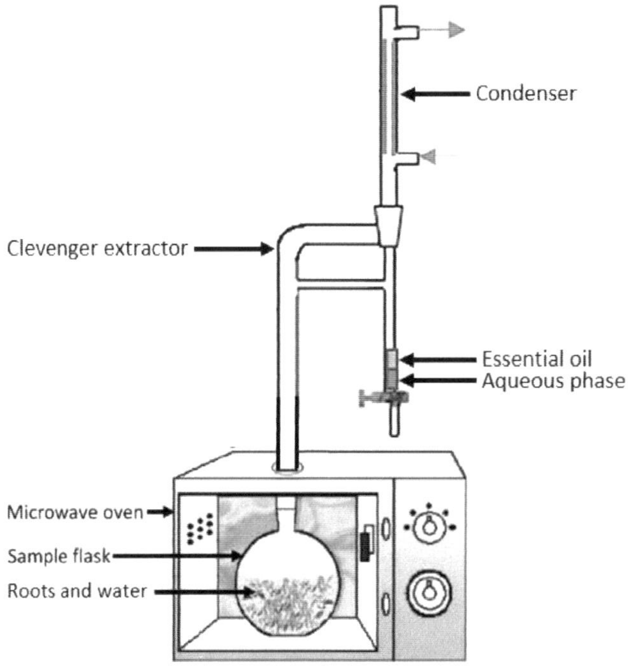

Figure 1. Schematic representation of the microwave hydrodistillation apparatus used in this study.

2.3. Chemical analysis of vetiver oil components
The components of the vetiver oil were identified by Gas Chromatography-Mass Spectrometry (GC–MS) analysis [6,10,11,12]. An Agilent 6980N gas chromatograph with an Agilent 5973 mass spectrometric detector was used for GC-MS. Gas chromatography-mass spectrometry was incorporated with a chromatography column HP-5.5% phenylmethylsiloxane, 30 m length, 0.32 mm film thickness, 0.25 μm internal diameter). The injection port and the interface were held at 230 and 280°C, respectively. The temperature was programmed from 50°C to 280°C at 10°C per minute with helium as the carrier gas. Most of the compounds were identified according to mass spectra (National Institute of Standards and Technology (NIST) and Wiley spectral library collection).

2.4. Electric consumption
The electric consumption of different extraction methods was calculated based on the influence of power consumption and extraction time. The general equation for the electric consumption described by (Eq. (2))

$$E_C = \frac{P.t}{3600000} \tag{2}$$

where E_C is electric consumption (kWh), P is power consumption (W) and t is time (s).
Additionally, the relative electric consumption of different extraction methods can be expressed by (Eq. (3))

$$E_C^* = \frac{E_C}{m} \tag{3}$$

where E_C^* is relative electric consumption (kWh/g) and m is mass of obtained essential oil (g).

2.5. CO_2 emission
The measurements of CO_2 emitted were carried out based on the procedures mentioned in the previous studies: to obtain 1 kWh of energy from coal or fossil fuels, 800 g of CO_2 will be released into the atmosphere during their combustion [6]. Thus CO_2 emission is described by (Eq. (4))

$$E_{CO_2} = \frac{E_C.800}{1000} \quad (4)$$

where E_{CO_2} is CO_2 emission (kg) and E_C is electric consumption (kWh).

For the relative CO_2 emission of different extraction methods was calculated according to (Eq. (5))

$$E^*_{CO_2} = \frac{E_{CO_2}}{m} \quad (5)$$

where $E^*_{CO_2}$ is relative CO_2 emission (kg/g) and m is mass of obtained essential oil (g).

3. Results and discussion

3.1. Effect of microwave power on the yield of vetiver oil

The microwave power exhibited significant effects on the extraction of vetiver oil (Fig. 2). By microwave hydrodistillation, when the microwave power increased from 300 W to 450 W, the yield of vetiver oil increased very fast. For example, when the microwave power increased from 300 W to 800 W for 180 min, the yield increased from 0.36% to 0.53%. Beyond 450 W, the yields of the vetiver oil did not increase.

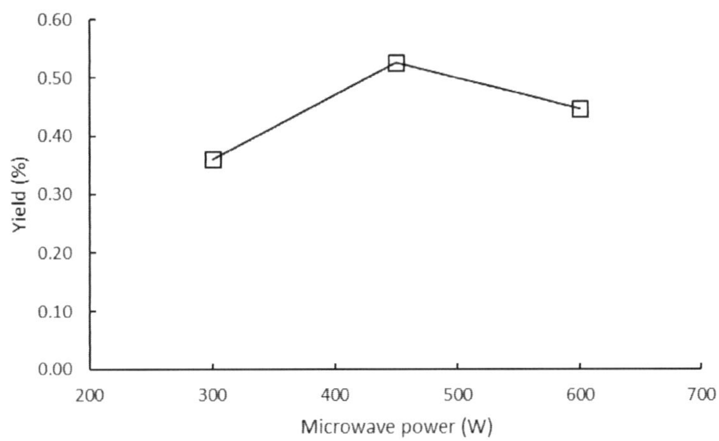

Figure 2. The effect of microwave power on the yield of vetiver oil (feed to solvent ratio of 0.50 g/mL).

The increase of vetiver oil yield in microwave hydrodistillation is probably due to the high pressure gradient formed inside the plant material (vetiver roots) [13]. Microwave absorption causes fast internal heating thus creating significantly high internal pressures which enhance the extraction of vetiver oil. The cell walls can swell and burst because of internal heating and thus further promotes the release of vetiver oil. Therefore an appropriate increase in microwave power led to rise in yield. Thus, the microwave power of 450 W was chosen as the output microwave power in microwave hydrodistillation.

3.2. Effect of feed to solvent (F/S) ratio on the yield of vetiver oil

As shown in Fig. 3, the yield of vetiver oil was found to increase with the increase of feed to solvent (F/S) ratio and then fall down at the optimum ratios. The smaller feed to solvent ratio means a larger concentration difference which favors mass transfer. But in microwave-assisted extraction, a higher solvent volume may give lower yield [14]. As found in Fig. 3, the yield of vetiver oil increased with the increase of feed before the feed to solvent (F/S) ratio reached 0.40 g/mL, and then they fell down. The phenomena were probably due to the inadequate stiring of the solvent when the microwave was applied at larger feeds. Hence, the feed to solvent (F/S) ratio of 0.40 g/mL seems to be appropriate for microwave hydrodistillation.

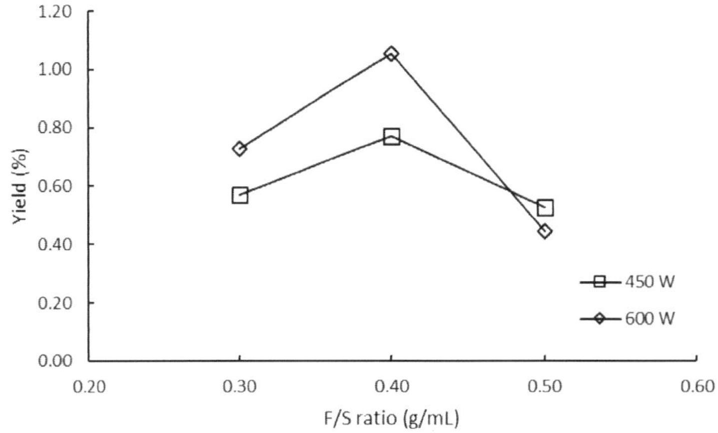

Figure 3. The effect of feed to solvent (F/S) ratio on the yield of vetiver oil.

3.3. Effect of extraction time on the yield of vetiver oil

Fig. 4 shows the effect of extraction time on the yield of vetiver oil over the range 60-180 min under a fix microwave power of 450W. From the Fig. 4, the amount of yield does not change significantly after 120 minutes where the yield obtained up to this period was 0.50%, 0.64% and 0.45% for feed to solvent ratio (F/S) of 0.30, 0.40 and 0.50 g/mL, respectively. The rate of extraction was high at the beginning of the extraction but get slow gradually by time. The effect of this parameter in microwave hydrodistillation has been studied by number researches on various plant materials [15,16,17] and almost all of them obtained the same configuration as illustrated in Fig. 4. These results confirmed the Fick's second law of diffusion which stated about the final equilibrium achieved by the solute concentrations in plant matrix and in the solvent after a certain time. This cause into no significant improvement in vetiver oil yield when prolonging the extraction time.

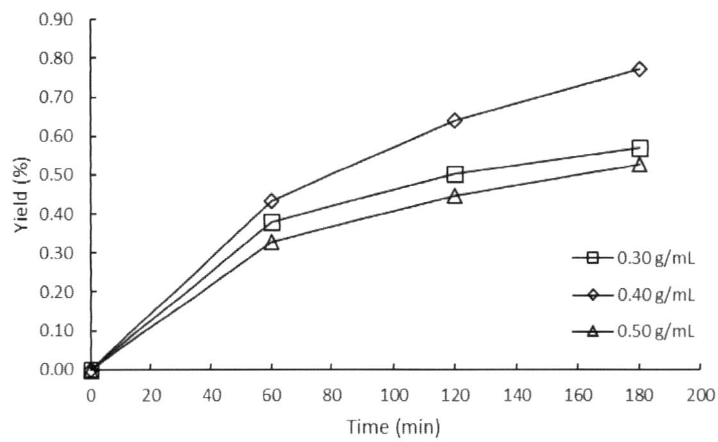

Figure 4. The effect of extraction time on the yield of vetiver oil (microwave power of 450 W).

3.4. Electric consumption and environmental considerations

The reduction of the extraction time is clearly advantageous for the proposed microwave hydrodistillation method in terms of cost and energy. In this study, a comparison between extraction of vetiver oil using hydrodistillation and microwave hydrodistillation methods has been performed. The extraction of vetiver oil using hydrodistillation method is based on previous research by Leite (2012) [5]. Leite (2012) [5] has done the extraction of vetiver oil with feed to solvent (F/S) ratio of 0.0167 g/mL for 24 hours using heating mantle (Glas Col STM1001 230 V/600 W) and obtained yield of 0.46%. While as a comparison then in this study selected the operating conditions from the extraction of vetiver oil using microwave hydrodistillation method that can be obtained yield almost equal to the extraction of vetiver oil using hydrodistillation method. So in this study for extraction of vetiver oil using microwave hydrodistillation method selected operating conditions as follows: microwave power of 300 W, feed to solvent (F/S) ratio of 0.40 g/mL and extraction time of 3 h. This is because in these operating conditions obtained yield almost equal to the extraction of vetiver oil using hydrodistillation method that is 0.49%.

The energy requirement needed to perform the extraction methods, based on the power consumptions of the hot plate (in hydrodistillation) and the microwave oven (in microwave hydrodistillation), considering the total periods of full extractions, was 14.40 kWh for hydrodistillation and 0.90 kWh for microwave hydrodistillation (Table 1). Kusuma and Mahfud (2017a) [6] showed that for the energy requirements, microwave hydrodistillation needs less than 1.30 kWh for normal performance but conventional methods expend more than 6.50 kWh. Relative electric consumption for the production of 1 g vetiver oil in hydrodistillation and microwave hydrodistillation was 294.63 and 2.72 kWh/g vetiver oil, respectively (Table 1). This indicates a substantial saving in the extraction cost when using microwave hydrodistillation instead of hydrodistillation. In addition Kusuma and Mahfud (2017a) [6] has also reported that the energy required for the extraction of essential oil from patchouli leaves (*Pogostemon cablin* Benth) was 6.55 kWh per gram of patchouli oil in hydrodistillation, but 1.62 kWh per gram of patchouli oil in microwave hydrodistillation.

Table 1. The electric consumption and environmental impact (CO_2 emissions) for the extraction of vetiver oil using hydrodistillation and microwave hydrodistillation methods

Extraction methods	Electric consumption (kWh)	Relative electric consumption (kWh/g vetiver oil)	CO_2 emission (kg)	Relative CO_2 emission (kg/g vetiver oil)
Hydrodistillation (Leite, 2012)	14.40	294.63	11.52	235.70
Microwave hydrodistillation	0.90	2.72	0.72	2.17

Regarding the environmental impact of pollution, the calculated quantity of CO_2 emitted in the atmosphere was higher in the case of hydrodistillation (11.52 kg CO_2) than microwave hydrodistillation (0.72 kg CO_2) (Table 1). Kusuma and Mahfud (2017a) [6] found that the amount of CO_2 which was released into the atmosphere was higher in hydrodistillation (5.36 kg CO_2) than microwave hydrodistillation (1.01 kg CO_2). Relative amounts of CO_2 emissions that result from the production of 1 g vetiver oil were higher in hydrodistillation (235.70 kg CO_2/g vetiver oil) than in microwave hydrodistillation (2.17 kg CO_2/g vetiver oil) (Table 1). This finding further indicated that there was a significant difference between hydrodistillation and microwave hydrodistillation in terms of the amount of CO_2 released into the atmosphere for the production of 1 g vetiver oil. Also, waste water was lower in microwave hydrodistillation than hydrodistillation. In addition Kusuma and Mahfud (2017a) [6] has also showed that the amount of CO_2 released into the atmosphere was dramatically higher in hydrodistillation (5.24 kg CO_2/g patchouli oil) than that in microwave hydrodistillation (1.30 g CO_2/g patchouli oil).

3.5. Gas Chromatography-Mass Spectrometry (GC-MS) analysis
To find out the components contained in vetiver oil analysis is done using Gas Chromatography-Mass Spectrometry (GC-MS). The results of the GC-MS analysis can be seen from the obtained chromatogram (Fig. 5). From the chromatogram can be seen that on the vetiver oil that has been extracted using microwave hydrodistillation method there are 19 components. Based on the results of GC-MS analysis also showed that the main components of vetiver oil that has been extracted using microwave hydrodistillation method were β-Gurjunene (30.12%), α-Vetivone (20.12%), 4-(1-cyclohexenyl)-2-trimethylsilylmethyl-1-buten-3-yne (13.52%) and δ-Selinene (7.27%). The results of the GC-MS analysis for vetiver oil that has been extracted using other methods showed different results. It can be seen from the results of the GC-MS analysis for vetiver oil that has been extracted using hydrodistillation method that has been done by Leite (2012) [5]. The main components of vetiver oil produced by hydrodistillation method were nootkatone (22.64%), khusimol (9.33%), pogostol (5.70%) and vetiselinenol (4.66%).

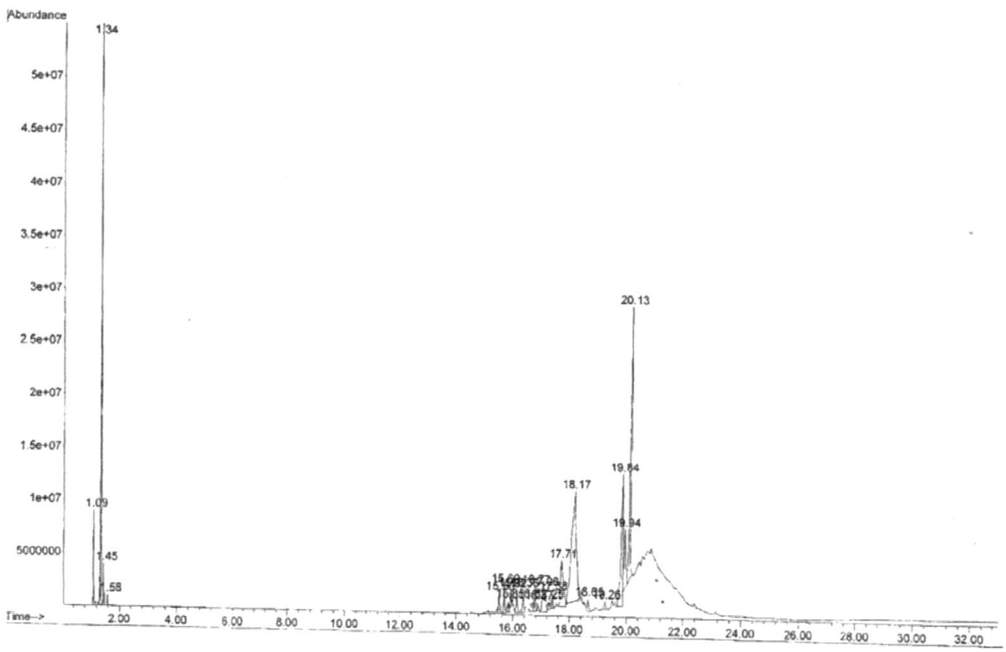

Figure 5. The chromatogram of vetiver oil that has been extracted using microwave hydrodistillation method.

4. Conclusion
The extraction of essential oil from vetiver roots (*Vetiveria zizanioides*) using microwave hydrodistillation offered important advantages over hydrodistillation: shorter extraction times (3 h vs. 24 h for hydrodistillation); better yields (with microwave hydrodistillation the obtainable vetiver oil quantity is higher than in case of hydrodistillation); it is environmentally friendly (the amount of CO_2 emission – a result of the extraction process of vetiver oil – was dramatically higher in hydrodistillation than that of microwave hydrodistillation); has reduced cost; and has less energy consuming compared to hydrodistillation. Therefore it can be used to confirm the efficiency of microwave hydrodistillation method and to explain how microwave hydrodistillation speed up the extraction process of vetiver oil.

In addition, based on the results of composition analysis for vetiver oil components using GC-MS, it can be seen that 19 components can be detected in the vetiver oil that has been extracted using microwave hydrodistillation method. The main components of vetiver oil that has been extracted using microwave hydrodistillation method were β-Gurjunene (30.12%), α-Vetivone (20.12%), 4-(1-cyclohexenyl)-2-trimethylsilylmethyl-1-buten-3-yne (13.52%) and δ-Selinene (7.27%).

References

[1] Kim H J, Chen F, Wang X, Chung H Y and Jin Z 2005 *J. Agric. Food Chem.* **53** 7691-7695
[2] Joy R J 2009 *Sunshine Vetiver Grass Chrysopogon Zizanioide* (L.) (Washington, D.C.: United States Department of Agriculture, Natural Resource Conservation Service, Available at: www.vetiver.org/USANRCS_Sunshine.pdf)
[3] Danh T 2007 *Development of Process for Purification of α and β-Vetivone from Vetiver Essential Oil and Investigation of Effects of Heavy Metals on Quality and Quantity of Extracted Vetiver Oil* (Sydney: University of New South Wales, PhD Thesis)
[4] Islam M P, Bhuiyan M K H and Hossain M Z 2008 *Agric. Eng. Int.: CIGR J.* **3**(5) 1-18
[5] Leite B 2012 *Extraction of Essential Oils from Vetiver* (*Vetiveria zizanioides*) *Grass* (KwaZulu-Natal: School of Chemical Engineering, University of KwaZuluNatal, Thesis)
[6] Kusuma H S and Mahfud M 2017a *RSC Adv.* **7** 1336-1347
[7] Jiménez-Carmona M M, Ubera J L and Luque de Castro, M D 1999 *J. Chromatogr. A* **855**(2) 625-632
[8] Khajeh M, Yamini Y, Sefidkon F and Bahramifar N 2004 *Food Chem.* **86**(4) 587-591
[9] Kusuma H S and Mahfud M 2017b *J. Appl. Res. Med. Aromat. Plants* **4** 55-65
[10] Li B, Zhang C, Peng L, Liang Z, Yan X, Zhu Y and Liu Y 2015 *Ind. Crops Prod.* **69** 329-334
[11] Wianowska D and Dawidowicz A L 2016 *Talanta* **151** 179-182
[12] Mahfud M, Putri D K Y, Dewi I E P and Kusuma H S 2017 *RASAYAN J. Chem.* **10**(1) 86-91
[13] Kusuma H S and Mahfud M 2017c *J. Appl. Res. Med. Aromat. Plants* **4** 46-54
[14] Eskilsson C S and Bjorklund E 2000 *J. Chromatogr. A* **902**(1) 227-250
[15] Asghari J, Touli C K and Mazaheritehrani M 2012 *J. Med. Plants Res.* **6**(28) 4475-4480
[16] Golmakani M-T and Rezaei K 2008 *Food Chem.* **109**(4) 925-930
[17] Hazwan H M, Hasfalina C M, Hishamuddin J and Zurina Z A 2012 *Int. J. Chem. Eng. Appl.* **3**(3) 173-177

The effect of microwave power on the production of biodiesel from nyamplung

L Qadariyah, F Mujaddid, Raka , S B Dhonny and M Mahfud

Department of Chemical Engineering Institut Teknologi Sepuluh Nopember, 60111 Surabaya, Indonesia

Email: lailatul_2008@yahoo.com

Abstract. Today, energy needs in Indonesia still rely on fossil energy sources that its availability in the world is increasingly depleted. Therefore, the research for alternative energy of petroleum must be developed, one of them is biodiesel. The use of microwave as energy source of biodiesel production can speed up the reaction time. So the microwave is considered more efficient. Seeds of nyamplung has an oil content of 71.4% (w/w) by weight. With the oil content of the nyamplung seeds has great potential when used as a raw material for biodiesel production. The aim of this research to study the effect of microwave power on the production of biodisel from nyamplung oil. Microwave power affects density, viscosity and yield of the product. The used of alkali catalyst, with higher the power , the lower the density and viscosity of the resulting product, but the resulting yield is 300 W. The power of more than 300 W is the opposite, resulting in the production of biodiesel using the optimum base catalyst at 300 W power.

1. Introduction

Along with economic growth, population, regional development and development from year to year, the need for energy needs of all sectors of the national energy users are also getting bigger. Books based Energy Security Indonesia (2014) states that fuel consumption increased rapidly from about 167.2 thousand kl/day (kl/d) in 2009 to 197.4 thousand kl / d, an increase of 18.1%, an increase in 5 years fuel consumption of 30.2 thousand kl / d. Increased consumption of fuel, especially gasoline and diesel due to the number of vehicles has increased every year.

In general reserves and the age of the national renewable energy resources are as follows:

Table 1. Age reserves and national energy sources

Name	Availability	Production	Remain (Year)
Crude Oil	4,7 billion barel	1,126 million barel/day	15
Natural Gas	9,3 TSCF	2,6 TSCF/tahun	35
Coal	4,968 billion ton	81,4 million ton/tahun	61

Source : Statistical Review of World Energy, 2005

Seeing the decreasing availability of petroleum and the increasing level of oil consumption coupled with the amount of imported fuel by Indonesia greater than exports, so it needs an alternative source of renewable energy so that the achievement of national energy security [1].

The potential of new and renewable energy is very significant, Indonesia has the potential of energy resources vary considerably.There are 75.091 MW geothermal, 29.164, mini / micro hydro 769.69 MW, biogass 2.3 million BOE, municipal waste 3,000 MW, solar power 480 kwh / m2 / day, wind 3-6 m / s and fuel vegetable 161.5 million BOE, (Director General of New and Renewable Energy and Energy Conservation EMR). But in use, need to pay attention to the concept of sustainable development is a development concept on the basis of value-adding resources. Therefore, it is necessary to develop Biofuel which is a government policy through presidential Instruction No. 1 of 2006 accompanied by Energy Minister Regulation No. 25 of 2013 on the use of biofuel. In the candy there is an obligation to use biofuels to fuel distributed to meet the needs of domestic energy such as biodiesel. The advantage of biodiesel is one of the alternative fuels that are environmentally friendly because biodiesel can reduce exhaust emissions of carbon monoxide and carbon dioxide and free sulfur content compared with that of other petroleum diesel [2].

Raw materials biodiesel are vegetable oils one of them is derived from seeds nyamplung plant. Nyamplung plant is a wild plant that grows in Indonesia, growing around the coast. This plant is usually only used for timber construction needs, furniture and others. Some of the advantages of biodiesel produced from nyamplung are nyamplung oil content is high compared to other types of plants (jatropha 40-60%, Sawit 46-54%, and Nyamplung 40-73%), some parameters have met Indonesian biodiesel quality standards, seed oil Nyamplung has a fuel two times longer than kerosene [3]. Nyamplung seed oil is a potential renewable energy resource as a basic material of biodiesel without having to compete with food needs [4]. Previous research results indicate that heating using microwave radiation in organic chemical synthesis takes a relatively short time compared to conventional heating [5].

Objective of research is to study the effect of microwave power on the production of biodisel from nyamplung oil. The transesterification reaction in this study was conducted in batch. Batch transesterification process is better than continuous process. The batch process is easy in controlling the reaction and does not require many equipments. In addition, the heating process of transesterification uses microwave radiation (microwave) which has different characteristics with conventional heating

2. Material and methods

2.1 Materials and chemicals
Nyamplung oil which will be used as feedstock in the manufacture of biodiesel is obtained from PT. Samino biofuel. Methanol is used as a binder compounds of non-lipid in the extraction process and as reagents for transesterification reactions with levels 98%, The catalyst KOH is a catalyst used during the transesterification process using microwave.

2.2 Transesterification Method
In employing solvent-free microwave extraction, we used a domestic microwave oven (EMM-2308X, Electrolux, maximum delivered power of 800 W) with wave frequency of 2450 MHz. The dimensions of the PTFE-coated cavity of the microwave oven were 48.5 cm × 37.0 cm × 29.25 cm.

IOP Conf. Series: Earth and Environmental Science **101** (2017) 012016 doi:10.1088/1755-1315/101/1/012016

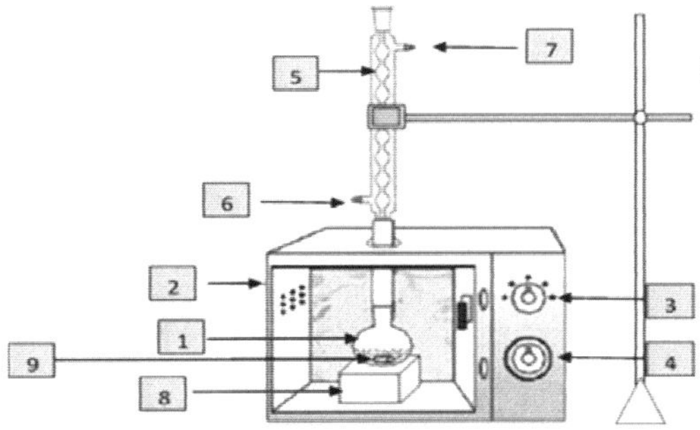

Figure 1. The series of transesterification tools

Transesterification is a major step in the process of making biodiesel as it aims to convert triglycerides contained in vegetable oils into Fatty Acid Methyl Ester (FAME) or biodiesel with glycerol byproducts.

First a mixture of methanol - a catalyst in according to variables are prepared, add acid catalyst with concentration according to variable , Running microwave radiation process in microwave oven with power and time of radiation according to variable are added with stirring using magnetic stirrer, Wash the solution using hot aquades (temperature 60 0C) inside the separator funnel, stop the heating and cooling the mixture separating the precipitate entering the reaction product into a separating funnel add the n-hexane to form three layers. Layer on FAME, the middle layer and bottom layer residual methanol in the form of glycerol, take the top screen and wash 3x with distilled water, biodiesel has been separated washed into the oven at 110 ° C.

3. Results and discussion

3.1 Effect of microwave power on yield

Biodiesel yield is a comparison between the mass of biodiesel (FAME) produced (transesterification product) with the mass of nyamplung seed oil prior to esterification. The data (biodiesel mass) obtained from the experiment is not purely biodiesel due to impurities during the process.

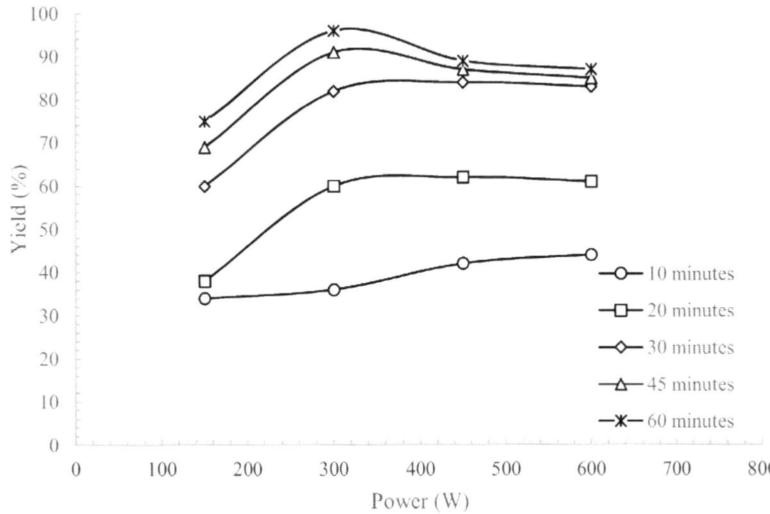

Figure 2. Effect of amount of catalyst to the biodiesel yield at 30 minutes.

Selection of 1% (w/w) base catalyst concentration as the best basic catalyst concentration (optimal) of the three catalyst concentration variables tested, since it yielded the largest yield in each radiation time. From the results of this trial which is represented on both images, the data pattern formed on the power and yield relationship, from 150 W to 300 W power occurs yield increases, whereas from 300 W to 600 W yields tend to decrease. At a power above 300 W, the reactant becomes hotter and methanol evaporates more frequently to meet the reflux condenser so that the contact between methanol and oil at a power above 300 W is less frequent compared to lower power. If using an alkaline catalyst, the use of power above 300 W, will cause a saponification reaction. As in some existing studies, on an alkaline catalyst, too high a temperature will decrease the yield of biodiesel due to a side reaction (saponification).

3.2 Effect of Microwave Power on Product Density

Density is the ratio of the mass amount of a substance to its volume at a certain temperature. The lower the temperature, the biodiesel will be higher and vice versa. The presence of glycerol in biodiesel affects the density of biodiesel because glycerol has a fairly high density (1.26 g/cm^3). So if glycerol does not separate well from biodiesel, then the biodiesel density will increase. The seed oil of nyamplung was measured using 5 ml pycnometer. Density limit permitted by SNI 04-7182-2006 (0.850-0.890 g /ml at 40 °C).

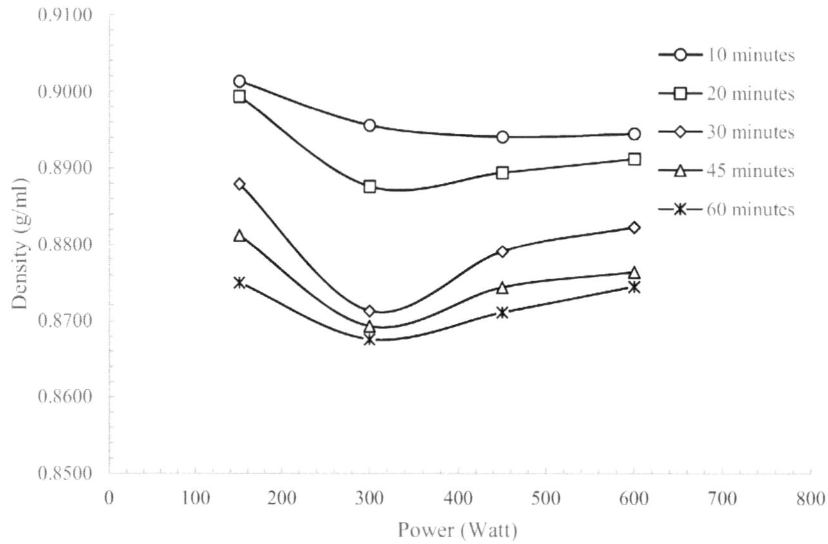

(a)

Figure 3. Effect of microwave power on product density with optimum catalyst Concentration 1% (w/w) oil with NaOH catalyst

Selection of 1% (w/w) base catalyst concentration as the optimum catalyst concentration of the three catalyst concentration variables tested. From these two figures, the data pattern formed on the power and density relationship, at 300 W power drops density of 150 W, whereas from 300 W to 600 W the biodiesel density increases. At 450 W and 600 W, the reactant faster to become hot and methanol more often evaporates to meet the reflux condenser so that the contact between methanol and oil on both power is less frequent than the lower power (150 W and 300 W). Since the contact between oil and methanol is rare, the conversion of triglycerides in oil to biodiesel is also low, resulting in increased density in transesterification products. In addition, the use of power above 300 W (high temperature), if using an alkaline catalyst, will cause saponification reaction. The presence of saponification reactions leads to an increase in product density. Ramadhas states that in the transesterification process of making biodiesel, will be produced methyl ester and glycerol. In the event of a saponification reaction in the transesterification process, the resulting biodiesel is more difficult to separate with glycerol. Glycerol, which is incorporated in biodiesel, causes the density in biodiesel to become larger, making it look more turbid [6].

3.3 Effect of Microwave Power on Product Viscosity

Nyamplung seed oil has a high kinematic viscosity, which is 53.4 cSt. (The Ministry of Forestry of the Republic of Indonesia (The Forestry Research and Development Agency.) 2008)). Given the transesterification of nyamplung seed oil into biodiesel, there will be a decrease in its kinematic viscosity. Where SNI 7182-2012 complies, the viscosity standard in biodiesel is 2.3-6.0 cSt. From the tests conducted on standard biodiesel (80% biodiesel and 20% diesel) the viscosity was 3.70 cSt. Where the value of this viscosity will be used as a reference in the analysis of this study.

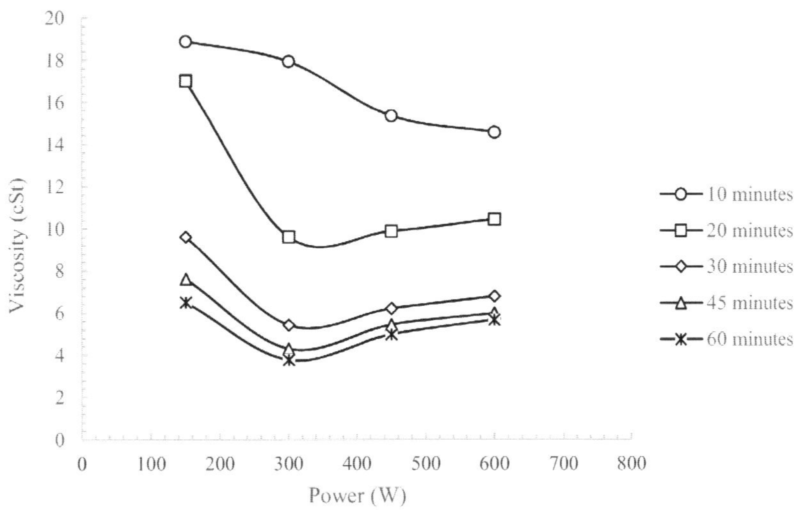

Figure 4. Effect of microwave power on product viscosity with optimum catalyst concentration 1% (w/w) oil with NaOH catalyst

Selection of 1% (w/w) base catalyst concentration as the best basic catalyst concentration (optimal) of the three catalyst concentration variables tested. From these two figures, the data pattern formed on power and density relationships, from 150 W to 300 W power, decreases viscosity, while from 300 W to 600 W the viscosity of biodiesel tends to increase. As in some existing studies, on an alkaline catalyst, excessively high temperatures will increase the viscosity of biodiesel due to side reactions (saponification). Ramadhas[6] stated that in the transesterification process of making biodiesel, will be produced methyl ester and glycerol. In the event of a saponification reaction in the transesterification process, the resulting biodiesel is more difficult to separate with glycerol. The glycerol in the biodiesel results in greater product viscosity, making it more turbid [6].

4. Conclusion

Microwave power affects the density, viscosity and yield of biodisel production. For an alkaline catalyst, the higher the power the lower the density and vicosity of the product, but the higher the power the higher yield of the product. The power of more than 300 W is drop, resulting in the production of biodiesel using the optimum base catalyst At 300 W power. The highest yield on 300 watts of power is 96 % with a reaction time of 60 minute.

References

[1] Soerawidjaja 2006 *Raw Material Aspect of Biodiesel Production in Indonesia* (AKIN Santo Paulus Semarang) pp 46-53
[2] S. H. Al-lwayzy T. Yusaf, R. A. Al-juboori, and B. Road 2014 *Biofuels from nabati* (Energies) pp 1996-1073
[3] Masyhud 2008 *Tanaman Nyamplung Berpotensi Sebagai Sumber Energi Biofuel* (Universitas Negeri Semarang) pp 20-2
[4] A. Prihanto 2008 *Peningkatan Yield Biodiesel dari minyak Biji Nyamplung Melalui Transesterifikasi Dua Tahap* (Universitas Maulana Malik) pp 8-44

[5] Kusmiyati 2008 *Reaksi katalitis esterifikasi asam oleat dan metanol menjadi biodiesel dengan metode distilasi reaktif* (Universitas Muhammadiyah Surakarta) chapter 12 pp 78-82
[6] Ramadhas 2010 *Biodiesel from nyamplung oil (AKPRIND Yogyakarta)* pp 130-6

Effect of Additional Suji Leaves and Turmeric Extract on Physicochemical Characteristic and Antioxidant Activity of Arenga-Canna Noodle

Miftakhussolikhah[1], D Ariani[1], ERN Herawati[1], A Nastiti[2], M Angwar[1], Y Pranoto[2]
[1]Research Unit for Natural Product Technology, Indonesian Institute of Sciences, Gading, Playen, Gunung Kidul, Yogyakarta, Indonesia
[2]Faculty of Agricultural Technology, Gadjah Mada University, Yogyakarta, Indonesia

E-mail: miftalipi@gmail.com

Abstract. Canna can be used as raw material for noodle but need a substitute material such as arenga starch. Arenga-canna noodle has dark appearance. Addition coloring agents from suji leaves and turmeric extract was done to improve product appearance and its functional characteristics. In this study, noodle was made with five variations of suji leaves and turmeric extract. Physical and chemical properties of noodle were analyzed. The results showed addition suji leaves extract and turmeric extract 0.4 g suji leaf/ml water and 0.06 g turmeric/ml water respectively, produce the best arenga-canna noodle quality. The addition of natural coloring agents increased antioxidant activity.

1. Introduction

Starch noodle (vermicelli) is a noodle that consumed by Indonesian people. Starch noodle is a flour-based product preferred by most Indonesians. Demand and consumption of noodle was significantly increase. However, it was not followed by raw materials availability, especially mung bean that remains to be imported for fulfillment. On the other hand, canna (*Canna discolor*) root is one of potential carbohydrate source but the utilization was still limited. So far canna utilization was still limited in starch form and the flour was not use optimally. Canna flour can be used as raw material in food processing such as starch noodle. We try to utilize flour rather than starch to explore canna utilization for noodle raw material.

Some research that investigate local material which can substitute noodle raw material has been conducted. According Li and Vasanthan [1] ideal starch for noodle raw material should have high amylose content, limited degree of swelling granules, low breakdown, and type C brabender curve. Canna root has relatively low amylose content so need a substitution material to improve canna noodle characteristics. Based on Alam [2], arenga starch has suitable characteristics for starch noodle, so arenga starch could be used for canna noodle substitute material.

Study about arenga-canna noodle characteristics has been conducted by Yosieto [3]. From the result, the best noodle produced from 75% arenga starch and 25% canna flour. The product has the closest physical properties to commercial products and the most preferred by the panelists. However, the appearance of noodle was less attractive and the panelist did not like with the color. Starch noodle from canna flour and arenga starch has dark color and ugly appearance. Therefore it is necessary to add dye toimprove physical appearance of arenga-canna noodle. Addition of natural coloring agents from suji leaves extract and turmeric extract was done to improve product appearance and its functional characteristics.

In this study, starch noodle was made with 75% arenga starch and 25% canna flour. Dye was added with five variations suji leaves extract at 0.4 g suji leaves/ml water; 0.6 g suji leaves/ml water 0.8 g suji leaves/ml water 1.0 g suji leaves/ml water and 1.2 g suji leaves/ml water and for turmeric extract at 0.06 g turmeric/ml water; 0.12 g turmeric/ml water; 0.18 g turmeric/ml water; 0.24 g turmeric/ml water and 0.30 g turmeric/ml water. Starch noodle without addition of natural coloring agents was used as control. Physical properties of noodle was analyzed. Chemical composition of selected starch noodle was then analyzed. The subjectives of this study are to determine the concentration of suji leaves and turmeric extract which produce best arenga-canna starch noodle quality and to improve its antioxidant activity.

2. Materials and Method

2.1. Materials

The main material used in this study was canna flour and arenga starch. Canna flour made from eight months old fresh canna bulbs. Arenga starch was bought from local industry. Suji leaves and turmeric was bought from local market. Maize starch noodle and rice noodle (Superior, PT Tiga Pilar Sejahtera) used as comparator commercial products.

2.2. Noodle with addition suji leaves and turmeric dye extract
Noodle was made from 75% arenga starch and 25% canna flour with dye extract addition.

Table 1. Variation of suji leaves and turmeric extract

Code	Addition of dye extract
GS 1	0,4 gram suji leaves / ml water
GS 2	0,6 gram suji leaves / ml water
GS 3	0,8 gram suji leaves / ml water
GS 4	1,0 gram suji leaves / ml water
GS 5	1,2 gram suji leaves / ml water
GK 1	0,06 gram kunyit / ml water
GK 2	0,12 gram kunyit / ml water
GK 3	0,18 gram kunyit / ml water
GK 4	0,24 gram kunyit / ml water
GK 5	0,30 gram kunyit / ml water

2.3. Determination of noodle physical properties before and after cooking,
Noodle strength was analysed based on Suryani [4], tensile strength and elongation of noodle was determined according to Chen *et al* [5].
2.4. Determination of noodle chemical characteristics
Water content, ash content, protein content and fat content determined based on AOAC method [6]. Antioxidant capacity was conducted according to Rohman and Riyanto [7].

3. Results

3.1. Physical properties arenga-canna noodle before and after cooking
Physical properties arenga-canna noodle with natural dye addition before and after cooking, including strength of raw noodle, tensile strength and elongation of cooking noodle were shown in Table 2.

Table 2. Physical properties arenga-canna noodle with natural dye addition

Code	Strength (N)	*Tensile Strength* (N)	Elongation (%)
GK 1	$3{,}21^{b}$	$0{,}23^{d}$	$21{,}16^{a}$
GK 2	$2{,}80^{b}$	$0{,}14^{c}$	$13{,}17^{a}$
GK 3	$2{,}65^{b}$	$0{,}08^{ab}$	$12{,}40^{a}$
GK 4	$1{,}71^{a}$	$0{,}05^{ab}$	$11{,}21^{a}$
GK 5	$1{,}16^{a}$	$0{,}03^{a}$	$10{,}84^{a}$
Control	$4{,}93^{c}$	$0{,}28^{d}$	$22{,}46^{a}$
Commercial A	$>400^{d}$	$0{,}10^{bc}$	$92{,}47^{b}$
Commercial B	$>400^{d}$	$0{,}05^{a}$	$20{,}84^{a}$
GS 1	$3{,}24^{d}$	$0{,}13^{c}$	$28{,}98^{b}$
GS 2	$2{,}57^{c}$	$0{,}12^{c}$	$19{,}05^{ab}$
GS 3	$2{,}32^{c}$	$0{,}10^{bc}$	$15{,}37^{ab}$
GS 4	$1{,}18^{b}$	$0{,}05^{ab}$	$12{,}37^{ab}$
GS 5	$0{,}63^{a}$	$0{,}02^{a}$	$7{,}54^{a}$
Control	$4{,}93^{e}$	$0{,}28^{d}$	$22{,}46^{b}$
Commercial A	$>400^{f}$	$0{,}10^{bc}$	$92{,}47^{c}$
Commercial B	$>400^{f}$	$0{,}05^{ab}$	$20{,}84^{b}$

The mean values within the same column followed by different superscript letters differ significantly ($p < 0.05$).

3.2. Chemical properties of arenga-canna noodle
Chemical properties arenga-canna noodle with natural dye addition was shown in Table 3.

Table 3. Chemical properties arenga-canna noodle

Chemical properties	Sample		
	GS 1 (%)	GK 1 (%)	Control (%)
Water content	11,60[b]	11,95[c]	11,77[bc]
Ash content	1,03[b]	0,88[a]	0,98[b]
Fat content	0,02[a]	0,05[a]	0,11[a]
Protein content	0,74[c]	0,53[ab]	0,44[a]
Carbohydrate content (*by difference*)	86,60[a]	86,59[a]	86,76[a]

The mean values within the same column followed by different superscript letters differ significantly ($p < 0.05$).

3.3. Antioxidant capacity of arenga-canna noodle

Antioxidant capacity of GS 1, GK 1, and controls noodle are presented in Table 4. Noodle with the addition of turmeric extract showed the highest antioxidant content.

Table 4. Antioxidant capacity of arenga-canna noodle

Sample	Antioxidant (%)
Control	0,03[ab]
GS 1	0,10[ab]
GK 1	0,58[c]
Arenga starch	0.00[a]

The mean values within the same column followed by different superscript letters differ significantly ($p < 0.05$).

4. Discussion

Statistical analysis (Table 2) showed that all canna-arenga noodle formula with addition suji leaves and turmeric resulting lower strength value than control and commercial product. More suji leaves and turmeric extract added to canna-arenga noodle, will produce weaker canna-arenga noodle, so resulting lower quality of noodle. The quality of noodle increase as the strength value (N) of noodle increase, because the noodle was more difficult to break. Strength of noodle influenced by bonds between molecules in the starch granules. The higher substitution of high amylose starch, more amylose molecules bond in the starch granules that causes the bonds between the starch molecules is getting stronger and not easily broken [4]. Thus, the addition of extract suji leaves and turmeric which does not contain starch in higher concentrations will interfere formation of bonds between amylose in starch granules, so arenga-canna noodle will be more easily broken.

Canna-arenga noodle with addition suji leaves and tumeric extract resulting different tensile strength compare to control products. Higher concentration of suji leaves and tumeric extract resulting lower tensile strength. Suji leaves and turmeric extract would add non-starch substances and will disrupt the bond between amylose molecules in crystalline form, so that the compactness of granular structure decreased, therefore lower force required to break the noodle.

However, higher concentration of turmeric and suji leaves extract resulting lower elongation. Elongation of arenga-canna noodle was proportional to the tensile strength. This is due to increased cohesiveness causing noodle elongation increased , so that the noodle will be more difficult to break [8]. Addition higher concentration of suji leaves and turmeric would add higher non-starch substances. Then it will disrupt the bond between amylose molecules in crystalline form, so that the compactness of granular structure will be decreased, therefore lower force required to break the noodle. As a result, the noodle quality decreases due the noodle would be easily broken when cooking.

Arenga-canna noodle with addition suji leaves and tumeric extract which selected based on physical characteristics then determinated their chemical characteristics, such as water content, ash content, protein content, fat content, carbohydrate content, amylose content, resistant starch (RS), and dietary fiber (Table 3). Chemical properties are properties that are important in determining the quality of arenga-canna noodle and to provide additional functional value.

Water content affect shelf life of food product. Higher moisture content of a food product affected greater degree of product deterioration during storage. The result showed that water content of noodle with addition natural dye extract from suji leaves and turmeric significantly different from controls. Water content of noodle affected by the type of material, the components contained, drying temperature, drying time, tools, material thickness and material size. Water content quality requirements for starch noodle based on SNI 01-3723-1995 was 14.5% (w / w). The water content of arenga canna noodle is still lower than SNI requirement, so all arenga canna noodle meets the requirements. The ash content indicates food inorganic substances levels in mineral elements form [9]. The presence of inorganic substances in noodle will affect the darker color of noodle caused by oxidation [10]. The ash content of arenga canna noodle GS1, GK1, and control were 1.03%; 0.88%; and 0.98% respectively. Maximum ash content according to SNI 01-3723-1995 for starch noodle is 0.5% (w / w). High levels of ash can be due to calcium, sodium, phosphorus and other minerals contained in the raw material of arenga-canna noodle. Fat content is not the quality parameters of noodle according to SNI 01-3723-1995. Low fat content of can prevent rancidity caused by oxidation of fatty acids [11].

Protein content is one of important parameter because it is directly related to the nutritional value. The amount of protein content will determine consumers acceptance, because protein is a nutritional compound which is also required by the human body. However, the protein content is not the quality parameters of noodle according to SNI 01-3723-1995. The protein values of noodle with natural pigments were significantly different and tend to be lower when compared with noodle control. This indicates that protein component in the tubers are a minor components. The protein content increased due to the addition of suji leaves and turmeric extract. Charles et al. [12] stated that the presence of proteins in various kinds of noodle will contribute to the texture properties of noodle.

Amylose content have very strong influence on the product characteristics. High amylose content on a product will simplify product encounter retrogradation. Starch with high amylose content was suitable for making noodle [13]. Extract natural dyes from plants that are used do not contain starch (non-starch). More non-starch ingredients were added, will made lower amylose content in starch noodle. Frei [14] classified amylose content as high amylose content (> 25%), moderate (20.1-25%), low (12.1-20%), very low (5.1 -12%) and glutinous (0-5%). Thus it can be said that the entire noodle products classified as high amylose.

The existence of resistant starch (RS) can be affected by various factors including processing, type of starch (amylose amylopectin ratio), physical characteristic of material (degree of hydration, particle size) and the presence of other components such as lipid, protein, etc [15]. Results of statistical analysis at significance level of 95% showed significantly different results between arenga canna noodle and control noodle. Arenga starch noodle containing 49.08% RS, while noodle control resulting lower RS content. Noodle control only contains 75% arenga starch so have lower amylose content. Higher RS content affected by high amylose content and due to processing.. It is also of the manufacturing processes in the manufacture of vermicelli done steaming, cooling. According Marsono [15], the process of heating or boiling, followed by other processes can improve RS content.

Dietary fiber is defined as a part of the plant that is not digested by human digestive tract enzymes including constituent of plant cell walls (cellulose, hemicellulose, pectin and lignin) and gum [16]. Dietary fiber content in arenga canna noodle at significance level of 95% have significantly different to noodle control. The fiber content is influenced by raw material used in noodle. Noodle from 100% arenga starch had lower dietary fiber because only contain arenga starch, while control noodle had higher dietary fiber because there was canna flour as substitute material. Processing could give an impact dietary fiber content in noodle. According Indriyani [17] longer heat treatment process will lead to more dietary fiber components damage. Thus the longer drying process, the more dietary fiber will be hydrolyzed.

Based on Table 4, noodle with the addition of turmeric extract showed the highest antioxidant content. It is caused by the content of antioxidants in turmeric is curcumin which resistant to heat treatment. Antioxidants contained in suji leaf is chlorophyll, which sensitive to heat. In drying process, noodle quality changes may occur such as degradation of antioxidant activity. Low antioxidant content of noodle with addition of suji leaf due to changes in chlorophyll into derivatives, such as feofitin and other chlorophyll derivatives that have antioxidant activity which is lower than the native chlorophyll [18]. The higher temperature, the higher rate of reaction, so antioxidant decreased faster. Antioxidants are compounds that can delay or prevent oxidation reactions of free radicals. Antioxidants react with free radicals that can inhibit free radicals to cause damage. In food stuffs, antioxidant found in many vegetables and tubers.

5. Conclusion

Addition suji leaves extract and turmeric extract 0.4 g suji leaf/ml water and 0.06 g turmeric/ml water respectively, produce the best arenga-canna noodle quality. Addition natural coloring agents increased antioxidant activity

Acknowledgements
This study was part of the research program of the Indonesian Institute of Sciences. The authors wish to thank

Professor Haryadi from the Gadjah Mada University for providing guidance.

References

[1] Li J H and Vasanthan T 2003 Hypochlorite Oxidation of Field Pea Starch and Its Suitability for Noodle Making Using an Extrusion Cooker *Journal of Food Research International* **36** 381-386
[2] Alam N 2008 Sifat Fisikokimia dan Sensoris Instant Starch Noodle Pati Aren yang Disubstitusi dengan Pati Tapioka *Journal Agroland* **15** 191–197
[3] Yosieto E 2014 Karakterisasi Sifat Fisik dan Sensoris Noodle Campuran Tepung Ganyong (Canna edulis Kerr.) dan Pati Aren (Arenga pinnata Merr.) *Skripsi* Universitas Gadjah Mada Yogyakarta
[4] Suryani C L 1999 Pemutihan dan Pengikatan Silang Pati Sagu untuk Substitusi Beras pada Pembuatan Bihun. *Thesis* Program Pasca Sarjana. Universitas Gadjah Mada Yogyakarta.
[5] Chen Z L, Sagis L, Legger A, Linssen J P H, Schols H A and Voragen A G J 2002 Evaluation of Starch Noodles Made from Three Typical Chinese Sweet Potato Starches *Journal of Food Science* **67** 3342-3347
[6] AOAC 1984 *Official Methods of Analysis* Washington Association of Official Analytical Chemists
[7] Rohman A and Riyanto S 2005 Daya Antioksidan Ekstrak Etanol Daun Kemuning (Murraya paniculata (L) Jack) secara in vitro *Majalah Farmasi Indonesia* **16** 136-140
[8] Fitriasari D 2013 Karakteristik Tepung Kentang Hitam Yang Dibuat Dengan Cara Pemarutan Dan Pengaruhnya Pada Subtitusi Pati Aren Pada Pembuatan Sohun *Skripsi* Fakultas Teknologi Pertanian. Universitas Gadjah Mada Yogyakarta.
[9] Winarno F G 2008 *Kimia Pangan dan Gizi* Gramedia Pustaka Utama Jakarta
[10] Hou G G 2011 *Asian Noodles: Sciences, Technology and Processing* John Wiley & Sons Inc, New Jersey
[11] Fennema O 1996 *Food Chemistry : Third Edition* Marcel Dekker Inc, New York
[12] Charles A L, Chang Y H, Ko W C, Sriroth K and Huang T C 2005 Influence Amylopectin Structure and Amylose Content on Gelling Properties of Five Cultivars of Cassava Starches *Journal of Agricultural and Food Chemistry* **53** 2717-2725.
[13] Hermann M 1996 Starch noodles from edible canna In: J. Janick (ed.) *Progress in new crops* ASHS Press, Arlington VA 507-508
[14] Frei M 2003 Studies on in vitro starch digestibility and the glycemic index of six different indigenous rice cultivars from the Philippines *Food Chemistry* **83** 395-402
[15] Marsono Y 1998 Review Resistant Starch: Pembentukan, Metabolisme dan Aspek Gizinya. *Agritech* **18** 29-35
[16] Spiller G A 2001 *Handbook of Dietary Fiber in Human Nutrition* CRC Press LLC.USA.
[17] Indriyani A 2007 Cookies Tepung Garut (*Maranata arundinaceae* L) Dengan Pengkayaan Serat Pangan. *Skripsi* Fakultas Teknologi Pertanian. Universitas Gadjah Mada Yogyakarta.
[18] Rufaida R 2008 Pembuatan Minuman Klorofil Daun Suji (*Pleomele angustifolia* N.E Brown) dan Evaluasi Mutunya Selama Penyimpanan *Skripsi* Fakultas Teknologi Pertanian. Institut Pertanian Bogor

IOP Conf. Series: Earth and Environmental Science **101** (2017) 012018 doi:10.1088/1755-1315/101/1/012018

Improving Properties of Arrowroot Starch (*Maranta arundinacea*)/PVA Blend Films by Using Citric Acid as Cross-linking Agent

Enny Sholichah[1,2], Bambang Purwono[2], Pramono Nugroho[1]

[1]Development Centre of Appropriate Technology Indonesian Institute of Sciences, Jl. KS. Tubun No. 5 Subang West Java, Indonesia, 41213
[2]Department of Chemistry, Faculty of Mathematics and Sciences Universitas Gadjah Mada, Sekip Utara Bulaksumur Yogyakarta Indonesia, 55281

E-mail address: enny.sh002@gmail.com

Abstract. This research studied the effect of PVA as organic polymer and citric acid as crosslinker agent in the arrowroot starch/PVA blend films. The properties of films were investigated by water uptake, water vapor permeability, mechanical properties, thermal stability, spectra of FTIR and XRD patterns. PVA used in this research influenced the film properties at the highest concentration. The cross-linkingsinter or intra molecules of arrowroot and PVA were developed as ester bonds which are formed from the reaction of hydroxyl groups consisting of starch and PVA with citric acid. The ester bond was confirmed by FTIR spectra. The increase of the amount of citric acid affected significantly on physical, chemical and mechanical properties, water uptake, WVP and crystallinity. Water barrier level was reduced by decreasing of water uptake and WVP succeeded significantly with increased crosslinking. Cross-linking impact the thermal stability of the films. The elasticity of the films also increases the production of citric acid as a plasticizer in the making of the films as a food packaging material.

1. Introduction

Tubers starch is a potential carbohydrate as the material of film preparation because of their capability forming a good film and its abundance in every region [1–5]. Starch consists two main components they are linear polymer defined as amylose and branched polymer known as amylopectin [6].Yulianti and Ginting [7] have made edible film using various type of tubers starch there were cassava, sweet potato, ganyong (*Canna edulis*) and arrowroot. They have found that the arrowroot film was the best mechanical properties. The weakness of the starch edible film is having high water vapor transmission rate because starch is hydrocolloid material has many hydroxyl groups which can make the hydrogen bonds with water molecules [8].

Modification process is required to improve the performance of starch edible film to implement it as food packaging material to improve the quality and shelf life of the product. Cross-linking modification on starch was done to produce a better thermal stability, mechanical properties and water resintance characteristic [9–12]. Furthermore, blending starch with organic polymer like polyvinyl alcohol (PVA) and cross-linking modification in film preparation was used by Negim et al. [13] and Liu et al. [14] to improve the quality of film especially in water barrier function. PVA is widely used for food packaging ingredient because it has good stability on oil or lipid, tensile strength, high elasticity [15–17] and biodegradable polymer [13,18].

Ether or esther bond is a covalent bond between linear or branched polymer that formed in a cross-linking process. Cross-linking modification gains the degree of polymerization and molecular mass, and also decrease the water solubility [19]. Many compounds can be applied as the cross-linker agents such as citric acid, sodium trimetaphosphat, sodium tripoliphosphat, epichlorohydrin, glutaraldehid, sodium tetraborat, dan $POCl_3$ [9], [14], and [20].

In our present work, citric acid was used as the crosslinker agent for arrowroot starch and PVA in film preparation. Citric acid was chosen because it shows high effectivity as a crosslinker agent at a low level. The other advantage is citric acid has lower hazardous level than glutaraldehyde, epicholohydrin and sodium hypophosphite [9]. The effect of concentration of PVA and citric acid on the physical and mechanical properties, water uptake, water vapor permeability, thermal stability and the structure of film matrix has been studied.

2. Experimental method

2.1. Materials

Arrowroot starch was obtained from UKM Sukabumi Yogyakarta, Polyvinyl alcohol (PVA) fully hydrolyzed was produced by Sigma Aldrich and citric acid by Merck.

2.2. Procedures

2.2.1. Film preparation

Arrowroot starch solution (5%) was heated and stirred until gelatinized (T: ± 70 °C), then citric acid was added. Heating and stirring were continuously proceeded at 80 °C for 10 minutes. This mix was called solution A. Solution B was prepared by dissolving PVA in water, heated and stirred until perfectly dissolved. Subsequently, citric acid was added into PVA solution. This mix was heated and stirred at 80 °C for 10 minutes. The amount of citric acid in both solutions was same. The Solution A and B were mixed at 80 °C for 10 minutes. At the end, the final solution was poured into Petri-dish and dried in the oven dryer at 50 °C. The films were stored in the tight plastic box at room temperature for at least 24 hours before characterization.

Table 1. Composition of films

Sample name	Arrowroot Starch (g)	PVA		Citric acid	
		(g)	(%)	(g)	(%)
P1C1	2	0.1	0.25	0	0
P1C2	2	0.1	0.25	0.1	0.25
P1C3	2	0.1	0.25	0.2	0.50
P1C4	2	0.1	0.25	0.3	0.75
P2C1	2	0.2	0.50	0	0
P2C2	2	0.2	0.50	0.1	0.25
P2C3	2	0.2	0.50	0.2	0.50
P2C4	2	0.2	0.50	0.3	0.75
P3C1	2	0.3	0.75	0	0
P3C2	2	0.3	0.75	0.1	0.25
P3C3	2	0.3	0.75	0.2	0.50
P3C4	2	0.3	0.75	0.3	0.75
P4C1	2	0.4	1	0	0
P4C2	2	0.4	1	0.1	0.25
P4C3	2	0.4	1	0.2	0.50
P4C4	2	0.4	1	0.3	0.75

The total amount of ingredient was 40g by adding water

2.2.2. Water Vapor Permeabiliity (WVP)

Water vapor permeability of films was determined based on the ASTM E96-00 [17]. The films were sealed on the weighing bottle (diameter ± 2.5cm) filled by silica gel (± 5 g). and then placed in the desiccator where the relative humidity (RH) was controlled at ±78% using saturated solution of NaCl. The weight of the bottles was measured every hour for 7 hours. The weight differences in the interval period (slope) were calculated to determine water vapor transmission rate(WVTR). The last step was calculation the WVP according to equation (1 & 2) below:

$$WVTR = \frac{slope}{A} = \frac{\Delta m}{\Delta t \; x \; A} \quad (1)$$

$$WVP = \frac{WVTR \; x \; L}{\Delta P} \quad (2)$$

Where L is the thickness of film (mm) and ΔP is the difference of Partial pressure of water vapor beetwen the inner and outer of film.

2.2.3. Water Uptake

The water uptake of films was measure according to the method described by Setiani et al. [22], the (2x2cm) films were dried in the oven dryer at 100-105 °C for 1 hours, then stored in the desiccator for 30 minutes and weight (initial weight). The next step was deeping the film in the water for 10 seconds, swabbing the remaining water on the film surface and weighing the films. This step was repeated until obtained the constant weight (final weight). The water uptake was calculated by equation (3).

$$Water\ uptake = \frac{\text{Final weight} - \text{Initial weight}}{\text{Initial weight}} x\ 100\% \quad (3)$$

2.2.4. Mechanical properties

The tensile strength and elongation were measured by using Texture Analyzer Zwick BL-GRS500N in accordance with the ASTM D 882 as used by [22]. The tensile strength and elongation were obtained by calculation using equations (4) dan (5).

$$\tau = \frac{F_{max}}{A} \quad (4)$$

$$\text{Elongation } (\%) = \frac{d_r}{d_0} \text{ x } 100\% \quad (5)$$

Where$\tau$is tensile strength (MPa); $F_{max}$is the maximum force A is the area of film cross-section (mm^2); d_r is the distance of rupture (mm) and d_0 is the distance onset of separation (mm).

2.2.5. Thermal Analysis

Thermal stability of film was analyzed usingLinseis–DTA PT1600. Sample (3 mg) was sealed on the aluminum pan and heated at 25 °C - 500 °C which the heating rate was 10°C/min.

2.2.6. Fourier Transform Infra Red (FTIR)

Identification of chemical structure of films modified cross-linking was carried out by FTIR Thermoscientific Nicolet iS- 10 in wave number 4000-500cm^{-1} and sensitivity 50cm^{-1}[3,23].

2.2.7. X-Ray Diffraction

X-ray diffractometer (RigakuSmartlab) scanned the films at 2θ: 5°- 40°, rate 2°/min. The instrument was set at 30 mA and 40 kV and use x-ray of Cu Kα with λ 1.54°A. The relative cristalinity was calculated according the equation (6) [24].

$$\text{Relative cristalinity} = \frac{\text{Cristaline Area}}{\text{Cristaline Area} + \text{Amorhous area}} \text{ x } 100\% \quad (6)$$

2.2.8. Statistical analysis

The effect of PVA and citric acid in the film properties was determined by Analysis of variance (ANOVA) and post-hoc (Duncan Test, P<0.05) test was performed to find out the difference between the variables. SPSS version 2016 programwas employed for Statistical analysis

3. Result and Discussion

3.1. Water Vapor Permeability (WVP)

WVP is a parameter of film properties that greatly affect the function of film as packaging material to increase the shelf life of food products. High water activity can accelerate the growth of microorganisms as contaminants that cause damage to a food product. Water vapor permeability describes the movement of water vapor from two different places or systems that are the environment inside and outside the film [25].

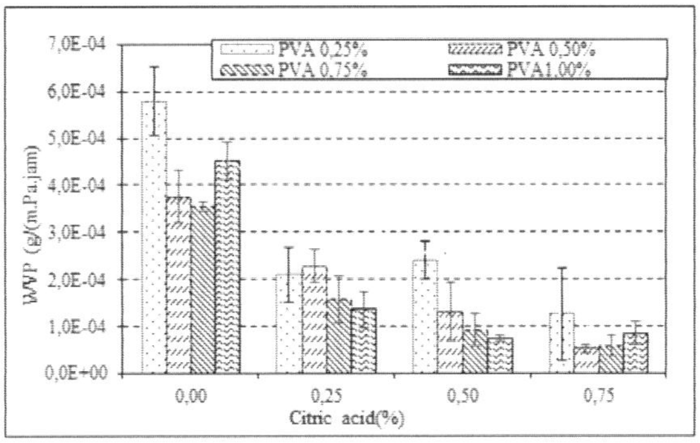

Figure 1. Water vapor permeability of films

PVA concentrations of less than 0.75% did not significantly influence the WVP but the addition of 0.25% citric acid was able to significantly decrease the WVP. Interaction of both also impact to the decreasing of the WVP of the films. Debeaufort and A. Voilley [25] confirmed that the material composition and film structure may affect the molecular transfer mechanism of water and the way to prevent the entry of water molecules. Citric acid served as in the formation of ester bonds intra and inter molecule starch and PVA causemolecules of starch and PVA increasingly compacted so that blocked the water molecule transfer. Furthermore, the ester bond formed reduces the hydroxyl group on starch and PVA where the hydroxyl groups are the active site of hydrogen bonds. Hydrogen bonding is the interaction between the water molecules and hydroxyl groups of starch and PVA (Figure 2).

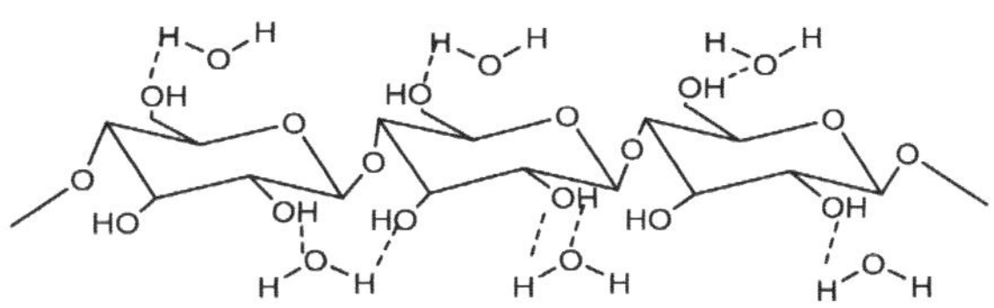

Figure 2 Hydrogen bonding between water molecule and starch

3.2. *Water Uptake*

The water uptake determination aimed to characterize the water resistance of the films. The small value of water uptake indicate the better level of the water resistance and vice versa [26].

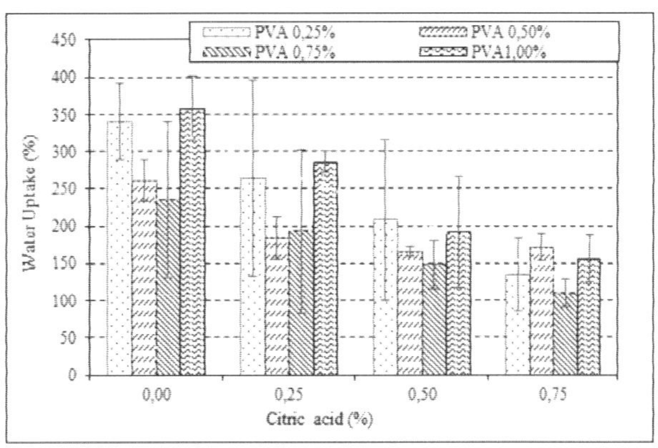

Figure 3. Water uptake of films

The concentration PVA (1%) significantly changed the water uptake but not in PVA 0.25-0.75%. While the increase of citric acid concentration influenced the water uptake of films significantly starts from 0.25%. It means that the higher the concentration of citric acid the smaller the water absorption. This result showed that cross-linking was successful to reduce the water absorption capacity of the film, it was equal with the study of Seligra et al. [23].

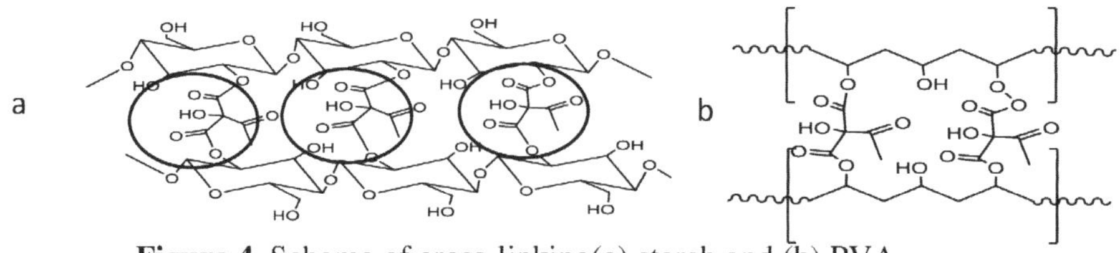

Figure 4. Scheme of cross-linking(a) starch and (b) PVA

3.3. Mechanical properties

Garcia et al. [27] stated that physical, chemical and temperature conditions are able to affect the mechanical properties of the film. Moreover, the interaction between the components of film impact the mechanical properties. Tensile strength is the maximum force given until the film is broken by the width of the cross-section area. while the elongation is the length of the strain until it is broken by the initial length.

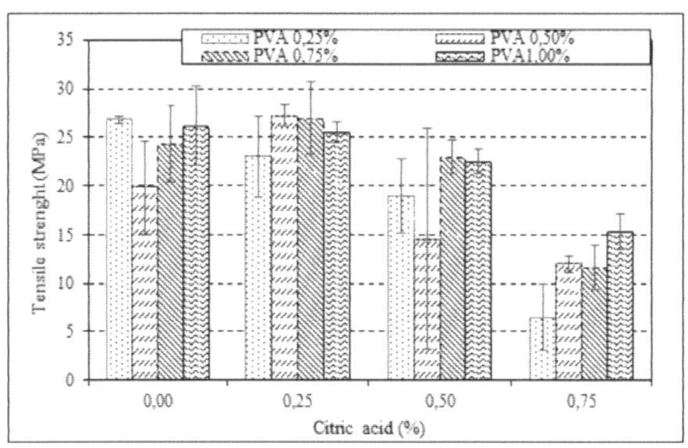

Figure 5. The tensile strength of films

The effect of PVA level is described in figure 5, where the concentrations of PVA 0.25-0.75% did not significantly impact the tensile strength of the films. The tensile strength of the film increased with the addition of 1% PVA. This result means that the PVA level less than 1% did not change the tensile strength of the films. It is equal to the results of Rapa et al. [18] and Azahari et al. [28] which the tensile strength of PVA films decreased with increasing amounts of corn starch addition. Gelatinized starch is at amorphous structure so that decreases the tensile strength of the film. Negim et al. [13] and Sadhu et al. [29] reported that every PVA monomer contains a hydroxyl group that can form hydrogen bonds so it is possible to increase the interaction strength in the matrix and produces higher tensile strength.

The use of citric acid as a crosslinking agent at concentrations of 0.5% and 0.75% reduced the tensile strength of the film significantly, this result is similar to [30] and [31]. The amount of citric acid used in film preparation greatly influences the mechanical properties of the film. According to the study of Reddy et al. [9], cross-linking between starch molecules that can increase the molecular weight of starch so that the tensile strength of the films is higher. The addition of citric acid less than 5% (dry basis) affects the increase of tensile strength of the film. According to Menzel et al.[10], the use of citric acid causes two reactions that are cross-linking and hydrolysis. At pH 2, the larger cross-linking as well as the hydrolysis process is indicated by the decreasing of the molecular weight of starch. Therefore. the decrease of tensile strength on the addition of citric acid 0.5 and 0.75% could be affected by the hydrolysis of glycoside bond in starch.

The elongation of the film did not change significantly with the addition of PVA (0.25-1%). The addition of citric acid (0-0.5%) did not significantly also affect the elongation of films. While concentration at 0.75%, the elongation increased significantly. This result is consistent with the study of Negim et al. [13] and Yoo et al. [30]. The hydrolysis reaction of the glycosidic bond in starch decreased the molecular weight of the starch which could increase the mobility of the starch molecules. Furthermore, the hydrolysis reaction also generated the amount of free OH groups that can form hydrogen bonds and proceed greater strain.

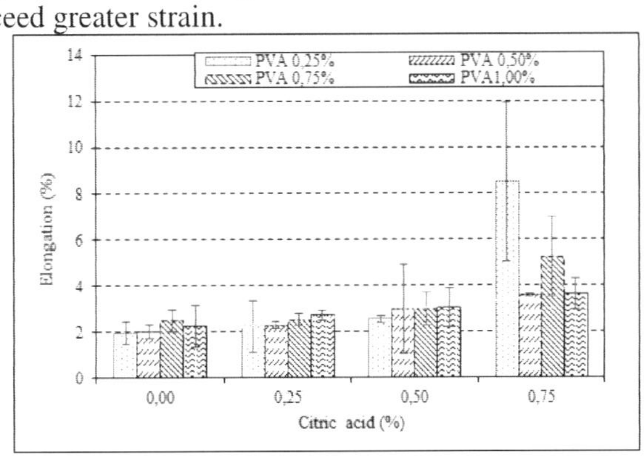

Figure 6. The elongation of film

3.4. Thermal analysis

DSC thermogram (Figure 7) shows that non-modified cross-linking film (P1C1) shows an endothermic peak at <100°C which is the temperature of gelatinizing arrowroot starch. These results were consistent with the thermogram of DSC starch arrowroot resulted by the study of Pratiwi [32] where the gelatinized temperature was at 69.17-80.03 °C, likewise Valencia et al [36] obtained the temperature of gelatinization of arrowroot starch is 65.5 ± 0.1°C. Inter and intra molecules interaction between the starch and PVA were developed by cross-linking due to the changes of structure of arrowroot starch so that there is no endothermic peak at that temperature.

Hemalatha [33] reported the DSC thermogram of pure PVA showed the dehydration temperature (Td), the glass transition temperature (Tg) and the melting temperature (Tm) were 30 °C, 114 °C and 222°C respectively. Figure 7 shows a wide peak at a temperature of 150-250 °C. It means that the glass transition process of the film occurs at this temperature, where crosslink modification leads to an increase in the glass transition temperature (Tg) of film. The melting process of the film occurs at a temperature of 243-335 °C (Table 2). The crosslink modification shifts the initial melting temperature (onset melting temperature) to lower and the end-point temperature becomes higher, this shifts due to cross-link stabilize the starch and PVA structure in film [9]. The widening endothermal peak corresponds to the crystallinity of the film. The widened peak indicates the crystallinity of the sample decreases [34]. The uncrosslinked films (P1C1) have a peak at a temperature of 244-267 °C which is possible because of the interaction between the components of arrowroot starch and PVA in the film matrix. This peak does not appear in the cross-linked films.

The enthalpy of melting (ΔHm) indicates the amount of energy required to melt the film matrix. The melting temperature (Tm) is the midpoint of the endothermic peak [34]. Table 2 shows that the greater the amount of citric acid used the value of ΔHm is increasing. This result related to the increase in the amount of the crosslinks formed so that in the melting process it takes more energy to break the bonds formed in the film matrix. This result is consistent with the study of Maiti et al. [35] which reported that the melting point of the starch and PVA film increased due to cross-linking.

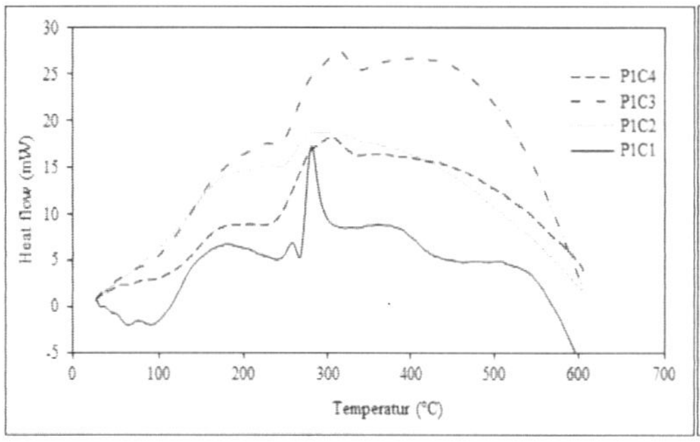

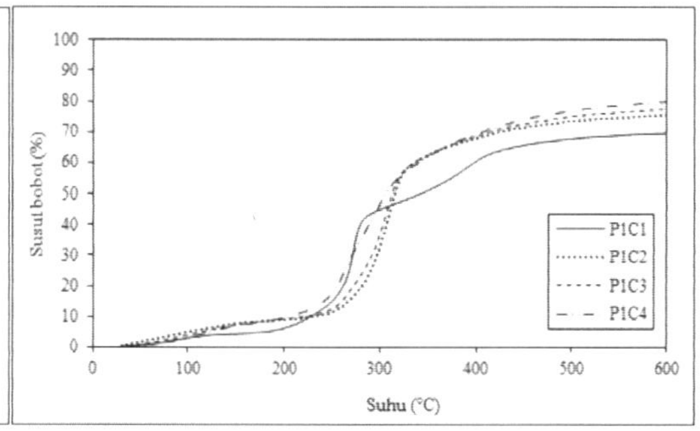

Figure 7 DSC Thermogram DSC of films **Figure 8**. TGA analysis of film

Table 2. The onset and end-point of melting temperature. melting temperature (Tm) and enthalpy of melting (ΔHm).

Sampel	Onset melting Temperature (°C)	End-point (°C)	ΔH_m (J/g)	T_m (°C)
P1C1	269.4	296.0	63.19	280.9
P1C2	252.0	335.8	96.37	280.5
P1C3	250.3	334.9	137.78	306.9
P1C4	246.0	331.1	147.05	282.9

ΔH_m (entalphyof melting); T_m (melting temperature)

The characteristics of film degradation due to heat are shown in figure 8. Uncrosslinked films (P1C1) undergo four stages of change: 206 °C, 289 °C, 405 °C and > 405 °C. Four stages are supported by the DSC thermogram, where the first stage is the water evaporation step (dehydration), then proceed with three stages of thermal decomposition of the film. The largest mass loss of uncrosslinked film occurs at temperatures of 200-250 °C and 250-350 °C for crosslinked films. The results of this study were equal with those reported by Hemalatha [33] stating that pure PVA has a melting point of 200-230 °C. The melting process is the beginning of the decomposition of a material. PVA decomposition through three stages. where the first step occurs at a temperature of 230-332 °C. The second stage can reach the temperature of 487 °C and the third stage at 487-645 °C. The mass loss in the second and third stages are relatively smaller than the first stage. The process of perfect decomposition is in the third stage. The degradation temperature of the crosslinked films shifted to be higher and only had 3 stages of thermal change. The temperature shift related to the presence of crosslinks between the arrowroot starch and PVA which could improve the thermal stability of the film as the results of the study [9].

3.5. FTIR

The FTIR spectra of the crosslinked film (Figure 9) appear peak at 1713 cm^{-1} that is attributed to the carbonyl group (C = O). This peak indicates the ester bonds between COOH groups of citric acid and OH groups on starch or PVA. Reddy et al. [9] and Seligra et al.[23], reported the similar result, where the formation of crosslinks by citric acid on starch showed the absorption bands at 1720 and 1724 cm^{-1}. The broad peak at 3300 cm^{-1} is characteristic of stretching O-H while the band 1148 cm^{-1} is a C-O group. The OH groups owned by starch and PVA molecules. The widespread form of the OH's peakis evident of the intermolecular interaction between starch and PVA hydrogen bonds [13]. The absorption bands at 935 and 890 cm^{-1} confirm the symmetric stretching of the C-O-C group [36].

The decrease of intensity of peak at 3300 cm^{-1} from 76.14% to 72.45% indicates the decrease of OH group due to the formation of ester bonds by citric acid. The intensity ratio of peaks at 3300 cm^{-1} and 1148 cm^{-1} (I_{3300} / I_{1148}) increased in the addition of citric acid from 1.12 to 1.31. It possibly due to the hydrolysis reaction on the glycoside bonds thus increasing the ratio of OH groups to C-O from C-O-C [23].

The peak at 1630 cm^{-1} was obtained in the uncrosslinked film spectra which indicates the presence of a strongly bonded water molecule in the film matrix. These results are consistent with the characteristics of the FTIR spectra of arrowroot starch of the study [36].

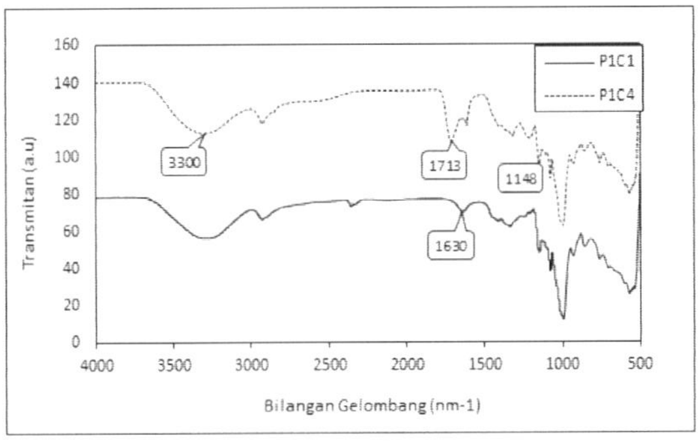

Figure 9. FTIR spectra of films

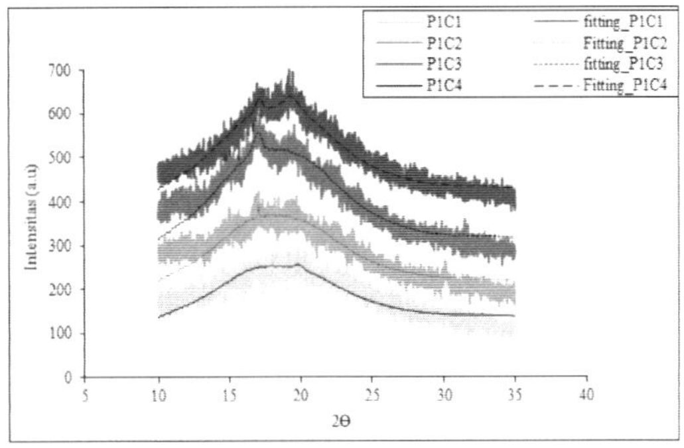

Figure10. XRD patterns of films

3.6. X-ray Diffraction

Diffractogram of uncrosslinked films shows no significant peak which means having an amorphous structure. The arrowroot starch is a native starch with a crystalline form of type A with the crystallinity is 20% [37]. Heating the arrowroot starch solution in the film preparation formed the gelatinized starch where the starch granules swell and adsorbed the water molecules thus destroying the crystalline structure of starch to form amorphous structure [6].

Table 3. The relative Cristalinity of films

Sample name	Relative Cristalinity [%]
P1C1	0.7
P1C2	0.2
P1C3	1.5
P1C4	6.1

Figure 10 shows the similarity of XRD patterns of the uncrosslinked and crosslinked films. The uncrosslinked film has a weak peak at 19.5°. While 0.25% of citric acid concentration (P1C2) appear a peak at 17°. The greater the citric acid concentration than the peak intensity at 17° becomes increased. At a 0.75% citric acid concentration indicates the presence of two peaks at 17° and 19.5°. According to the results of Faridah et al. [37], arrowroot starch is starch type crystalline A with peaks at 15°, 17°, 20° and 23°. Peaks at 17° and 19.5° indicate the reassociation of linear starch components to form a regular or crystalline structure. The reassociation is possible due to the formation of crosslinks reduce the distance between molecules of starch and PVA so that the greater the chance of forming intermolecular hydrogen bonds.

The crystallinity of the film increases with the addition of citric acid concentration (Table 3). This result is related to the increasing of the intermolecular interaction between arrowroot starch and PVA so that its crystallinity increases. According to Herlina [38], cross-link modification on starch increases the interaction of starch with polyfunctional compounds that can react with the -OH group on the amylose or amylopectin structure which can strengthen the hydrogen bonds in the starch chain.

4. Conclusion

The arrowroot/PVA blend film has been prepared by adding citric acid as a crosslinker agent. PVA used in this research influenced the film properties at the highest concentration. Citric acid greatly affected the properties of the films. The cross-linkings between arrowroot and PVA were developed as ester bonds which are formed from the reaction of hydroxyl groups consisting of starch and PVA with citric acid. The ester bond was confirmed by FTIR spectra. The increase of the amount of citric acid affected significantly on physical, chemical and mechanical properties, water uptake, WVP and crystallinity properties. Water barrier level was reduced by decreasing of water uptake and WVP succeeded significantly with increased crosslinking. The elasticity of the films also increases the production of citric acid as a plasticizer in the making of the films as a food packaging material.

Acknowledgment
The authors greatly thank those have supported this research especially to Indonesian Institute of Sciences, The Ministry of Research, Technology and Higher Education of the Republic of Indonesiafor funding this research and Synchrotron Light Research Institute – Thailand for providing SAXSIT software for XRD analysis.

References
[1] Ardiansyah Ryan 2011 Pemanfaatan Pati Umbi Garut Untuk Pembuatan Plastik Biodegradable 99
[2] Sarifudin A 2013 *Pencirian Bioplastik Komposit Tepung Singkong dan Natrium Alginat dengan Aditif* Limonen*{Bibliography}a Kulit Jeruk*
[3] Hernandez D P, Medina-jaramillo C, Lopez-Cordoba A and Goyanes S 2017 Edible cassava starch fi lms carrying rosemary antioxidant extracts for potential use as active food packaging *Food Hydrocoll.***63** 488–95
[4] Caribé L, Reis B, Oliveira C, Souza D, Betânia J, Cezar A, Larroza I and Izabel J 2014 Food and Bioproducts Processing Active biocomposites of cassava starch : The effect of yerba mate extract and mango pulp as antioxidant additives on the properties and the stability of a packaged product *Food Bioprod. Process.***94** 382–91
[5] Santacruz S, Rivadeneira C and Castro M 2015 Edible films based on starch and chitosan. Effect of starch source and concentration, plasticizer, surfactant's hydrophobic tail and mechanical treatment *Food Hydrocoll.***49** 89–94
[6] Wurzburg O . and Szymanski C . 1970 Modified Starches for the Food Industry *J. Agric. Food Chem.***18** 997–1001
[7] Yulianti R and Ginting E 2012 Perbedaan Karakteristik Fisik Edible Film dari Umbi-umbian yang Dibuat dengan Penambahan Plasticizer *Balai Penelit. Tanam. Kacang-kacangan dan Umbi-umbian* 131–6
[8] Wijayanti A and Harijono H 2014 Pemanfaatan Tepung Garut (Marantha arundinaceae l) sebagai Bahan *Pembuatan* Edible Paper dengan Penambahan Sorbitol *J. Pangan dan Agroindustri***3** 1367–74
[9] Reddy N and Yang Y 2010 Citric acid cross-linking of starch films *Food Chem.***118** 702–11
[10] Menzel C, Olsson E, Plivelic T S, Andersson R, Johansson C, Kuktaite R, Järnström L and Koch K 2013 Molecular structure of citric acid cross-linked starch films *Carbohydr. Polym.***96** 270–6
[11] Santoso B, Pratama F, Hamzah B and Pambayun R 2015 Karakteristik Fisik dan Kimia Pati Ganyong dan Gadung Termodifikasi Metode Ikatan Silang *Agritech***35** 273–9
[12] Mengatto L, Graciela M, Rubiolo A, Rintoul I and Luna J 2013 Hydrophilic and hydrophobic interactions in cross-linked chitosan membranes *Mater. Chem. Phys.***139** 181–6
[13] Negim E S M, Rakhmetullayeva R K, Zh Y G and Urkimbaeva P I 2014 Improving biodegradability of polyvinyl alcohol / starch blend films for packaging applications **3** 263–73
[14] Liu B, Xu H, Zhao H, Liu W, Zhao L and Li Y 2017 Preparation and characterization of intelligent starch/PVA films for simultaneous colorimetric indication and antimicrobial activity for food packaging applications *Carbohydr. Polym.***157** 842–9
[15] Pereira V A, de Arruda I N Q and Stefani R 2015 Active chitosan/PVA films with anthocyanins from Brassica oleraceae (Red Cabbage) as Time–Temperature Indicators for application in intelligent food packaging *Food Hydrocoll.***43** 180–8
[16] Liu B, Xu H, Zhao H, Liu W, Zhao L and Li Y 2017 Preparation and characterization of intelligent starch / PVA films for simultaneous colorimetric indication and antimicrobial activity for food packaging applications *Carbohydr. Polym.***157** 842–9
[17] Monjazeb Marvdashti L, Koocheki A and Yavarmanesh M 2017 Alyssum homolocarpum seed

gum-polyvinyl alcohol biodegradable composite film: Physicochemical, mechanical, thermal and barrier properties *Carbohydr. Polym.***155** 280–93
[18] Râpă M, Grosu E, Andreica M and Hetvary M 2014 Polyvinyl alcohol and starch blends : properties and biodegradation behavior Journal of Environmental Research and Protection **11** 34–42
[19] Alcázar-Alay S C and Meireles M A A 2015 Physicochemical properties, modifications and applications of starches from different botanical sources *Food Sci. Technol.***35** 215–36
[20] Hirsch J B and Kokini J L 2016 Understanding the Mechanism of Cross-Linking Agents (POCl 3 , STMP , and EPI) Through Swelling Behavior and Pasting Properties of Cross-Linked Waxy Maize Starches *Cereal Chem***79** 102–7
[21] Luchese C L, Frick J M, Patzer V L, Spada J C and Tessaro I C 2015 Synthesis and characterization of biofilms using native and modified pinhao starch *Food Hydrocoll.***45** 203–10
[22] Setiani W, Sudiarti T and Rahmidar L 2013 Preparasi Dan Karakterisasi Edible Film Dari Poliblend Pati Sukun-Kitosan *Valensi***3** 100–9
[23] Seligra P G, Medina Jaramillo C, Famá L and Goyanes S 2016 Biodegradable and non-retrogradable eco-films based on starch–glycerol with citric acid as crosslinking agent *Carbohydr. Polym.***138** 66–74
[24] Frost K, Kaminski D, Kirwan G, Lascaris E and Shanks R 2009 Crystallinity and structure of starch using wide angle X-ray scattering *Carbohydr. Polym.***78** 543–8
[25] Debeaufort F and Voilley A 2015 Lipid-Based Edible Films and Coatings *Edible Films and Coating for Food Aplications* (Springer Science+Business Media) pp 135–210
[26] Darni Y and Utami H 2010 Studi Pembuatan dan Karakteristik Sifat Mekanik dan Hidrofobisitas Bioplastik dari Pati Sorgum *J. Rekayasa Kim. dan Lingkung.***7** 88–93
[27] Garcia M A, Pinotti A, Martino M N and Zaritzky N E 2009 Characterization of Starch and Composite Edible Films dan Coatings *Edible Films and Coating for Food Aplications* pp 169–209
[28] Azahari N A, Othman N and Ismail H 2011 Biodegradation Studies of Polyvinyl Alcohol / Corn Starch Blend Films in Solid and Solution Media **22** 15–31
[29] Sadhu S D, Soni A and Varmani S G 2014 Preparation of Starch-Poly Vinyl Alcohol (PVA) Blend Using Potato and Study of Its Mechanical Properties Meenakshi Garg **3** 33–7
[30] Yoon S, Chough S and Park H 2005 Properties of Starch-Based Blend Films Using Citric Acid as Additive . II
[31] Jiugao Y and Ning W 2005 The Effects of Citric Acid on the Properties of Thermoplastic Starch Plasticized by Glycerol 494–504
[32] Pratiwi M 2016 *Structural Changes of Arrowroot Starch (Marantha arundinaceae L.) as The Impact of Multiple Treatment by Acid Hydrolysis, Debranching, Autoclaving-Cooling, and Heat Moisture Treatment (HMT)*
[33] Hemalatha K ., Parvatikar N and Rukmani K 2015 Influence of ZnO nanoparticles on thermal behavior of Poly Vinyl Alcohol films *Int. J. Sci. Tech. Res.***5** 106–15
[34] Salleh M S N, Nor N N M, Mohd N and Draman S F S 2017 Water Resistance and Thermal Properties of Polyvinyl Alcohol-Starch Fiber Blend Film **20045**
[35] Maiti S, Ray D and Mitra D 2012 Role of Crosslinker on the Biodegradation Behavior of Starch /Polyvinylalcohol Blend Films 749–59
[36] Valencia G A, Moraes Iz C F, Lourenco R V, Habitante M Q B and Sobral P J do A 2014 Maranta (Maranta arundinacea L) Starch Properties *International Conference on Food properties*
[37] Faridah D N, Fardiaz D, Andarwulan N and Sunarti T C 2014 Karakteristik Sifat Fisikokimia Pati Garut (*Maranta arundinaceae*) *Agritech***34** 14–21
[38] Herlina 2010 Karakterisasi Sifat Fisika, Kimia, dan Fungsional Bahan Pati Umbi Gembili (*Dioscorea esculenta* L.) Termodifikasi secara Ikatan Silang dengan Natrium Tripolifosfat *Agrotek***4** 60–7

Physical, mechanical, and barrier properties of sodium alginate/gelatin emulsion based-films incorporated with canola oil

A Syarifuddin*, Hasmiyani, A Dirpan, M Mahendradatta

Department of Food Science and Technology, Hasanuddin University, Makassar

E-mail address : adiansyah@agri.unhas.ac.id

Abstract. The development of mixed emulsion-based films formed by sodium alginate/gelatin incorporated with canola oil can offer particular properties such as water vapor barrier properties. The different ratios of sodium alginate/gelatin and sodium alginate/gelatin emulsion-based films incorporated with canola oil were developed and their effects on films' physical, mechanical and barrier properties were assessed. Here we set out to examine whether canola oil addition and different ratio of sodium alginate/gelatin modified physical, mechanical, and barrier properties of films. To do so, the films were prepared by vary the ratio of sodium alginate/gelatin (2.5, 1, 0.5). Canola oil addition induced changes in moisture content, thickness, solubility, water vapor transmission rate (WVTR), percent elongation at break ($p<0.05$). In addition, it is apparent that varying ratio of sodium alginate to gelatin induced change the mechanical properties of films. The reduction of sodium alginate to gelatin decreased the tensile strength of both films. Improved values of WVTR, tensile strength and solubility at break were observed when the ratio of sodium alginate/gelatin emulsion film incorporated with canola oil was 2.5. Therefore, different ratio of sodium alginate/gelatin incorporated with canola oil can be used to tailor emulsion films with enhanced water vapor barrier and mechanical properties

1. Introduction

The prevalence of novel techniques to maintain the quality of food without the requirement of food preservatives has encouraged food manufactures to improve edible film properties particularly linked in food preservation. One approach to further improve barrier properties for water vapor transfer is to design composite films by dissolving a lipid or oil into hydrocolloid film-forming solutions to make emulsion based-films. Emulsions are thermodynamically unstable systems and are prone to destabilize (e.g., coalescence, creaming) during food processing, transportation or storage [1] and emulsified films show better mechanical properties in comparison with pure lipid layers [2]. Many of the most important properties of emulsion-based films are determined by the right selection of the components used in the polymer matrix, compatability and preparation techniques [2].

Polysaccharides and proteins are usually applied to form films with good mechanical properties and can impart a barrier properties to the oxygen and carbon dioxide but their inherent hydrophilic nature makes them a poor barrier for moisture loss. In order to have better funcionality properties, most composite films were formed by dissolving hydrophobic compounds into hydrocolloid aqueous solution. Hydrocolloids (polysaccharides and proteins) form a continuous and cohesive network and hydrophobic compounds provide the moisture barrier properties [3,4]. For example, oil addition over the hydrocolloids matrix forms bilayer films or dispersed oil into the hydrocolloid matrix forms emulsion (emulsified films).

Sodium alginate, a copolymer consisting of D-mannuronic and L-guluronic acid monomers [5], has been known as a pH and electric fiel-responsive polymers. Films prepared with hydrocolloids such as sodium alginate generally provide a strong films but remain exceptionally poor to water because of their hydrophilic nature [6,7]. In addition, film-forming potential of protein such as gelatin is mainly used because edible films made from gelatin are commonly characterized very transparent and form an excellent barrier to oxygen and carbondioxide but the mechanical properties of these films are vulnerable to higher water content. Several researchers have tried to improve the quality of gelatin films by adding corn starch [8], alginate [9], pectin [10]. The use of oils in order to improve moisture barrier properties of films has been well documented [11,12,13]. Canola oil has potential to be incorporated into hydrocolloids matrix since 18 carbon fatty acids account for about 95% of canola's total fatty acids [14].

Sodium alginate and gelatin have been known for their ability to increase viscoelastic of gels [15] and to form good film-forming properties [16]. However film properties of different ratios of sodium alginate/gelatin with added canola oil have not been investigated.

The aims of the present work were to assess the effect of different ratios of sodium alginate/gelatin on films's properties. Physical and mechanical properties changes of sodium alginate/gelatin emulsion-based due to canola oil incorporation were also investigated. This evaluation was carried out by measuring the physical, mechanical and barrier properties to films.

2. Materials and Method

2.1. Materials

Sodium alginate (SA) and gelatin (G) purchased from Nura Jaya (Surabaya, Indonesia) were used as film-forming components of the hydrophilic continous phase for emulsion-based films. Canola oil (CO) purchased from local market was used as hydrophobic disperse phase, and glycerol purchased from local chemical store was added as plasticizer. Span 80 and tween 80 employed as emulsifiers were purchased from local chemical store.

2.2. Methods

2.2.1. Film formation

Films were produced by using casting technique in two stages. First, polysaccharides films were produced in different ratios of sodium alginate and gelatin in order to find out the formulation that resulted in films with better performance. This formulation then was used in the second stage of the study along with canola oil. In first stage, the effect of different ratios of sodium alginate/gelatin on physical, mechanical and barrier properties was evaluated. The effect of the canola oil addition was also evaluated in the second stage of the study.

Films were produced by using casting technique. First, polysaccharides films were prepared by mixing amounts of κ-carrageenan and gelatin to achieve a final polysaccharides concentration of 2.25% (w/w).

For sodium alginate/gelatin films-based films (SA/G), sodium alginate, gelatin and glycerol aqueous solutions were prepared (Table 1). Sodium alginate was dissolved in 200 ml of warmed distilled water (60^{0}C) for 15 min using magnetic stirrer. The gelatin was succesively added under constant stirring for 15 min. Then, glycerol was added under constant stirring for 15 min. After 45 min stirring, the solution was kept in waterbath for 2 h. The obtained film solution were cooled at room temperature. 20 ml of film-forming solutions were spread onto petri dishes and they were subsequently dried in an oven at 50^0C for 18 h. Finally, dried film were peeled off from the dishes and kept in dessicators at room temperature.

For sodium alginate/gelatin emulsion-based films incorporated with canola oil (SA/G CO), the formulations above were used in addition with canola oil (40% of polysaccharides) (Table 2). After 2 h in waterbath, another step is neccesarry before pouring of film-forming solution in order to incorporate blends of emulsifiers and canola oil. The amount of span 80 and tween 80 required were pre-solubilised in canola oil. Then, all of these components were homogenized with Ultra-Turrax at 24.000 rpm during 2 min. 20 ml of film-forming solutions were poured onto petri dishes and they were dried in an oven at 50^0C for 18 h. Finally, dried film were peeled off from the dishes and kept in dessicators at room temperature.

Table 1. Experimental design of edible films with sodium alginate/gelatin (SA/G)

Ratio Sodium alginate/gelatin	Sodium alginate (% w/w)	Gelatin (% w/w)	Glycerol (% w/w)
2.5	1.61	0.64	30
1	1.125	1.125	30
0.5	0.75	1.5	30

Table 2. Experimental design of sodium alginate/gelatin emulsion-based films incorporated with canola oil (SA/G CO)

Ratio Sodium alginate/gelatin	Sodium alginate (% w/w)	Gelatin (% w/w)	Glycerol (% w/w)	Canola oil (% w/w)
2.5	1.61	0.64	30	40
1	1.125	1.125	30	40
0.5	0.75	1.5	30	40

2.2.2. Film characterization

Thickness

Film thickness was measured by using a digital caliper (KRISBOW KW06-422) with 200 mm x 8" and 0.01 mm accuracy. The values showed represent the means of five measurements randomly taken during each evaluated samples.

Water Vapour Transmisson Rate (WVTR)

The water vapour transmisson rate was determined using gravimetric dessicant method [17] that was modified. Small cups with lids were prepared and filled with silica gel. Then, the films were placed on the mouth of cup then sealed with wax. After sealed, the cups were weighed then placed in dessicator containing NaCl with concentration 70% for interval time 0, 8, 24, 32, 48 h. Then, the weights of cup were recorded. Data obtained then made linear regression equation in order to obtain the slope of weight of cup. Water vapor transmission rate is expressed by the slope of weight of cup (g/h) divided by the area of the film tested (cm^2). The mean value of three measurement replications was reported for each sample.

Tensile Strength (TS) and Percent Elongation at Break

TS and EB were measured using Universal Testing Machine (Hung Ta, HT-2010). Film strips (10 cm x 2 cm) were prepared and mounted to a separate grip of Universal Testing Machine. Initial grip separation was set at 60 mm and the cross head-speed was set at low speed. The tensile strength of the specimen was calculated by dividing the maximum load to rupture (F) and the area of films (A).

$$\text{TS} = \frac{F}{A}$$

Percentage elongation at break was calculated by dividing the film elongation at the moment of rupture of the specimen occured (b) by initial length of the specimen (a) and multiplying by 100.

$$EAB = \frac{b}{a} \times 100$$

Solubility in water [18]
The solubility of samples was expressed as the percentage of the film dry mass which is soluble after 24h immersion in distilled water. First, 3x2 cm of samples were dried in an oven at 105^{0}C for 24 h and weighed (W1). After 24 h drying, each samples was immersed into a 50 ml tube containing 10 ml of distilled water. After 24 h immersion in distilled water, the solution was filtered using filter paper (Whatman no.1), and finally the samples remained on the filter paper were oven dried at 105^{0}C during 24 h after which the samples were weighed to determine the dried remnant insoluble mass (W2). Solubilty was then calculated using the formula:

$$\text{Solubility (\%)} = \frac{W1 - W2}{W1} \times 100$$

Moisture content (MC)[19]
The moisture content of samples was determined by calculating the loss of film weight after drying in an oven at 105^{0}C for 24 h.

3. Statistical analysis

Data analyses were performed using R software (R-3.3.1 release). T-test was used to determine significant differences among the mean of physical and mechanical values calculated in SA/G and SA/G CO films. Three replications were used to determine physical and mechanical properties. For all data analyses, the effects were considered significant when $P<0.05$. The relationship between physical, mechanical and barrier properties was studied using explorative principal component analysis (PCA) with R software (R-3.3.1.*release*).

4. Results and Discussion

Table 3. Summaries of mean physical, mechanical and barrier properties of SA/G and SA/G CO films

	Parameters					
Films	Moisture Content (%)	Thickness (mm)	Water Solubility (%)	WVTR (g/h.cm^2)	Tensile Strength (MPa)	Elongation at break (%)
SA/G films						
2.5	26.85±1.31	0.12±0.005	27.98±2.27	0.091±0.000	10.68±2.94	24.23±3.26
1	27.91±0.22	0.13±0.003	45.44±18.72	0.090±0.000	9.71±1.79	21.63±3.86
0.5	24.00±0.27	0.13±0.002	36.81±3.08	0.091±0.000	6.79±1.22	20.97±0.42
SA/G CO films						
2.5	22.53±0.86	0.26±0.03	22.66±11.88	0.027±0.006	11.35±1.19	53.60±3.08
1	27.10±0.28	0.25±0.02	30.13±0.23	0.031±0.005	8.99±2.16	103.53±19.08
0.5	16.14±0.97	0.20±0.02	34.76±3.95	0.033±0.004	7.49±1.98	130.22±14.19

Values were mean ± standard deviation

Table 4. Summaries of t-test of physical and mechanical properties between SA/G and SA/G CO films

Films	Moisture Content (%)	Thickness (mm)	Solubility (%)	WVTR ($g/h.cm^2$)	Elongation at break (%)	Tensile Strength (MPa)
SA/G	26.25 ± 1.88^a	0.13 ± 0.00^a	36.74 ± 12.18^a	0.091 ± 0.000^a	22.28 ± 2.94^a	9.06 ± 2.53^a
SA/G CO	$21.93\pm.4.81^b$	0.24 ± 0.03^b	29.185 ± 8.20^b	0.030 ± 0.005^b	95.78 ± 35.75^b	9.27 ± 2.31^a

Values were mean ± standard deviation. Different superscripts within the same column indicate significant differences ($P<0.05$)

Moisture content

Table 3 showed the water content of SA/G films and SA/G CO films at different ratios (2.5, 1, 0.5). Results show that the reduction of sodium alginate (and the corresponding increase of gelatin) lead to a decrease of water content on both films. Possibly, the different ratios of SA/G presented the moisture content values (24.00-27.91%) higher than SA/G CO (16.14-27.10%). In this study we also did t-test to determine significant differences among moisture content values calculated in SA/G and SA/G CO. T-test revealed that SA/G CO films showed lower moisture content in comparing with SA/G films. From these results, it is apparent that varying the concentration of gelatin and canola oil addition can result in modification of the network of structure of the films, which is reflected in the moisture content.

Thickness

Thickness of both films (SA/G and SA/G CO) at different ratios was shown in Table 3. Films made with SA/G particularly with the highest ratio (ratio 2.5) showed the lowest thickness. Conversely, the highest thickness (ratio 2.5) was observed from films made with SA/G CO (ratio 2.5). The possible reasoning for this may be because of the sodium alginate increased. Increasing the amount of sodium alginate used leading to increased immiscibility of the blend resulting in increasing thickness obtained. These results was also supported by t-test on thickness values calculated in SA/G and SA/G CO films incorporated with canola oil. T-test revealed that SA/G CO showed a thicker films in comparing with SA/G. The fact that thickness of that SA/G CO were higher than that of SA/G allowed us to postulate that thickness is inversely related to canola oil addition. Some authors reported that edible films containing hydrophobic substances such as waxes and oils can form thicker films [7,20].

Water Solubility

The water solubility of both films at different ratio was shown in Table 3. Addition of canola oil to films containing different ratio of sodium alginate/gelatin (SA/G CO) resulted in a lower water solubility than those observed for films containing SA/G only. Both ratio 1 of film SA/G and ratio 0.5 of SA/G CO showed the highest values of water solubility indicating influence of gelatin used. These results similar with observed by some authors who found that the addition of gelatin significantly increased solubility in water [8]. The addition of canola oil at different ratios of sodium alginate/gelatin was also observed. T-test revealed that the presence of canola oil within the lowest sodium alginate and the highest gelatin concentration (ratio SA/G 0.5) led to blocking of leaching out these polysaccharides into water.

Water vapor transmission rate

The WVTR of both films at different ratio was shown in Table 3. The WVTR values were lower for all ratio of SA/G CO (0.027-0.033 g/h.cm^2) compared with SA/G films (0.090-0.091 g/h.cm^2). T-test results showed that the presence of canola oil reduced the WVTR values by 30% relative to the values of sodium alginate/gelatin films. Incorporations of canola oil at different ratios of sodium alginate/gelatin showed lower the WVTR values due to the formation of an inter-connecting lipid network within the film matrix, which provides hydrophobicity and thus reduces the adsorption of water molecules, as has been observed in emulsified films by other authors [21,22]. As a result, the water vapor transmission through the SA/G CO would be different with SA/G films.

Mechanical properties

In our study, the properties of tensile strength and elongation at break of both films at different ratio can be seen in Table 3. In this study, a higher amount of sodium alginate used (and the corresponding decrease of gelatin), the highest tensile strength values obtained. It may be due to the presence of intermolecular interactions between polymers. These results were similar with results obtained by some authors who found that high tensile strength were observed for films produced with low gelatin and high starch concentration [23].

On the other hand, the elongation at break of SA/G films demonstrated an intense decrease from 24.23% to 20.97%. In films made with SA/G only the elongation values decreased as gelatin concentration increased (and the corresponding decrease of sodium alginate). A similar results were also observed by some authors [24]. Conversely, an intense increase were observed on SA/G CO. This could be due to the plasticizing effect of canola oil. By t-test we found that the elongation values of SA/G CO significant increased.

PCA analysis

In our study, a PCA was carried out to select the best ratio of films among SA/G and SA/G CO with respect to their physical, mechanical and barrier properties. Fig. 1 showed the PCA of the first two principal components of both films and also showed how the different ratios of sodium alginate/gelatin affected physical, mechanical and barrier properties. The first two components (PC1 and PC2) together explained 84.27% of the variance, with PC1 and PC2 explaining 54.63% and 29,64% of the variance, respectively (Fig.1A). It was clearly separate between SA/G films and SA/G CO films (Fig.1B). PCA1 reflects the effect of film solubility and WVTR and showed mainly a positive correlation with solubility and WVTR and negative correlation with elongation. PCA2 is positive correlated with thickness, TS and MC and negative correlated with elongation and solubility.

PCA showed that thickness and TS of SA/G CO ratio 2.5 and SA/G CO ratio 1films (Fig.1B) were correlated since they are plotted to each other. This is in accordance to the previous results where films containing hydrophobic substances such as waxes and oils can form thicker films. In addition, a higher sodium alginate with the presence canola oil (SA/G CO ratio 2.5) make a strong intermolecular interactions between polymers (and the corresponding higher TS). Whereas MC of SA/G films ratio 2.5 versus elongation of SA/G CO films ratio 0.5 loadings, plotted on opposite side of the PC vector, were negatively correlated between each other. Solubility of SA/G film ratio 0.5 loading which is placed orthogonally against elongation of SA/G CO films ratio 0.5 do not correlate: in fact, it is known that increasing amount of gelatin used, higher values of elongation were obtained.

PCA showed that films composed of ratio 2.5 or 1 of SA/G CO were the best ratio due to good combination of tensile strength, thickness and WVTR. However, when the properties of those two ratios (2.5 and 1) were compared, ratio 2.5 present better mechanical and barrier properties (higher TS, low solubility and low WVTR). Thus, the films composed of sodium alginate/gelatin incorporated with canola oil with ratio 2.5 (SA/G CO 2.5) were elected as a better films.

(A)

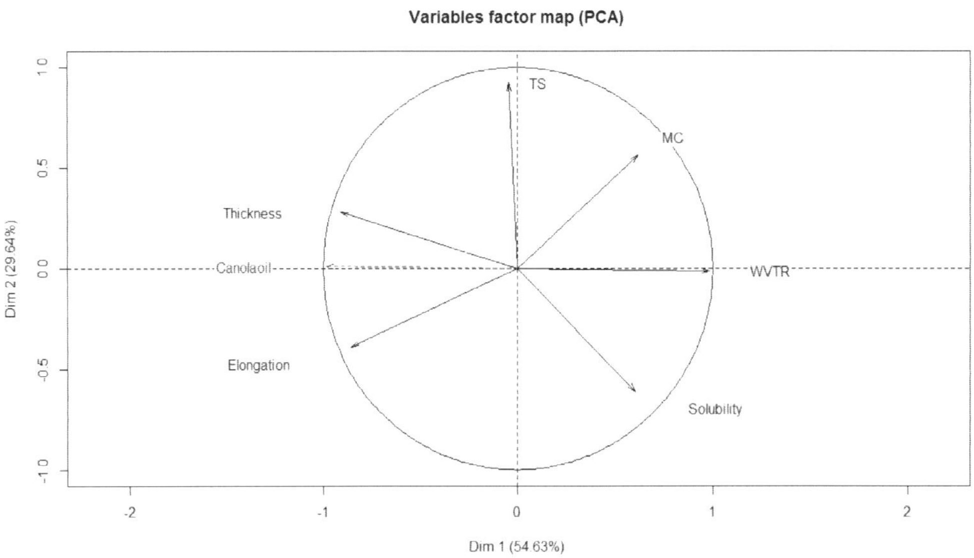

(B)

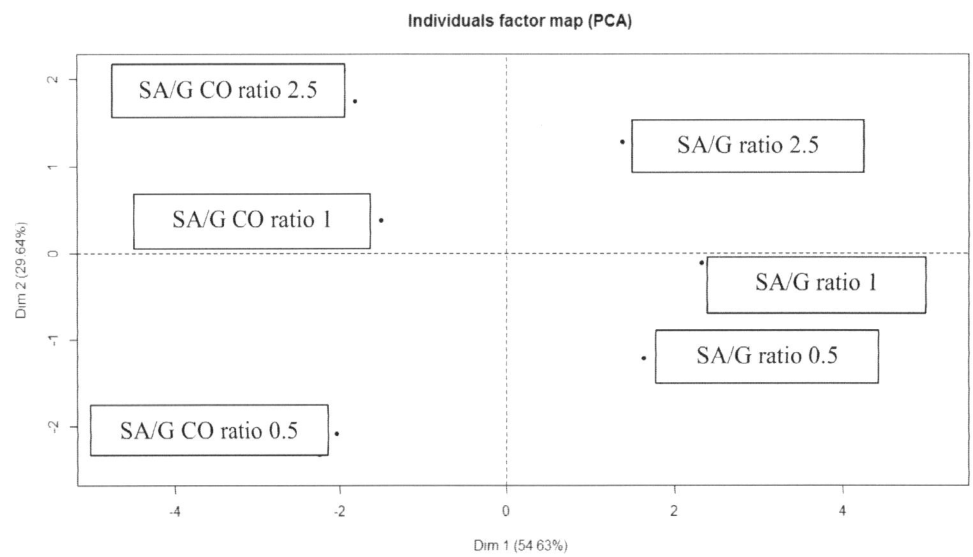

Fig.1A and B. PCA of different ratio of SA/G and SA/G CO films.

5. Conclusions

The incorporation of canola oil into films composed with sodium alginate and gelatin affects theirs basic functional properties. Canola oil incorporation significantly affect moisture content, thickness, WVTR, solubility in water and elongation at break. For all sodium alginate/gelatin emulsion film incorporated with canola oil, the 2.5 ratio films showed the higher tensile strength, lower solubility and lower WVTR.

Acknowledgments

We gratefully acknowledge support of Hasanuddin University via PPI-WCU 2016.

References

[1] Mao, L., Roos, Y.H., O'Callaghan, D.J. and Miao, S. 2013. Volatile release from whey protein isolate-pectin multilayer stabilized emulsions: Effect of pH, salt and artificial salivas. Journal of Agricultural and Food Chemistry **61**(26): 6231-6239..

[2] Galus, S. and Kadzinska, J. 2015. Food applications of emulsion-based edible films and coatings. Trends in Food Science & Technology **45**(2): 273-283.

[3] Kamper, S.L. and Fennema, O. (1984a) Water vapor permeability of an edible, fatty acid, bilayer film. Journal of Food Science, **49**(6), 1482-5.

[4] Kamper, S.L. and Fennema, O. (1984b) Water vapor permeability of edible bilayer films. Journal of Food Science, **49**(6), 1478-81.

[5] Borchard, W., Kenning A., Kapp A., Mayer C. 2005. Phase diagram of the system sodium alginate/water: a model for biofilms. Inte J Biol Macromol **35**(5): 247-56

[6] Guilbert, S. (1986) Technology and application of edible protective films. In:Food packaging and preservation, M. Mathlouthi (ed.), Elsevier Applied Science Publishers, New York, pp. 371-94.

[7] Kester, J. J. and Fennema, O. R. 1986. Edible films and coatings: A review. Food Technology **40** (12): 47-59.

[8] Fakhouri, F.M., Martelli, S.M., Caon, T., Velasco, J.I. and Mei, L.H.I. 2015. Edible films and coatings based on starch/gelatin: Film properties and effect of coatings on quality of refrigerated Red Crimson grapes. Postharvest Biology and Technology, **109**: 57-64.

[9] Dong, Z., Wang, Q and Du, Y. 2006. Alginate/gelatin blend films and their properties for drug controlled released. Journal of Membran Science, **280**(1-2): 37-44.

[10] Rezvanian, M., Ahmad, N., Amin, M.C.I.M., Shiow-Fern, Ng. Optimization, characterization, and in vitro assesment of alginate-pectin ionic cross-linked hydrogel film for wound dressing applications. International Journal of Biological Macromolecules **97**: 131-140.

[11] Cerquira, M.A., Souza, B.W.S., Teixera, J.A., Vicente, A.A. 2012. Effect of glycerol and corn oil on physicochemical properties of polysaccharides films-A comparative study. Food Hydrocolloids **27**:175-184.

[12] Ghasemlou, M., Aliheidadri, N., Fahmi, R., Shojaee-Aliabadi, S., Keshavarz, B., Cran, M.J., Khaksar, R. 2013. Physical, mechanical and barrier properties of corn starch films incorporated with plant essential oils. Carbohydrate Polymers, **98**(1): 1117-1126.

[13] Zukas, B.G. and Gupta, N.R. 2016. Improved water barrier properties of calcium alginate capsules modified by silicone oil. Gels, **2**, 14.

[14] Ackman, R.G. 1990. Canola fatty acids-an ideal mixturefor health, nutrition, and food use, in Canola and Rapeseed. Production, Chemistry, Nutrition and Processing Technology (ed F. Sahidi). Avi Book, Van Nostrand Reinhold, New York.pp. 81-98.

[15] Derkach, S.R., Ilyin, S.O., Maklakova, A.A., Kulichikhin, A.G. and Malki, A.Y. 2015. The rheology of gelatin hydrogels modified by κ-carrageenan. LWT-Food Science and Technology **63** (1): 612-619.

[16] Pranoto, Y., Lee, C.M. and Park, H.J. 2006. Characterizations of fish gelatin films added with gellan and κ-carrageenan. LWT-Food Science and Technology **40** (5): 766-774.

[17] Poeloengasih, C.D., dan Marseno, D.W. 2003. Karakterisasi edible film komposit protein biji kecipir dan tapioka. Jurnal Teknologi dan Industri Pangan **XIV**(3): 224-232.

[18] Ahmad, M., Benjakul, S., Prodpran, T. and Agustini, T.W. 2012. Physico-mechanical and antimicrobial properties of gelatin film from the skin of unicorn leatherjacket incorporated with essentials oils. Food Hydrocolloids **28**: 189-199.

[19] Zahedi, Y., Ghanbarzadeh, B. and Sedaghat, N. 2010. Physical properties of edible emulsified films based on pistachio globulin protein and fatty acids. Journal of Food Engineering **100** (1): 102-108.

[20] Greener, I. K. and Fennema, O. 1989. Evaluation of edible, bilayer films for use as moisture barrier for food. Journal of Food Science **54**(6): 1400-1406.

[21] Fabra, M., Pérez-Masiá, R., Talens, P., & Chiralt, A. (2011). Influence of the homogenization conditions and lipid self-association on properties of sodium caseinate based films containing oleic and stearic acids. Food Hydrocolloids, **25**, 1112-1121

[22] Shellhammer, T., & Krochta, J. (1997). Whey protein emulsion film performance as affected by lipid type and amount. Journal of Food Science, **62** (2), 390-394.

[23] Zhong, Q-P.; Xia, W-S. 2008. Physicochemical properties of edible and preservative films from chitosan/cassava starch/gelatin blend plasticized with glycerol. Food Technology and Biotechnology, **46**, p. 262-269

[24] Xiao, C., Lu, Y., Gao, S., and Zhang, L. 2001. Characterization of konjac glucomannan-gelatin blend films. Journal of Applied Polymer Science, **79**:1596-1602.

IOP Conf. Series: Earth and Environmental Science **101** (2017) 012020 doi:10.1088/1755-1315/101/1/012020

Sensory and Textural Characteristics of Noodle Made of Ganyong Flour (*Canna edulis Kerr.*) and Arenga Starch (*Arenga pinnata Merr.)*

ERN Herawati[1], D Ariani[1], Miftakhussolikhah[1], E Yosieto[2],M Angwar[1], Y Pranoto[2]
[1]Research Unit for Natural Product Technology, Indonesian Institute of Sciences, Gading, Playen, Gunung Kidul, Yogyakarta, Indonesia
[2]Faculty of Agricultural Technology, Gadjah Mada University, Yogyakarta, Indonesia

E-mail: ervika.rnh@gmail.com

Abstract. Ganyong (Canna edulis Kerr) is a local tuber which highest amount of starch content, but has not been fully utilized well at present. One way to improve the usefulness of canna is to process it into noodle, but it needs arenga starch which has high amylose content. The aim of this research was to study the sensory and textural properties of noodle made from canna flour and arenga starch. Research methodologies consist of: (i) characterization of canna flour and arenga starch, (ii) noodle production, and (iii) characterization sensory and textural properties of the noodle. Noodle was made with five ratio variations of canna flour and arenga starch, i.e. 100:0; 75:25; 50:50; 25:75; and 0:100. Sensory analysis was done by hedonic scoring method with attributes : color, stickiness, elasticity, firmness, surface smoothness and overall liking. Textural properties analyses consist of tensile strength, elongation, and stickiness measurements. The results showed that canna flour and arenga starch can be used in noodle making process. Noodle with 25% of canna flour was the most favored product and has the best textural properties. Factors which affect textural properties of product are the amylose and amylopectin amount in each starch. Tensile strength, elongation, and stickiness measurements of noodle with 25% of canna flour were 0,13 N; 41,61%; and 0,0115N respectively.

1. Introduction

Indonesia has a great potential in agricultural product, including tubers. One of the local tubers that can be developed is ganyong (Canna edulis Kerr). Ganyong contain high amount of carbohydrates. It can be consumed as a source of energy supply for the body. Processsing ganyong tuber into flour and starch can be highly prospective in terms of functional properties and chemical composition [1]. Ganyong is a potential agricultural product that can be used for food diversification. Therefore, it is necessary to increase the utilization and consumption of ganyong. Processing ganyong/canna tuber into noodles can increase the consumption of ganyong, but it need other ingredient because amylose content of canna tubers relatively low. Local ingredient that can be used is arenga starch which have high amount of amylose content. Arenga starch has been used for raw material of noodle because it has similar character with green bean starch, which contain high amylose, has a limited degree of bubbling granules, and has a viscosity Branbender curve type C [2].

Arenga starch has physical properties, chemical properties and functionally similar to the green bean starch that commonly used as a raw material in the manufacture of starch noodle and instant noodle [3]. Amylose content in arenga starch is high >35% [4]. By using canna flour and arenga starch in noodle-making, it is expected to develop a noodle made from local materials with good sensory, physical properties and acceptable by society. Some researcher reported use of canna flour and wheat flour in noodle processing [5, 6]. Srikaeo et al [7] and Wande et al [8] has been reported use of canna starch in a rice noodle product. There is no report about sensory and textural properties noddle made from canna flour and arenga starch. The aim of this research was to study the sensory and textural properties of noodle made from canna flour and arenga starch. By conducting this research, it is expected that the value of canna as local farm product can be elevated, i.e. by processing it as raw material for noodle making.

2. Materials and Method

2.1. Materials

Canna flour that used in this research was produced by Indonesian Institute of Sciences (LIPI) in Yogyakarta. The other ingredient was arenga starch from Klaten, Yogyakarta. The tools used for noodle making are noodle making machine (Food Extruder PD-45N, La Pramigiana). UTM (Universal Testing Machine Zwick 0.5-Lloyd's Universal Testing Instrument) used for texture analysis.

2.2. Noodle process

Making noodles is based on the common practice in a large scale [9] with some modification to be performed in a laboratory scale. Noodle was made with five ratio variations of composition between canna flour and arenga starch, i.e. 100:0; 75:25; 50:50; 25:75; and 0:100. The first step was weighing a mixture of canna flour and arenga starch with a total of 300 grams. Starch flour is blended until homogeneous, then water is added to flour mixture. Stirring the flour and water is performed with a mixer, so that the dough can be hydrated evenly. The mixture is then molded into pellets with a length of 3-5 cm using the extruder with a diameter of 15 mm. Making pellets aims to expand the surface of the dough so as to facilitate the process of gelatinization when steamed. Pellets then steamed for ± 3 minutes until the surface becomes shiny. Steaming should not be too long because it is desired to gelatinization partially on the surface of the pellet. While pellets were still hot, put immediately into the extruder to be molded into noodle threads. The mold used in this process has a diameter of 0.7 mm. The obtained raw noodle is then hung on a cart and steamed for ± 20 minutes until the color of noodle becomes transparent. The cooked transparent noodle is allowed to stand until it reaches room temperature, then separated and dryed in a cabinet dryer at 55°C for 6 hours until the moisture content is about 10%.

2.3. Noodle analysis

Characteristic of canna flour and arenga starch were analysis with three replicates. Sensory analysis of noodle using hedonic scoring method with attributes: color, stickiness, elasticity, firmness, surface smoothness and overall liking to 25 panelist, with 1-7 grading scale (1=lowest preference, 7=highest preference). Dried noodle cooked using boiling water with a long time in accordance with the optimum cooking time, then allowed to stand no more than 30 minutes in a closed container. Samples are presented in colored bowl, and each panelist get a bowl of the same color for each sample. Textural properties analysis of noodle consist of tensile strength, elongation, and stickiness measurements [10] with three replicates. The design of the experiment used a completely randomized design (CRD). Statistical data analysis using the software of SPSS version 16 with One Way Anova method with a significant level of 5% in the comparison of means using Duncan method.

3. Results

3.1. Characteristic of canna flour and arenga starch

The characteristic of canna flour including water content, ash content, fat content, amylosa content, crude protein content, carbohydrate content, and starch content can be shown in Table 1.

Table 1. Characteristic of canna flour*

Component	Content	
	% Wet basis (*wb*)	% Dry basis (*db*)
Water content	6.27	
Ash content	2.98	3.18
Fat content	0.55	0.59
Amylosa content	32.10	34.37
Crude protein content	2.85	3.04
Carbohydrate content	87.35	93.19
Starch content	53.0	56.55

*three replicates

The characteristic of arenga starch including water content, ash content, fat content, amylosa content, crude protein content, carbohydrate content, and starch content can be shown in Table 2.

Table 2. Characteristic of arenga starch*

Component	Content	
	% Wet basis (*wb*)	% Dry basis (*db*)
Water content	15.27	
Ash content	0.11	0.13
Fat content	0.01	0.01
Amylosa content	37.86	44.68
Crude protein content	0.06	0.07
Carbohidrat content	84.58	99.82
Starch content	77.69	91.69

*three replicates

3.2. Noodle process
Canna flour and arenga starch used as raw materials in noodle processing as shown in Figure 1.

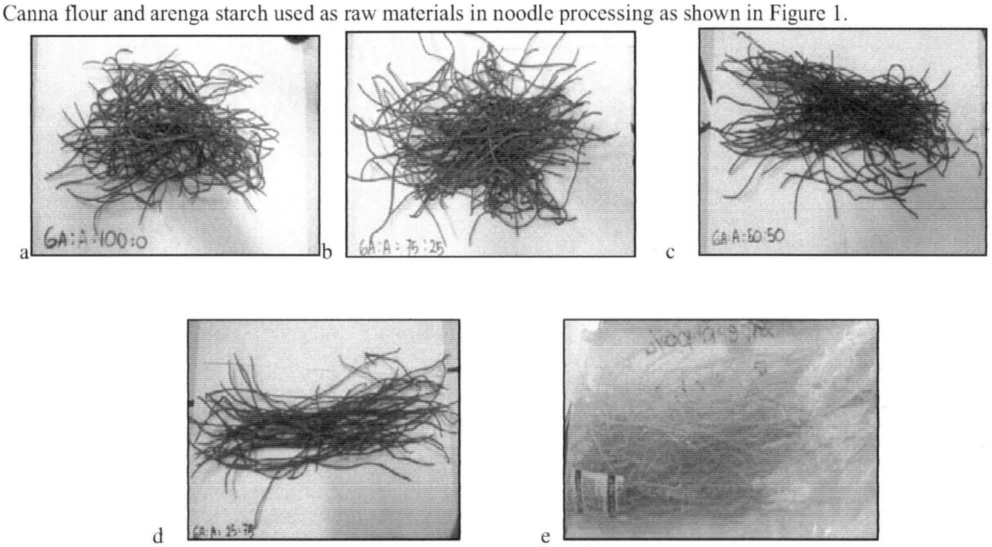

Figure 1. (a) noodle made from canna flour:arenga starch = 100:0; (b) noodle made from canna flour:arenga starch = 75:25; (c) noodle made from canna flour:arenga starch = 50:50; (d) noodle made from canna flour:arenga starch = 25:75;
(e) noodle made from canna flour:arenga starch = 0:100

3.3. Sensory Preference
Sensory analysis was performed with hedonic scoring method to identify which product is the most preferred by the panellist using attributes such as : color, stickiness, elasticity, firmness, surface smoothness and overall liking. Sensory analysis on noodle products aims to determine which products are most preferred by the panellist to reflect acceptance of products which is currently marketed. Sensory characteristics of noodle made from canna flour and arenga starch after cooking (cooked) can be seen in Table 3.

Table 3. Sensory preference characteristic of cooked noodle*

Sample (Canna flour: arenga starch)	Sensory preference					
	Color	Stickiness	Elasticity	Firmness	Surface smoothness	Overall liking
100:0	3.04^{ab}	2.12^{a}	2.04^{a}	2.40^{a}	2.28^{a}	2.16^{a}
75:25	2.76^{a}	2.08^{a}	2.00^{a}	2.56^{a}	2.24^{a}	2.12^{a}
50:50	3.40^{ab}	3.44^{b}	3.08^{b}	3.72^{b}	3.44^{b}	3.52^{b}
25:75	3.76^{bc}	4.64^{c}	4.28^{c}	4.16^{b}	4.16^{c}	4.56^{c}
0:100	4.24^{c}	3.68^{b}	3.44^{b}	3.72^{b}	4.00^{bc}	3.6^{b}

*25 replicates as represent of 25 panelist
Note : The same superscript symbol in the same column indicate that sample are not significantly different at a significance level of 95%
Scale : 1=lowest preference, 7=highest preference

3.4. Textural characteristic of cooked noodle
Textural characteristics of noodle made from canna flour and arenga starch after cooking (cooked) can be seen in Table 4. Textural properties include tensile strength, elongation, and stickiness measurements.

Table 4. Textural characteristic of cooked noodle*

Sample (canna flour: arenga starch)	Textural properties		
	Tensile strength (N)	Elongation (%)	Stickiness (N)
100:0	0.03^{a}	13.85^{a}	0.0056^{a}
75:25	0.06^{b}	17.35^{a}	0.0109^{b}
50:50	0.07^{b}	20.69^{a}	0.0148^{c}
25:75	0.13^{d}	41.61^{b}	0.0115^{b}
0:100	0.18^{e}	37.87^{b}	0.0059^{a}

*three replicates
Note : The same superscript symbol in the same column indicate that sample are not significantly different at a significance level of 95%.

4. Discussion

As shown in Table 1, the water content of canna flour was 6,27% w.b. The ash content of canna flour was 2,98% (wb) or 3,18% (db). The result was higher than the maximum ash content from garut flour standard (SNI) which is 0,5% (b/b). The high ash content of canna flour can be caused by residue of metallic material left during the process of flouring. The fat content of canna flour was 0,59%(db). The low fat content in canna flour was related with the low fat content of ganyong tubers. Protein content of canna flour was 3,04%(db). Amylose content of canna flour was 34,37% (db). Aprianita et al [11] reported the amylose content from canna flour was 233.2%. Chansri et al. [12] reported that ideal starch as raw material for noodle is starch with high amylose content (21-28%).

Arenga starch contain high amount of carbohydrate, starch, and amylose content. As shown in Table 2, arenga starch contain high amount of amylose which is 44.68% (db). Arumdinari [13] also reported the high amylose content of arenga starch was 40,78% (db). It caused arenga starch commonly used as a raw material in the manufacture of starch noodle and instant noodle [3, 4].

As shown in Figure 1, the noodle color were brown, which is influenced by the color of canna flour used. Differences in the proportion / ration of canna flour used affects the color of the products, especially the brightness of the products. The more canna flour used, the brightness of the product will be reduced. Noodle made from 100% canna flour has high brightness values because the product does not have a smooth surface and still contain ungelatinized flour perfectly.

Preference for color attributes in the noodle made from canna flour 100%, 75%, and 50% was not significantly different. The highest level of color preference is noodle made from arenga starch 100%. This could be due to visibility of noodle-based canna which has a brownish color (Figure 1) that is rarely found in the market, in contrast to noodle from arenga starch 100% which has clear visibility like commercial product. Previous research reported differences in the proportion of canna flour used affects the color of the noodle, especially the brightness of the noodle [14]. The more canna flour used, the brightness of the noodle product will be reduced.

Attributes stickiness judged by the ease of the product to disintegrate before being consumed. The desired product is a solid product that is not easily destroyed when scooped. While the elastic properties is one of the textural properties of the most frequently assessed when eating noodle products. Noodle is the preferred product that is elastic, but easily chewed. As shown in Table 3, stickiness and elasticity preference were increased with lower addition of canna flour. Previous study reported that use of canna flour will increase the noodle cooking loss [14]. Higher cooking loss leads to undesirable noodles characteristics such as, high starch solubility and poor cooking resistance or tolerance that causes stickiness in texture of noodles. Thus, the more canna flour is used, the lower of stickiness and elasticity preference, due to the higher of cooking loss value. The highest level of preference for stickiness and elasticity of the noodle is product noodle made from canna flour:arenga starch = 75:25. It appears that the stickiness and elasticity in this product differs markedly from other noodle mixture. Noodle product is said to be sticky if it is attached to the teeth and palate when eaten.

Firmness preference of noodle made from canna flour 100% and 75% were not significantly different, while the other product (noodle made from canna flour 50%, 25%, and arenga starch 100%) also not significantly different. Qazi *et al* [15] reported the effect of canna starch in the stretch firmness of rice noodle. Firmness preference correlated with stability ratio, as variation between amylose and amylopectin content. Firmness of noodle also correlated with hydration index and swelling index. Previous study [14] reported that value rehydration noodle made from 100% canna flour and noodle made from 75% canna flour are quite large, so it cause the lower firmness preference. The best firmness preference was noodle with canna flour 25%, but not significantly different with noodle made from canna flour 50% and noodle made from arenga starch 100%.

Noodle surface smoothness can be assessed through visibility and sensations in the mouth of panelists. Products that have the highest level of preference in this case is the noodle made from canna flour 25%, but not significantly different to the products made from canna flour 50% and arenga starch 100%. Noodle products which were overall preferred by panelists is noodle made from canna flour:arenga starch = 25:75. This product has the highest scale, significantly different with other canna flour and arenga starch noodle. The highest scale of noodle made from canna flour:arenga starch = 25:75 was positively correlated with the other atributes, i.e. stickiness, elasticity, firmness, and surface smoothness.

Tensile strength and elongation are important textural properties in noodle. Noodle quality which customers want is a product with a value of tensile strength and high elongation. Both of these properties are correlated with each other positively. The more compact the structure of the noodle, the elongation will increase, and it will be more difficult to break up when pulled. Amylose content in the noodle will affect the strength of the product when stretched [2, 12]. Noodle products from mung bean starch has a tensile strength values higher than the product of canna starch [12]. From the results of statistical tests with $p < 5\%$, it was obtained that the tensile strength of the sample noodle significantly different at each variation of the ratio, except the noodle made from canna flour:arenga starch = 75:25 and 50:50. Tensile strength and elongation values increasing as more arenga starch added in the mix due to an increase in the amylose crystalline clusters noodle threads. Noodle made from high amylose material has a tensile yield strength and strong texture / compact [3]. High amylose content which causes retrogradation perfect so that the structure of starch instant noodle produced more rigid, robust and compact [3]. Qazi *et al* [15] reported the effect of canna flour that decrease the tensile strength of noodle. Meanwhile, the elongation noodle made from canna flour 100%, 75% and 50% were not significantly different. Noodle made from canna flour 25% and arenga starch 100% were not significantly different. Elongation values increasing as more arenga starch added in the mix due to an increase in the amylose crystalline clusters noodle threads. The more compact the structure of the noodle will increase the elongation.

Stickiness is also a determinant factor of noodle quality, because higher level of stickiness will cause the strands of noodle difficult to separate after the cooking process. Sticky nature of the noodle product is not favored by consumers, so it is expected that noodle have lower levels of stickiness. Sticky nature commonly found in products with a high amylopectin content, since amylopectin is able to absorb water but has a less compact structure so that water can not be absorbed properly restrained in starch granules [16, 17, 18]. Stickiness of noodle product can be seen in Table 4. It shows that there is a trend, i.e. if more arenga starch is added, the value of the product stickiness is greater. But the results of statistical tests with $P < 0.05$ indicates that the stickiness of the product made from canna flour 100%, arenga starch 100% were not significantly different. The stickiness level of noodle made from canna flour 75% and 25% were not significantly different.

5. Conclusion

The ratio between canna flour and arenga starch affect the sensory and textural characteristic of noodle. Preference for color attributes in the fifth sample of canna noodle was not significantly different. The highest level of preference for stickiness and elasticity of the noodle is noodle made from canna with composition of canna flour:arenga starch = 75:25. Noodle products which were overall preferred by panellists is noodle made from canna with composition of canna flour:arenga starch = 25:75. Factors which affect textural properties of product are the amylose and amylopectin amount in each starch. The value of tensile strength, elongation, and stickiness will increase with more addition of arenga starch. The best textural properties is noodle with 25% of canna flour. Tensile strength, elongation, and stickiness measurements of noodle with 25% of canna flour were 0,13 N; 41,61%; and 0,0115N respectively.

Acknowledgements
This study was part of the research program of the Indonesian Institute of Sciences. The authors wish to thank Professor Haryadi from the Gadjah Mada University for providing guidance.

References

[1] Caroline and Ilmi 2016 Production of Indonesian Canna edulis type IV resistant starch through acetylation modification *International Food Research Journal* **23** 491-497
[2] Lii C Y and Chang Y H 1986 The Properties of Some Modified Starches and Their Applications. *Role of Chemistry in the Quality of Processed Food* Food and Nutrition Press Inc Westport Connecticut USA
[3] Rahim A 2008 The Effect of Noodle Treatment on Physicochemical Properties of Instant Noodle from Arenga starch (in Indonesian language) *Journal Agroland* **15** 101–105
[4] Alam N and Saleh M 2009 Charateristics of Palm Sugar Tree at Various Growth Phases (in Indonesian language) *Journal Agroland* **16** 199 – 205
[5] Santacruz S, Pennanen M and Ruales J 2009 Protein enrichment of oriental noodles based on Canna edulis starch *Asian Journal of Food and Agro-Industry* **2** 521-538
[6] Ilmi I L 2012 Instant Noodle Formulation based Canna Flour with Addition of Red Palm Oil (RPO) as Carbohydrate Food and Pro-Vitamin A (in Indonesian language) *Thesis* Bogor Agriculture University
[7] Srikaeo K, Mingyai S and Sopade P 2011 Physicochemical properties, resistant starch content and enzymatic digestibility of unripe banana, edible canna, taro flours and their rice noodle products *International Journal of Food Science & Technology* **46** 2111-2117
[8] Wandee Y, Uttapap D, Puncha-arnon S, Puttanlek C, Rungsardhtong V, and Wetprasit N 2015 Quality assessment of noodles made from blends of rice flour and canna starch *Food Chemistry* 179 85-93
[9] Haryadi 2006 *Rice Processing Technology* (in Indonesian language) Yogyakarta UGM Press
[10] Chen Z L, Sagis L, Legger A, Linssen J P H, Schols H A, and Voragen A G J 2002 Evaluation of Starch Noodles Made from Three Typical Chinese Sweet Potato Starches *Journal of Food Science* **67** 3342-3347.
[11] Aprianita, Vasiljevi C, Bannikova A and Stefan 2014 Physicochemical properties of flours and starches derived from traditional Indonesian tubers and roots. *Journal Food Science Technology* **51** 3669–3679
[12] Chansri R, Puttanlek C, Rungsadthogy V and Uttapap D 2006 Characteristics of Clear Noodles Prepared from Edible Canna Starches *Journal of Food Science* **70** 337-342
[13] Arumdinari R 2012 The Substitution Effect of Cassava Flour On Arenga starch Noodle Characterisation (in Indonesian language) *Thesis* Gadjah Mada University Yogyakarta Indonesia
[14] Herawati E R N, Ariani D, Miftakhussolikhah, Angwar M, Pranoto Y 2015 Physical Properties Characterization of Noodle Made From Ganyong Flour (Canna edulis Kerr.) and Starch Sugar (Arenga pinnata Merr.) *Proceeding International Conference Food for a Quality Life*. SEAFAST Center, Bogor Agriculture University.
[15] Qazi I M, Rakshit S K, Tran T, Javidullah, Khan M Z 2014 Effect of Blending Selected Tropical Starches with Rice Flour on The Cooking Quality and Texture of Rice Based Noodles. *Sarhad J. Agric.* **30** 257-264
[16] Huang Y C and Lai H M 2010 Noodle Quality Affected by Different Cereal Starches *Journal of Food Engineering* **97** 135-143
[17] Sandhu K S and Kaur M 2010 Studies on noodles quality of potato and rice starches and their blends in relation to their physicochemical, pasting and gel textural properties *Food Science and Technology* **43** 1289–1293
[18] Kruger J E, Matsuko R B and Dick J W 1996 *Pasta and Noodle Technology* American Association of Cereal Chemist Inc St Paul Minnesota USA

Quality assessment of dry noodles made from blend of mocaf flour, rice flour and corn flour

N Afifah[1] and L Ratnawati[1]

[1] Development Center for Appropriate Technology, Indonesian Institute of Sciences Subang, Indonesia

E-mail: Syabiljafa2008@gmail.com

Abstract. Mocaf flour, rice flour, and corn flour in different concentrations were used to produce dry noodles. The aims of this study were to investigate physicochemical properties of the flours and the quality characteristics of dry noodles made from these flours. Dry noodles were prepared by gelatinization of blending flours followed by feeding the dough to extruder and drying overnight in room temperature. Flours were analyzed for chemical and pasting properties and noodle samples were evaluated for chemical, cooking, and texture properties. The results showed there were significant differences in protein content and elongation between noodles made from blending mocaf and rice flour with that of blending mocaf, rice, and corn flour. The moisture, ash, and protein content of noodle samples ranged from 10.98 to 14.18%, 1.23 to 1.39%, and 4.09 to 5.58%, respectively. Values of noodle cooking quality were ranging from 12.0 – 13.8 minutes, 10.6 to 14.3%, and 204 to 248%, respectively for cooking time, cooking loss, and cooking weight. The elongation, hardness, and adhesiveness of noodles ranged from 276 to 374%, 3,523 to 10,478 gf, and -81.99 to -52.49 g.sec.

1. Introduction

The primary source of carbohydrate in Indonesia now leads in rice and flour-based foods including noodle. It needs to be aware because most of the types of noodle are made of wheat flour. Indonesia is not wheat-producing country, so that the direction of these changes can lead to dependence on imported food that endangers national food. Dependence of food against another country can affect the susceptibility to foreign intervention in economic and political [1]. In order to reduce imports of wheat flour, a lot of efforts have been made to find a replacement. Besides that, there are recently growing consumer demands for gluten-free products in the food industry including noodles [2]. One source of carbohydrates that can replace wheat flour is canna starch, rice flour, corn starch, and buck wheat flour [3].

Rice, cassava and corn are three types of crops with the highest production number in Indonesia. In 2015, rice, cassava, and corn production was 75,397,841 tons, 21,801,415 tons, and 19,612,435 tons respectively [4]. Many Indonesian cuisines use rice flour as an ingredient. The physical and chemical attributes of food during processing are affected the physicochemical properties of rice. Mocaf is a product derived from cassava that uses the principle of modifying cassava cells in fermentation. Advantage mocaf has aroma and flavor better than regular cassava flour, white has more color than usual cassava flour [1]. Corn as raw material of noodles has advantages such as having a natural yellow color produced by β-carotene, lutein, and zeaxanthin [5].

The heating processing for pre-gelatinization is critical step for maintaining noodle strands because gluten-free flour is not able to form a cohesive dough structure [2]. Generally, there are two different ways to produce gluten-free noodles [6]. The first gluten-free flour is mixed with water, then the dough is extruded through a die and the extruded strands are dropped into hot water. The other way is that gluten-free flour batter is prepared, coated onto a rotating heated drum, transferred into a steaming tunnel, and then dried before cutting.

Exploration of non-wheat carbohydrate source for the manufacture of noodles has been done, ranging from the partial substitution up to full substitution without using wheat flour. Yadaf, et al. [7] have substituted wheat flour by 25 percent of sweet potato, colocasia and water chestnut flours to manufacture noodles. Wandee, et al. [8] reported the substitution of rice flour with canna cross-linked starch has improved the tensile strength and elongation of noodles. Meanwhile, a composite of corn flour and cassava flour has been done by Pato et al. [5] and showed that on ratio of 55% cassava flour and 45% corn flour resulted in instant noodle which met the Indonesian quality standard of Instant Noodle (SNI 01-3551-2000) except for protein content of less than 4%. The addition of selected tropical starches such as mung bean starch, canna starch and sweet potato starch to rice flour had significantly modified cooking quality and texture properties of noodles [9]. The physicochemical properties of starch have an essential role in both starch noodle processing and the final starch noodle quality [10].

However, very little information is available on the blending of mocaf, rice, and corn flours for noodle production. Our study was meant to investigate the chemical and pasting properties of mocaf, rice, and corn flours. The applicability their blends in relation to their dry noodle preparation was also examined by chemical, textural, and cooking analysis.

2. Materials and Methods

2.1. Raw materials

Commercial mocaf flour was purchased from small medium enterprise "Harapan Jaya" Subang Regency, Indonesia. The corn flour and rice flour was produced in Food Laboratory, Development Center for Appropriate Technology. Rice and corn seed were collected from local market in Subang, Indonesia. The corn seed was crushed using the chopper machine, and then the corn husk was separated by floating process. The rice corn (grits) soaked in water for 4 hours, dried until moisture content less than 10% (wb), then refined using a disc mill. Rice was soaked in water for 2 hours and drained well. Fine milled to make flour using disc mill machine, and dried flour rice till moisture content less than 10% (wb). All of flour was sifted using a sieve size of 40 meshes.

2.2. Dried noodle preparation

Noodles were produced by mixing 1 kg of blending flour appropiated with treatment and additional ingredients salt (1%) and water (40%) based on flour. Mix all of materials lasted 15 minutes using commercial laboratory mixer and steamed for 30 minutes. The resulting dough was fed in the small sacle single-screw extruder (constructed by Development Center for Appropriate Technology) at temperature setting of 60 ^{o}C and speed 50 rpm. Noodles strand was dried overnight in room temperature. The treatment of blending flour as follows:

MR1 = 50% the mocaf flour and 50% rice flour
MR2 = 60% the mocaf flour and 40% rice flour
MR3 = 70% the mocaf flour and 30% rice flour
MRC1 = 40% the mocaf flour, 30% rice flour, and 30% corn flour
MRC2 = 50% the mocaf flour, 30% rice flour, and 20% corn flour
MRC3 = 60% the mocaf flour, 30% rice flour, and 10% corn flour

2.3. Analyses of noodles

Noodle samples were analyzed for chemical, cooking, and texture properties.

2.3.1. Chemical analysis. The procedures for analysis of moisture and ash content are as outlined by the Indonesian National Standard (SNI 01-2891-1992) about analytical methods for food and beverages. Protein content was measured according to the Dumas combustion method by using DuMaster Buchi D-480, Switzerland.

2.3.2. Pasting Analysis. The pasting properties of flours were evaluated using a Rapid Visco Analyzer (RVA-TecMaster, Macquarie Park, Australia). Mocaf, rice, and corn flours were analyzed following the procedure of RVA^{TM} General Pasting Method STD2. Flour samples, weighing 3.5 g (14% moisture content), were mixed with 25 g of distilled water in disposable aluminum RVA canisters. The sample was spinned (160 rpm) at 50°C for 1 min, heated to reach 95°C in 7.5 min, and then held at 95°C for 5 min. Afterwards, it was cooled back to 50°C in 7.5 min, and then held at 50°C for 2 min.

2.3.3. Cooking quality analysis. Cooking time, cooking weight, and cooking loss of dry noodles was measured according to the method reported by Wandee et al. [8]. Dried noodles (about 5 g) were cut into lengths of 4-5 cm and cooked in 200 ml boiling distilled water in a covered beaker. The optimum cooking time was evaluated by observing the time of disapperance of white core of the noodle strands every 30 second, by pressing the cooked noodle between two transparent glass slides. Dry noodles (about 1 g) were cut into small pieces lenghts of 3–5 cm, and boiled in 30 ml water boiling water until completely cooked according result of cookint time test. The cooked noodles were then placed on a nylon screen, rinsed with distilled water, drained for 1 min. The wet noodle immediately was weighed as cooking weight. The cooking water and rinse water in a pre-weighed glass was evaporated and dried in a hot-air oven at 105 °C to a constant weight. Cooking weight of noodles was expressed as the percentage of g cooked noodle/g dried noodle, while cooking loss was expressed as the percentage of dry matter loss during cooking to dry sample weight. Cooking weight and cooking loss were calculated by equation (1) and (2).

$$Cooking\ weight\ (\%) = \frac{W_1 - W_2 \times (1 - M)}{W_2 \times (1 - M)} \times 100\% \qquad (1)$$

$$Cooking\ loss\ (\%) = \frac{W_3}{W_2 \times (1 - M)} \times 100\% \qquad (2)$$

Where, W_1 is the weight of noodles after cooking (g); W_2 is the weight of noodles before cooking (g); W_3 is the weight of dry matter (g) and M is the moisture content of noodles before cooking (%).

2.3.4. Textural Profile Analysis. Textural profile of noodles was expressed as elongation, hardness and adhesiveness and measued using a texture analyzer (TAXT-Plus, Stable Micro Systems, Surrey, UK). Strip of noodles was cooked appropriate to time cooking test, drained for about 2 minutes at room temperature. Analysis of noodle texture followed Stable Micro System guidelines and [11] with some modification. Elongation of cooked noodles was measured by using spaghetti tensile grips (A/SPR) rig at pre-test speed of 1 mm/s, test speed of 3 mm/s, post-test speed of 10 mm/s, and initial distance between clamps of 2 cm. Hardness and adhesiveness of cooked noodles was examined by placing a set of noodle strands on the flat metal plate under the probe. Test was done by using a cylinder P/36R probe at mode; trigger type, auto 0.5 g; pre-test speed 2 mm/s, test speed 2 mm/s; post-test speed 10 mm/s, and strain of 60%. Five replicates of cooked noodles at each level of treatment were determined.

2.3.5. Statistical analysis. Variance analysis of data of each treatments was carried out on a statistical package SPSS. The significant difference among mean values was analyses using one-way analysis of variance (ANOVA) followed by Duncan's test at a significance level of ($P < 0.05$).

3. Result and Discussion

Mocaf, rice, and corn flour contain no gluten, the quality of noodles made from blending these flours depend solely on the physicochemical properties of their starch. Table 1 presented the chemical and pasting properties of these flours.

Table 1. Chemical and pasting properties of mocaf, rice, and corn flour

Parameter	Mocaf flour	Rice flour	Corn flour
Moisture content, %	10.62	5.05	6.05
Ash content, %	1.29	0.27	0.48
Protein content, %	1.77	10.07	7.07
Amylose content, %	27.83	26.61	27.98
Amylopectin content, %	36.04	40.33	32.92
Peak Viscosity, cP	4,755	4,510	1,999
Hot Paste Viscosity, cP	2,775	1,835	1,770
Breakdown Viscosity, cP	1,976	2,675	229
Final Viscosity, cP	3,246	3,193	4,034
Setback Viscosity, cP	471	1,358	2,264
Pasting Temperature, ^{o}C	69.5	79.7	79.7

The flours used in this study were enough dry with moisture content below 11%. Ash content of mocaf flour was higher than rice and corn flour, but the lowest in protein content. Amylose content is an important factor that will affect noodle quality. Corn flour had the higher amylose content than rice and mocaf flour. Table 1 showed that peak viscosity (PV) and hot paste viscosity (HPV) of mocaf flour was highest, following by rice flour and corn flour. The high PV indicates the ability of starch to swell freely before their physical breakdown [12]. HPV gives an indication of the the ability of starch to withstand heating and shear stress. The measure of the susceptibility of cooked starch granules for disintegration can be assessed from breakdown viscosity value. Final viscosity (FV) shows the ability of the material to form a viscous paste or gel after cooking and cooling. The tendency of starch retrogradation can be predicted by using setback viscosity value. Corn flour had the highest final viscosity and setback viscosity value, indicating that the starch was easier to retrograde. Mocaf flour had a lower pasting temperature than corn flour and rice flour. Amylose contents, lipids contents and by branch chain-length distribution of amylopectin may attribute to pasting properties of starch [13].

The moisture content, ash content, and protein content of dry noodles made from mocaf, rice, and corn flour are presented in Table 2.

Table 2. Chemical analysis of dry noodles made from mocaf, rice, and corn flour

Treatments	Moisture, %	Ash, %	Protein, %
MR1	14.18^{a}	1.23^{a}	5.55^{a}
MR2	10.98^{b}	1.24^{a}	4.89^{b}
MR3	13.65^{ac}	1.39^{b}	4.09^{c}
MRC1	13.13^{c}	1.29^{ac}	5.58^{a}
MRC2	13.33^{ac}	1.25^{a}	5.09^{d}
MRC3	13.74^{ac}	1.33^{c}	4.52^{e}

Same capital letters indicate no significant differences among treatments (p > 0.05).

The increasing moisture content of products will result the reducing their shelf life, hence proper drying of the noodles will be required [14]. The moisture content of noodle samples ranged from 10.98% to 14.18%, and were not significantly different (p>0.05), except the sample MR2 with 60% of mocaf flour and 40% rice flour was significantly different (P<0.05) from other samples. The final moisture content of dried noodles is usually less than 14% [6].

The ash content of dry noodles ranged from 1.23 to 1.39%. There were no significant difference (p>0.05) between all samples except for samples MR3 (contained 70% mocaf flour and 30% rice flour) and MRC3 (contained 60% mocaf flour, 30% rice flour, and 10% corn flour). Compositional differences of the flour blends resulted in the differences of noodle ash content. The ash content of noodles increased with addition of mocaf flour. In this study, the ash content of mocaf flour, corn flour, and rice flour were 1.29%, 0.48%, and 0.27% respectively.

The results showed the protein content of noodles ranged from 4.09% to 5.58% with samples MRC1 and MR3 having the highest and lowest values respectively. Samples MR1 and MRC1 were not significantly different (p>0.05), but differed significantly (p<0.05) from other samples. The addition of mocaf flour decreased the protein content of noodles. The low protein content of noodles was as a result of the low protein content of mocaf flour for 1.77%, whereas corn flour and rice flour had protein content of 7.07% and 10.07% respectively. Pato et.al [5] showed that protein content of instant noodles decreased when addition of tapioca flour increased and corn flour decreased.

The cooking test can be influenced by the nature and quality of raw materials, the shape of the final product, the processing system, drying and the cooking system itself [15]. Figure 1 – 3 presented cooking tests which included cooking time, cooking loss, and cooking weight of dry noodles made from mocaf, rice, and corn flour.

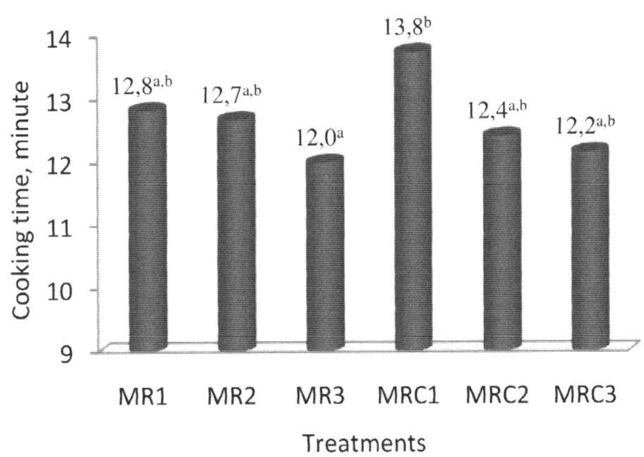

Figure 1. Cooking time of dry noodles made from mocaf, rice, and corn flour

The cooking time of the different samples ranged from 12.0 – 13.8 minutes, with samples MR3 (70% mocaf flour, 30% rice flour) and MRC1 (40% mocaf flour, 30% rice flour, and 30% corn flour) having the lowest and highest values, respectively. The results showed that there were significant differences (p<0.05) in the cooking time between sample MR3 and MRC1, but not were significant differences from the other samples. The difference in cooking time of noodles obtained in this work may indeed be due to difference in their gelatinization properties. The low pasting temperature can decrease cooking time of noodles [16]. Table 1 showed that mocaf flour had the lowest pasting temperature, followed by rice flour and corn flour. Therefore the pasting temperature of flour mixtures was increased, as the ratio of mocaf flour in the blending flour decreased. Furthermore amylose-lipids complex in corn flour might have delayed the swelling of individual granules within the noodle causing delay in cooking [16]. Omeire et al. [17] studied noodle made with 80% of wheat flour, 10% of cassava flour, 10% of defatted flour, and reported the cooking time of the noodle samples ranging from 10 to 16 minutes. Cooking time of noodles prepared from blending of rice flour with canna starch, mung bean starch, and sweet potato starch ranging from 7.2 to 9.2 minute [9].

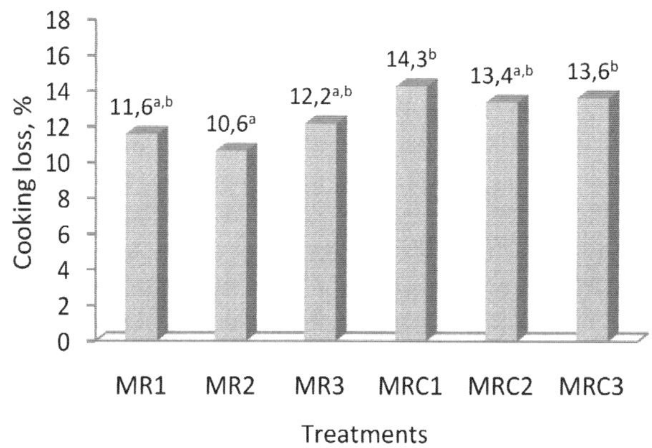

Figure 2. Cooking loss of dry noodles made from mocaf, rice, and corn flour

Cooking loss show the amount of solids (dry matter) which dissolve in water during cooking. The result in the cooking loss ranged from 10.6 to 14.3%. There were significant differences between sample MR2 with MRC1 and MRC3, but not were not significant differences from the other samples. Amylose content correlated negatively with cooking loss [12] and the higher peak viscosity, cooking loss of noodles tended to be higher [18]. Purwandari et al. [19] reported cooking loss of gluten-free noodle ranging from 11.20% to 27.38%. These noodles made from konjac flour, breadfruit and pumpkin flour with texturing agent fermented cassava flour or tapioca flour.

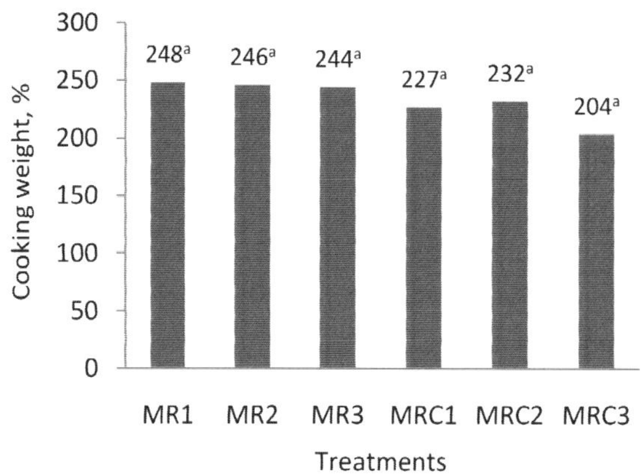

Figure 3. Cooking weight of dry noodles made from mocaf, rice, and corn flour

Cooking weight showed the weight gain of the noodles during the cooking and it is an indication of the amount of water was absorbed by noodles during cooking. Excess water absorption usually results soft and sticky noodles, but insufficient water absorption produces noodles with a coarse and hard texture [20]. The cooking weight of the noodle samples were ranging from 204% to 248%, and not significantly different (p>0.05) among all samples. The high cooking weight would be related to the high starch and protein content of the noodle samples [14]. Starch content of rice, mocaf, and corn flour were 66.94%, 63.87%, and 60.90%, respectively. Qazi et al. [9] reported noodles made form pure rice flour, mung bean starch, canna starch, and sweet potato starch or blends of rice flour with one of this starch were ranged from 168.5 % to 188.9 %. Similarly with the results from [21, 9] that the variation in cooking weight of these samples might also be due to the differences in the swelling power and pasting properties of flour and starch used to made the noodles.

The texture properties (elongation, hardness, and adhesiveness) of dry noodles made from mocaf, rice, and corn flour are shown in Table 3.

Table 3. The texture properties of dry noodles made from mocaf, rice, and corn flour

Treatments	Elongation (%)	Hardness (gf)	Adhesiveness (g.sec)
MR1	275.72^{a}	3,523^{a}	-54.73^{ab}
MR2	292.07^{a}	5,125^{ab}	-52.49^{a}
MR3	302.17^{ab}	5,953^{ab}	-57.03^{ab}
MRC1	374.20^{c}	10,478^{c}	-55.85^{ab}
MRC2	362.45^{c}	7,507^{b}	-77.09^{ab}
MRC3	335.79^{bc}	5,183^{ab}	-81.99^{b}

Same capital letters indicate no significant differences among treatments (p > 0.05).

Elongation shows the distance to break the noodle strands. Elongation is the measure for extensibility and correlates well with tensile strength. The elongation of noodles ranged from 275% to 374%. There were significantly different (p<0.05) between noodle samples made from blending mocaf and rice flour (MR1, MR2, MR3) with noodle samples made from blending mocaf, rice, and corn flour (MRC1, MRC2, MRC3). The highest elongation found in sample MRC1 (noodle made from mocaf, rice, and corn flour of 40%, 30%, 30%, respectively). Noodles prepared from blending of mocaf flour 50% and rice flour 50% (MR1) required shortest time to break, which indicated the lowest extensibility. Increased elongation was found in noodles made from high amylose flours. The same results reported by [12] that the increase of amylose content of rice flour resulted noodles which more difficult to stretch and break. There was positive correlation between amylose content of rice flour with tensile strength and extensibility of noodles. Relative extension (elongation) of noodles prepared from blending of rice flour with canna starch, mung bean starch, and sweet potato starch ranging from 101% to 239% [9].

The hardness of noodles ranged from 3,523 gf to 10,478 gf with samples MRC1 (40% of mocaf flour, 30% rice flour, and 30% corn flour) and MR1 (50% of mocaf flour and 50% rice flour) having the highest and lowest values, respectively. High hardness of noodles may be attributed to the highest amylose content. Amylose of starch will bind to each other with a bonding matrix, will increase the hardness of noodles. Low gel hardness of starch may be attributed to lower amylose content of starch which caused less retrogradation of the starch during gel formation and consequently weaker gel structure [21].

Adhesiveness is a measure of the stickiness of the noodle. The high amylose in the mixed flour decreased the adhesiveness of noodles. The lower amylose content formed the weak gel structure, resulting in greater soluble solids and so higher adhesiveness. In addition, Retrogradation amylopectin required longer times than amylose and amylopectin crystals are less stable than amylose crystals [22]. The higher amylopectin content caused the noodle dough too sticky due to amylopectin was difficult to retrograde to maintain the noodle structure. Gluten-free noodle made from konjac flour, breadfruit and pumpkin flour had hardness ranging 2,237 to 4,954 gf and adhesiveness ranging -158 to -1,097g [19].

Assessing the best treatment used effective index method [23]. Parameters fixed in this calculation were chemical quality, cooking quality and texture quality. The result showed that combination 40% mocaf flour, 30% rice flour, and 30% corn flour (sample MRC1) had the first grade.

The advantages for consumers and also noodles industry if they consume or use three ingredients of mocaf, rice and corn flour are the availability of abundant raw materials at affordable prices, no addition of synthetic dyes due to the natural yellow color of corn flour, and noodles contain high dietary fiber. Noodles made from mung bean starch are preferred for its desired appearance and excellent texture because high amylose content. Canna noodles also have much better to extrusion noodles made from sweet potato. Pure sweet potato starch is generally considered inferior for the noodles production compared with other starches such as mung bean, canna, and corn starch but sweet potato starch plays a role in substitution for expensive those starches [3]. But the availability of mung bean and canna in Indonesia is not as abundant as cassava, rice or corn.

4. Conclusion

Physicochemical properties of flours, especially the amylose and protein content, associated with the quality of noodle produced. Noodles produced from blending mocaf, rice, and corn flour had protein content, cooking loss, elongation, and hardness higher than that of blending mocaf and rice flour. Protein content of noodle increased as decrease in percentage of mocaf flour in the noodle. Noodle made from 40% mocaf flour, 30% rice flour, and 30% corn flour had the best quality mainly in term protein content and texture properties, even though the cooking quality was less. This noodle had protein content, ash content, elongation, hardness, adhesiveness, cooking time, cooking loss, and cooking weight of 5.58%, 1.29%, 374.20%, 10,478 gf, -55.85 g.sec, 13.75 minutes, 14.28%, and 226.91%, respectively.

References

[1] Budiarto 2011 Opportunity of modified cassava flour (mocaf) as wheat flour substitute an alternative materials to support food security *Proc. Int. Conf. on Natural Resources Climate Change and Food Security in Developing Countries (Surabaya)* p 837-44

[2] Jeong S, Kim H W and Lee S 2017 Rheological and secondary structural characterization of rice flour-zein composites for noodles slit from gluten-free sheeted dough. *Food Chem.* **221** 1539–45

[3] Tan H Z, Li Z G and Tan B 2009 Starch noodle: history, classification, materials, processing, structure, nutrition, quality evaluating and improving. *Food Res. Int.* **42** 551-76.

[4] Indonesian Statistics Agency 2017 https://www.bps.go.id/site/resultTab

[5] Pato U, Yusuf Y, Isnaini R F and Dira D M 2016 The quality of instant noodle made from local corn flour and tapioca flour. *J. Adv. Agr. Tech.* **3** 118-23

[6] Fu, B X 2008 Asian noodles: history, classification, raw materials, and processing. *Food Res. Int.* **41** 888–902.

[7] Yadaf BS, Yadaf R B, Kumari M and Khatkar B S 2014 Studies on suitability of wheat flour blends with sweet potato, colocasia and water chestnut flours for noodle making. *LWT Food Sci. Technol.* **57** 352-58

[8] Wandee, Y, Uttapap D, Puncha-arnon S, Puttanlek C, Rungsardthong V and Wetprasit N 2015 Quality assessment of noodles made from blends of rice flour and canna starch. *Food Chem* **179** 85–93

[9] Qazi, I M, Rakshit S K, Tran T, Javidullah and Khan M Z 2014 Effect of blending selected tropical starches with rice flour on the cooking quality and texture of rice based noodles *Sarhad J. Agr.* **30 (2)** 257-64

[10] Chen Z, Schols H A and Voragen A G J 2003. Physicochemical properties of starches obtained from three different varieties of Chinese sweet potatoes. *J. Food Sci.* **68** 431-37.

[11] Impaprasert R, Piyarat S, Sophontanakij N, Sakulnate N, Paengkanya S. Borompichaichartkul, C. and Srzednicki, G 2017 Rehydration and textural properties of dried konjac noodles: effect of alkaline and some gelling agents. *Horticulturae* **3 (20)** 1-10

[12] Fari M J M, Rajapaksa D and Ranaweera K K D S 2011 Quality characteristics of noodles made from selected varieties of Sri Lankan rice with different physicochemical characteristics. *J.Natn.Sci.Foundation Sri Lanka* **39 (1)** 53- 60

[13] Singh N, Kaur L, Sandhu K S, Kaur J, Nishinari K 2006 Relationships between physicochemical, morphological, thermal, rheological properties of rice starches. *Food Hydrocolloids* **20** 532–42

[14] Omeire G C, Kabuo N O, Nwosu J N, Peter-Ikechukwu A and Nwosu M O 2015 Enrichment of wheat/cassava noodles with partially defatted protein-rich flours. *IOSR-JESTFT* **9 (5)** 121-25

[15] Garcia L G C, e Silva A H S, de Castro Cunha P and Damiani C 2016 Preparation of gluten-free noodles incorporated of jabuticaba peel flour *J. Food and Nutr. Res.* **4 (2)** 82-7

[16] Singh N, Singh J, Sodhi N S 2002 Morphological, thermal, rheological and noodle-making properties of potato and corn starch. *J. Sci. Food Agric.* 82:1376–1383

[17] Omeire G C, Nwosu J N, Kabuo N O, Nwosu M O and Obasi N E 2015 Cooking properties and sensory evaluation of enriched cassava/wheat noodles. *Int. J. Innovative Res. Technol. Sci. (IJIRTS).* **3 (2)** 46-50

[18] Lase V A, Julianti E and Lubis L M 2013 Bihon type noodles from heat moisture treated starch of four varieties of sweet potato. *J. Teknologi dan Industri Pangan.* **24 (1)** 89-96

[19] Purwandari U, Khoiri A, Muchlis M, Noriandita B, Zeni N F, Lisdayana N and Fauziyah E 2014 Textural, cooking quality, and sensory evaluation of gluten-free noodle made from breadfruit, konjac, or pumpkin flour. *Int. Food Res. J.* **21 (4)** 1623-27

[20] Li P H, Lu W C, Hsieh C W, Li T C and Huang D W 2015 Rheological properties of dough and quality of salted noodles supplemented with djulis (*Chenopodium formosanum* Koidz.) Flour. *J. Agr. Sci.* **7 (6)** 84-92

[21] Sandhu K S, Kaur M and Mukesh 2010 Studies on noodles quality of potato and rice starches and their blends in relation to their physicochemical, pasting and gel textural properties. *LWT Food Sci. Technol.* **43 (8)** 1289–93.

[22] Tam L M, Corke H, Tan W T, Li J and Collado L S. 2004. Production of *bihon*-type noodles from maize starch differing in amylose content *Cereal Chem.* **81 (4)** 475–80

[23] De Garmo E P, Sullivan W G and Canada J R 1984 *Engineering Economy* (New York : MacMillan Publishing Company).

Correlation of concentration of modified cassava flour for banana fritter flour using simple linear regression

A Herminiati[1], T Rahman[1], E Turmala[2] and C G Fitriany[2]

[1]Center for Appropriate Technology Development - Indonesian Institute of Sciences
Jl. KS Tubun No. 5 Subang, West Java, Indonesia.
[2]Department of Food Technology - Pasundan University
Jl. Setiabudhi No. 193 Bandung, West Java, Indonesia

E-mail address: herminiati@yahoo.com

Abstract. The purpose of this study was to determine the correlation of different concentrations of modified cassava flour that was processed for banana fritter flour. The research method consists of two stages: (1) to determine the different types of flour: cassava flour, modified cassava flour-A (using the method of the lactid acid bacteria), and modified cassava flour-B (using the method of the autoclaving cooling cycle), then conducted on organoleptic test and physicochemical analysis; (2) to determine the correlation of concentration of modified cassava flour for banana fritter flour, by design was used simple linear regression. The factors were used different concentrations of modified cassava flour-B (y1) 40%, (y2) 50%, and (y3) 60%. The response in the study includes physical analysis (whiteness of flour, water holding capacity-WHC, oil holding capacity-OHC), chemical analysis (moisture content, ash content, crude fiber content, starch content), and organoleptic (color, aroma, taste, texture). The results showed that the type of flour selected from the organoleptic test was modified cassava flour-B. Analysis results of modified cassava flour-B component containing whiteness of flour 60.42%; WHC 41.17%; OHC 21.15%; moisture content 4.4%; ash content 1.75%; crude fiber content 1.86%; starch content 67.31%. The different concentrations of modified cassava flour-B with the results of the analysis provides correlation to the whiteness of flour, WHC, OHC, moisture content, ash content, crude fiber content, and starch content. The different concentrations of modified cassava flour-B does not affect the color, aroma, taste, and texture.

Keywords—banana fritter flour; concentration; modified cassava flour.

1. Introduction

Indonesia is one of the tropical countries. The country of Indonesia has abundant natural wealth, especially on local food crops of tubers. One of the most recognized tuber crops is cassava. Utilization of cassava is still very limited. Therefore, a series of studies are required to increase the potential of cassava as an alternative food source of carbohydrate tubers that are in demand by the community.

Program and activity of agricultural development of food crops in West Java Province in 2015-2019 is "Pajale" is one program that aims to develop cultivation of crops, one of which is cassava. Development sites will be conducted in several areas such as in Bandung, Bogor, Cianjur, Subang, Sukabumi, Sumedang, and Tasikmalaya.

Subang regency produces superior varieties of cassava. One of them is in Gandasoli Village, Tanjung Siang District. This region is one of the potential areas of cassava producers, and one of the most sought after varieties is manggu. The average of Tanjung Siang-Subang district's produce reaches 114.400 quintals per year [1].

Utilization of cassava into flour can increase the economic value of cassava. Cassava flour can be processed into a variety of processed food products one of them as a mixture of coated flour or seasoned flour for fried products. The National Standarization Agency defines seasoned flour as a food ingredient in the form of a mixture of flour and spices with or without the addition of other foodstuffs and allowable food additives [2].

The modified starch is a particular treated flour to provide better properties for improving the prior nature, particularly its physicochemical and functional properties or to alter some other properties [3]. The food industry has now begun to utilize the use of modified flour as a food aid for certain food products. The addition of modified flour to food products can enhance the superiority of the quality of food products. The advantage of modified flour that can make the product more crispy, better in terms of mouthfeel, color and flavor when compared with traditional ingredient products such as insoluble fiber [4].

The purpose of this research is to know the different types of processed cassava flour (cassava flour, modified cassava flour-A, and modified cassava flour-B) and the exact modified cassava flour of concentration against banana fritter flour which are expected to have color, aroma, taste, and texture favored by consumers.

2. Research methods

2.1. Materials, chemicals, and tools

The study is conducted at the Laboratory of Post-Harvest Development, Center for Appropriate Technology Development, Indonesian Institute of Sciences, Subang. The materials used in this study are cassava flour, modified cassava flour-A (using the method of the lactid acid bacteria), and modified cassava flour-B (using the method of the autoclaving cooling cycle), derived from cassava Manggu variety with harvesting age of 9 months, obtained from cassava farmer in the village district of Gandasoli, Tanjung Siang, Subang district. Supporting ingredients include: sago flour, corn starch, sugar, salt, baking soda, gum arabic, vanilla, kepok banana, and coconut oil. The chemicals used for chemical analysis are 1.25% H_2SO_4, 3.25% NaOH, 95% ethanol, distilled water, Luff Schoorl's Solution, HCl 9.5 N, concentrated HCl, universal pH indicator, H_2SO_4 6 N, KI, $Na_2S_2O_3$, and starch.

The tools used in the processing of banana fritter flour are a digital balance, slicer, cabinet dryer, vibrator screen, plastic packaging, frying pan, and spoon. The tools used for physicochemical analysis include: cup, spatula, digital balance sheet, eksikator, tang krus, oven, furnace, filter paper, funnel, erlenmeyer 500 mL, stirring rod, bath, porcelain cup, centrifuge tube, centrifuge, Kett Electric Laboratory C-100-3 Whitenessmeter, burette, 500 mL dilution flask, clamp, stative, dropper, stirer, and measuring pipette.

2.2. Formulation of banana fritter flour

Preliminary research is the determination of the types of flour (50% cassava flour, 50% modified cassava flour-A, and 50% modified cassava flour-B), other ingredients (20% sago flour, 15% corn starch, 10% sugar, 1.2% salt, 0.7% baking soda, 0.9% gum arabic, and 2.2% vanilla). Selected flour is done by organoleptic test, then the selected flour is analyzed to whiteness of flour, WHC, OHC, moisture content, ash content, crude fiber content, and starch content [5].

The process of making for banana fritter flour through the stages: weighing the material according to the formulation (50 g types of flour, 20 g sago flour, 15 g corn starch, 10 g sugar, 1.2 g salt, 0.7 g baking soda, 0.9 g gum arabic, 2.2 g vanilla), mixing ingredients, applying to fried bananas, mixing with water, banana coating, frying, and incision.

Main research is the determination of different concentrations of the selected flour to be used as banana fritter flour. Then performed the analysis of WHC, OHC, whiteness of flour, moisture content, ash content, crude fiber content, starch content [5], and organoleptic test [6] applied to banana fritter flour.

The process of making for banana fritter flour through the stages: weighing the material according to the formulation (selected flour 40%; 50%; 60%), sago flour 20%, corn starch 15%, sugar 10%, salt 1.2%, baking soda 0.7%, gum arabic 0.9%, vanilla 2.2%), mixing of ingredients, applying to fried bananas, mixing with water, banana coating, frying, and incision.

2.3. *Treatment and data analysis*

The treatment design used consisted of two variables, those are the independent variable and the dependent variable. The independent variable (X) of this experiment consisted of flour concentration with three levels: (X_1) 40%, (X_2) 50%, and (X_3) 60%. The dependent variable (Y) of this experiment consisted of seven treatment methods (Y_1) WHC, (Y_2) OHC, (Y_3) whiteness of flour, (Y_4) moisture content, (Y_5) ash content, (Y_6) crude fiber content, and (Y_7) starch content.

The research design model used is simple linear regression with two variables measured, ie selected flour concentration to: WHC, OHC, whiteness of flour, moisture content, ash content, crude fiber content, and starch content. Furthermore each data obtained results plotted to the curve, then performed a simple linear regression analysis to determine the relationship between variables measured by processing methods, the equation is: $Y = a + bX$

Y = Concentration of selected flour
a = Constants
b = Regression coefficient (increase if positive value or decrease if negative value)
X = Measured variable (WHC, OHC, whiteness of flour, moisture content, ash content, crude fiber content, and starch content).

Physical responses include: water holding capacity (WHC), oil holding capacity (OHC), and whiteness of flour. Chemical responses include: moisture content, ash content, crude fiber content, and starch content. Organoleptic response using hedonic method [6]. Assessment is given to the color, aroma, taste, and crisp texture of 30 somewhat trained panelists. Quantitative qualitative data obtained are performed using numerical scales: (1) strongly dislikes, (2) somewhat dislikes, (3) dislikes, (4) rather likes, (5) likes, (6) very likes.

3. Results and discussion

3.1. *Preliminary research result on organoleptic response*

Banana fritter flour is made from different types of flour but uses the same concentration, then organoleptic testing on color, aroma, taste, and texture. Results on organoleptic response from the panelists were presented in Table 1.

Table 1. Results on organoleptic response

Sample	Organoleptic response			
	Color	**Aroma**	**Taste**	**Texture**
1. Cassava flour (50%)	4.39	4.95	4.28	4.30
2. Modified cassava flour-A (50%)	4.34	5.09	4.92	5.05
3. Modified cassava flour-B (50%)	4.76	5.22	4.98	5.19

The result of the test of the color of banana fritter flour get a high value of 4.76 from sample modified cassava flour-B, with the response from the panelist stated rather like. The color produced from the banana fritter flour as a coating is influenced by the banana color used, resulting in a non enzymatic browning reaction with maillard reaction. When aldose /ketose is exposed to heat and reacts with the amino group, there is the production of various components such as flavour and dark-colored polymers [7]. According to [8], the reaction of carbohydrates, especially reducing sugar with the primary amino group, produces a brown material, for example in the process of frying banana fritter flour.

The result of organoleptic tests on the aroma of banana fritter flour showed a high value of 5.22 from sample modified cassava flour-B, with a response from the panelists states like. The aroma contained in the banana fritter flour is affected by the distinctive aroma of vanilla addition. Aroma of food is caused by the formation of volatile compounds [9] from vanilla.

The result of organoleptic test on banana fritter flour taste showed a high value 4.98 from sample modified cassava flour-B, with criterion of response from panelist states rather like. The taste found in banana fritter flour is influenced by the addition of sugar, which provides a savory taste and the use of oil in the frying process. The taste is influenced by aromas, foodstuffs, crispness, and food maturity levels [9].

The result of organoleptic test on banana fritter flour texture shows a value of 5.19 from sample modified cassava flour-B, with criteria of response from panelists express likes. Crispy texture on bananas fritter flour is influenced by the addition of corn starch, so banana fritter flour tend to be more crunchy and easily broken when bitten. In line with the results of research [10] that in the process of making wet noodles by using modified cassava flour, need the addition of corn flour to get a good noodle texture.

The result of organoleptic testing showed that the selected type of flour is modified cassava flour-B (using the method of the autoclaving cooling cycle), the results of physicochemical analysis of modified cassava flour-B are presented in Table 2 below.

Table 2. The results of physicochemical analysis of modified cassava flour-B

Parameter	Analysis results	SNI No. 01-2997-1992 for cassava flour	Explanation
Whiteness of flour	60.42%	Min 85%	Not yet
Water Holding Capacity	41.17%	-	-
Oil Holding Capacity	21.15%	-	-
Moisture content	4.40%	Max 12%	Fulfill
Ash content	1.75%	Max 1.5%	Not yet
Crude fiber content	1.90%	Max 1.5%	Not yet
Starch content	68.56%	Min 75%	Not yet

The result of the analysis of modified cassava flour-B refers to Indonesian National Standard (SNI) No. 01-2997-1992 [11] for cassava flour. Moisture content of modified cassava flour-B already meets SNI, but for whiteness of flour, ash content, crude fiber content, and starch content have not fulfilled Indonesian National Standard.

3.2. Main research results against physicochemical response of banana fritter flour

The results of physicochemical analysis of different concentrations of modified cassava flour-B to be used as banana fritter flour are presented in Table 3 below.

Table 3. Results of physicochemical analysis of banana fritter flour

Parameter	Concentration of modified cassava flour-B (%)		
	40%	50%	60%
Water Holding Capacity	22.06	25.47	31.20
Oil Holding Capacity	18.05	17.39	16.33
Whiteness of flour	84.80	81.60	76.10
Moisture content	9.46	8.41	7.98
Ash content	2.75	2.52	1.86
Crude fiber content	1.66	2.14	2.61
Starch content	66.40	67.36	70.37

3.2.1. Water holding capacity. WHC is used to measure the ability of flour to retain water absorbed, the WHC value is influenced by the water content in the food. Water added to flour for doughs of coating, will affect physical properties and processing. The absorption of water in flour is also influenced by the size and structure of the starch granules, wherein the smaller starch granules will increase solubility and increase water absorption [12]. An increase in WHC value in the use of more concentration of flour (60%), because with the increasing use of modified cassava flour, the moisture content of banana fritter flour will be lower, causing the flour to absorb more water so that the water holding capacity also higher.

The ability of flour to absorb and retain water is not only influenced by the moisture content in the ingredients, it is also influenced by several factors, those are the amylose content, the size of the starch granules, and the fat content of the ingredients. Water absorbed in starch molecules is caused by the physical properties of the granules as well as intramolecularly bound [13].

The result of linear regression analysis shows the correlation of the difference of flour concentration to WHC from banana fritter flour, which can be seen in Figure 1. The correlation between different flour concentration on banana fritter flour processing, shown by **r** value of linear regression equation, Y = 3.41+(0.45x) with correction coefficient (r) is 0.9894. This shows that between different concentrations of flour and WHC on banana fritter flour have a very strong correlation, so the increase of flour concentration from banana fritter flour affects the WHC produced, because if the correlation coefficient value of 0.8 - 1 indicates very strong correlation.

Perfect linear relationship with positive slope, this is indicated by a positive slope of 0.9894 means that the higher the concentration of flour the higher the value of WHC produced.The result of the 40% flour concentration analysis method resulted in lower average WHC compared with a 60% concentration resulting in a high WHC average.

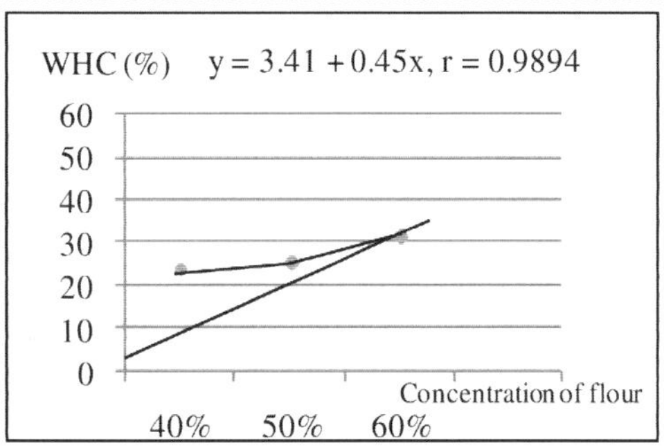

Figure 1. Graph of linear regression method of treatment of water holding capacity

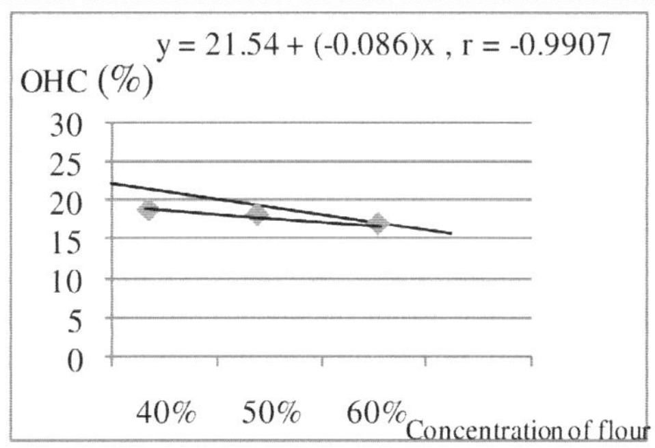

Figure 2. Graph of linear regression method of treatment of oil holding capacity

3.2.2. Oil holding capacity. OHC is used to measure the ability of flour to retain the oil it absorbs. This ability is determined by the presence of fat and fiber content. Fat can form a layer that is hydrophobic on the surface of the fiber network while the fiber has the ability to absorb oil. The low fat content in the flour will make the flour absorb more oil from the outside. High fiber content in flour will make the flour has the ability to absorb and retain more oil.

Oil holding capacity is due to oil trapped in a porous starch matrix or in the structure of the amylose helix or amylopectin due to the formation of the amylose-lipid complex. The changes of hydrophobic groups are more amyloid-lipid complex due to high temperature [14]. The decline in OHC value in the use of more concentration of flour (60%), this occurs because the ability of banana fritter flour to absorb and retain oil is affected by the frying process. Flour that has a greater OHC value will absorb more oil and hold the oil used for frying, this causes the cooking oil used quickly run out.

The result of linear regression analysis showed the correlation of different concentration of flour to OHC from banana fritter flour, which can be seen in Figure 2. The correlation between different flour concentration on banana fritter flour processing, shown by **r** value of linear regression equation, Y = 21.54+(-0.086x) with correction coefficient (r) is 0.9907. This shows that between different concentrations of flour with OHC on banana fritter flour has a very strong correlation, so the decrease of flour concentration from banana fritter flour has an effect on the resulting OHC, because if the correlation coefficient value of 0.8 - 1 indicates very strong correlation .

The perfect linear relationship with slope is indicated by a negative slope (-0.9907), meaning that the higher the flour concentration, the lower the resulting OHC value. The results of the 60% flour concentration analysis method resulted in lower average OHC value compared with 40% concentration resulting in a high average OHC. According to [15 and 16] the ability to absorb high oils on flour shows the flour has a part that is lipophilic.

3.2.3. Whiteness of banana fritter flour. The whiteness of flour is strongly influenced by the process of starch extraction, the more pure starch extraction process then the resulting flour will be white [17]. The white density of modified cassava flour as raw material is 60.42%, after going through the formulation process with the supporting material to become banana fritter flour has increased the whiteness of flour, which can be seen in Table 3.

The result of linear regression analysis shows the correlation of the difference of flour concentration to the whiteness of banana fritter flour, which can be seen in Figure 3. The correlation between different flour concentration on banana fritter flour processing, shown by **r** value of linear regression equation, is Y = 102.5833-(-0.043x) with correction coefficient (r) is 0.9885. This shows that between different flour concentration and whiteness on banana fritter flour have a very strong correlation, so that the concentration of flour from the banana fritter flour affects the degree of white produced, because if the correlation coefficient value of 0.8 - 1 indicates correlation very strong.

The perfect linear relationship with slope is indicated by a negative slope of -0.9885 meaning that the higher the flour concentration the lower the whiteness of the resulting flour. The result of the 60% flour concentration analysis method yields a low average whiteness of flour value compared to the 40% concentration resulting in a high average white degree. This is influenced by the addition of sago flour with 89.9% white grade, on banana fritter flour formulation, the higher concentration of modified cassava flour (60%), less concentration of sago flour, compared to modified cassava flour concentration (40%), then the concentration of sago flour more.

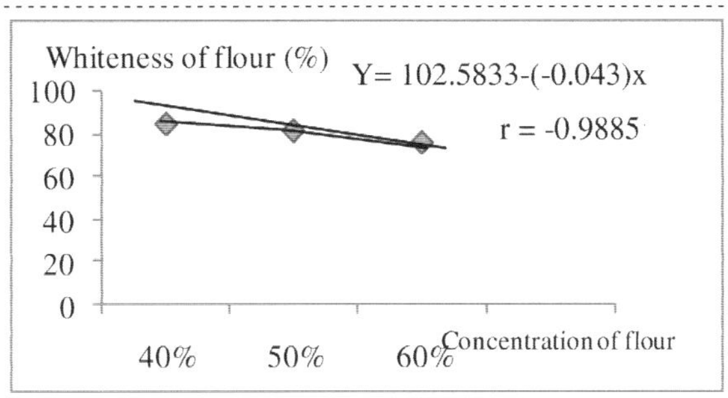

Figure 3. Graph of linear regression method of treatment of whiteness of flour

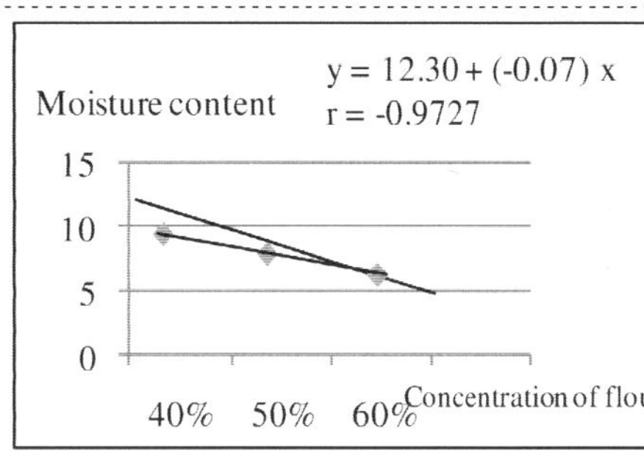

Figure 4. Graph of linear regression method of treatment of moisture content

3.2.4. Moisture content. Water is a major component of food, which plays an important role in determining the various reactions and quality of food. The moisture content in food is related to food damage by microorganisms [18]. High moisture content in flour product will greatly disrupt the stability of the product, which may result in flour agglomeration during storage [19]. The decrease in banana fritter flour moisture content in the use of more flour concentration (60%), due to the influence of moisture content of modified cassava flour (4.4%) added to the formulation. The increasing concentration of modified cassava flour used, the lower the resulting moisture content. The result of linear regression analysis shows the correlation of flour concentration to moisture content of banana fritter flour, which can be seen in Figure 4.

The correlation between different flour concentration on banana fritter flour processing, shown by **r** value of linear regression equation, is Y = 12.30+(-0.07x) with correction coefficient (r) is 0.9727.

This shows that between different concentrations of flour and moisture content in banana fritter flour has a very strong correlation, so that the concentration of flour from banana fritter flour affects the moisture content produced, because if the correlation coefficient value of 0.8-1 indicates correlation very strong. The perfect linear relationship with slope is indicated by a negative slope of -0.9727 meaning that the higher the concentration of flour the lower the moisture content of the flour produced.

The result of the 60% flour concentration analysis method resulted in average low moisture value compared with a concentration of 40% resulting in a high average moisture content. This is influenced by the addition of modified cassava flour (4.4% moisture content) added to the formulation. The increasing concentration of modified cassava flour used, the lower the moisture content of the banana fritter flour formulation. According to [20] that increasing of moisture content can cause the product loss of crispness.

3.2.5. Ash content. The content of starch ash is related to the mineral content in it. Ash content is strongly influenced by the type of materials, age of materials, and others. According to [21], the content of flour ash varies from 0.30 - 1.40% (ww). The greater the ash content of a food, the greater the mineral content contained in the food. In the formulation of banana fritter flour is added salt and baking soda which is inorganic salt, so it is still left as ash when it is done. The result of linear regression analysis shows the correlation of flour concentration to ash content of banana fritter flour, which can be seen in Figure 5.

The correlation between different flour concentration on banana fritter flour processing, shown by **r** value of linear regression equation, is Y = 4.59+(-0.043)x with correction coefficient (r) is 0.9613. This shows that between different flour concentration and ash content on banana fritter flour have a very strong correlation, so that the concentration of flour from banana fritter flour affects the content of ash produced, because if the correlation coefficient value of 0.8-1 indicates correlation very strong.

The perfect linear relationship with slope is indicated by a negative slope of -0.9613 meaning that the higher the concentration of flour the lower the ash content of the flour produced. The result of the 60% flour concentration analysis method yielded an average low ash content value compared to a concentration of 40% resulting in a high average ash content. This is influenced by different concentrations in the addition of modified cassava flour (ash content of 1.75%). The addition of salt and baking soda which is inorganic salt does not give a real effect, because it is added at the same concentration. The addition of sago flour containing ash content of 4.86% in banana fritter flour formulation has an effect on ash content. Due to the lower concentration of modified cassava flour, the addition of sago flour is higher.

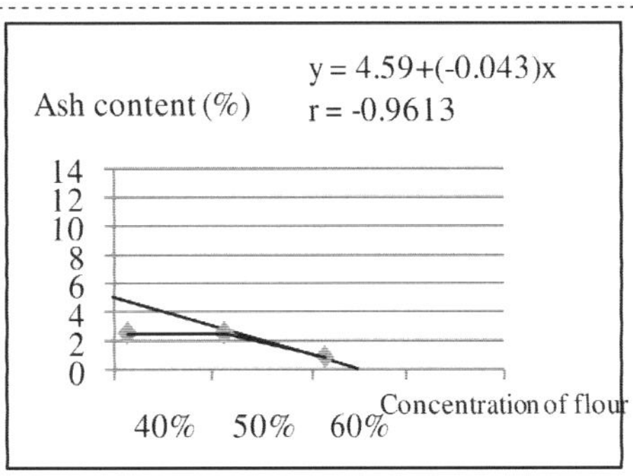

Figure 5. Graph of linear regression method of treatment of ash content

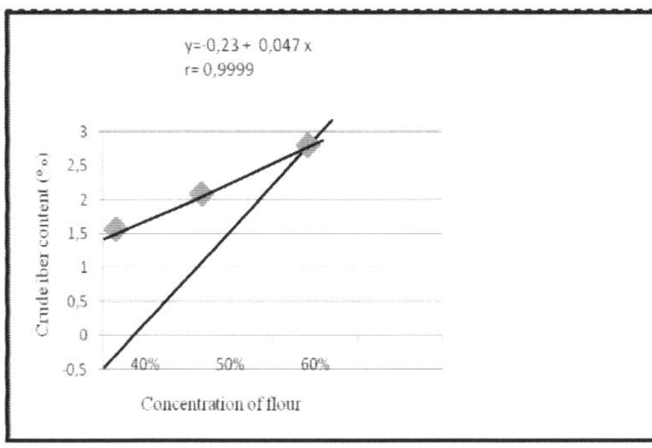

Figure 6. Graph of linear regression method of treatment of crude fiber content

3.2.6. Crude fiber content. Fiber is a component of plant tissue that is resistant to hydrolysis by enzymes in the stomach and small intestine [22]. The crude fiber is determined from the residue after the food is treated with acids and strong bases. The result shows that with the increasing concentration of modified cassava flour, the crude fiber content is increasing. The result of linear regression analysis shows the correlation of the difference of flour concentration to the crude fiber content of banana fritter flour, which can be seen in Figure 6.

The correlation between different flour concentration on banana fritter flour processing, shown by **r** value of linear regression equation, ie Y = (0.23+0.047x) with correction coefficient (r) is 0.9999. This shows that between different flour concentration and crude fiber content in banana fritter flour has a very strong correlation, so that the concentration of flour from the banana fritter flour affects the crude fiber content produced, because if the correlation coefficient value is 0.8-1 shows very strong correlation.

The perfect linear relationship with slope is indicated by a positive slope of 0.9999 meaning that the higher the concentration of flour the higher the crude fiber content produced. The result of the 60% flour concentration analysis method yielded a mean value of high crude fiber content compared with a concentration of 40% yielding a low average crude fiber content. This is influenced by different concentrations in the addition of modified cassava starch (1.90% fiber content).

3.2.7. Starch content. The composition of starch is an important factor in determining the texture and characteristics of the coating flour. Starch is composed of two main fractions, those are amylose and amylopectin [23]. Amylose plays a role in the absorption of oil during frying, high flour amylose content can be used to reduce oil absorption due to its ability to form film [24].

The increase of banana fritter flour starch content in the use of more flour concentration (60%), due to the effect of modified cassava flour concentration added to the formulation. The increasing concentration of modified cassava flour used, the higher the resulting starch content. The result of linear regression analysis shows the correlation of flour concentration difference to starch content of banana fritter flour, which can be seen in Figure 7 below.

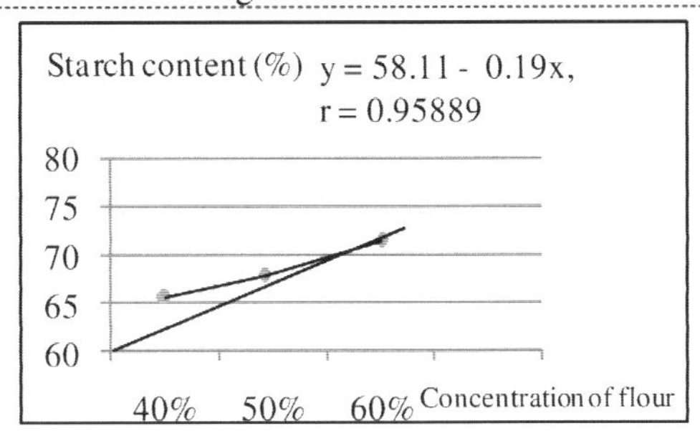

Figure 7. Graph of linear regression method of treatment of starch content

The correlation between different flour concentration on banana fritter flour processing, shown by **r** value of linear regression equation, is Y = 58.11–0.19x with correction coefficient (r) is 0.9588. This shows that between different flour concentration and starch content in banana fritter flour have a very strong correlation, so that the increase of flour concentration of banana fritter flour has an effect on the starch content produced, because if the correlation coefficient value of 0.8-1 indicates correlation very strong.

The perfect linear relationship with slope is shown with a positive slope of 0.9588 means the higher the concentration of flour, the higher the starch content produced. The result of the 60% flour concentration analysis method yielded an average high starch content value compared to a concentration of 40% resulting in a low average starch content. This is influenced by different concentrations in the addition of modified cassava flour (67.31%), and addition of corn starch containing 88% starch content. According to [25] starch with high amylopectin content tend to give fragile and crispness product characters, while amylose give of texture and violence more resistant.

3.3. Main research result on organoleptic response

Banana fritter flour from modified cassava flour-B with a concentration of 60%; 15% sago flour; 10% corn starch; 10% sugar; 1.2% salt; 0.7% baking soda; 0.9% gum arabic; and 2.2% vanilla, selected from the simple linear regression based on WHC, OHC, whiteness of flour, moisture content, ash content, crude fiber content, and starch content.

Banana fritter flour color after the coating process on banana of kepok and frying process has a golden yellow color. The result of organoleptic testing on the color of banana fritter flour from sample modified cassava flour B (60%), gets value of 4.49, with the response from the panelists say rather like. Different concentrations of modified cassava flour have no effect on color parameters. According to [26] decreasing of brightness and brown color in product due to the frying process.

The aroma found in banana fritter flour is influenced by the distinctive aroma that arises from banana and the addition vanilla of banana fritter flour formulation. The result of organoleptic testing on the aroma of banana fritter flour from sample modified cassava flour B (60%), gets value of 5.22, with the response from the panelists say like. Different concentrations of modified cassava flour have no effect on flavour parameters. According to [27] vanilla aroma is included into a group of volatile compounds that are not heat resistant.

The result of organoleptic testing on the taste of banana fritter flour from sample modified cassava flour B (60%) gets value of 4.33, with the response from the panelists say rather like. Different concentrations of modified cassava flour have no effect on taste parameters. Sweet and savory of banana fritter flour flavors are influenced by the addition of sugar in the formulation of banana fritter flour and the use of cooking oil in the frying process. When flour is fried then the water molecules will evaporate and be replaced by oil that makes air cavities in the food, so resulting in the product of swelling and crunchy [28]

The result of organoleptic testing on the texture of banana fritter flour from sample modified cassava flour B (60%) gets the value of 4.41, with the response from the panelists say rather like. Different concentrations of modified cassava flour have no effect on texture parameters. The crunchy texture of banana fritter flour is influenced by the addition of corn starch and baking soda ($NaHCO_3$) to banana fritter flour formulation. According to [29] corn starch when the frying process provides a more crunchy texture and easily broken when bitten, corn starch is recommended for use as a coating material (fried products). The addition of baking soda will make the banana fritter flour expand and increase the crispness, because at the time of heating/frying, baking soda will release carbon to form a fragile and crunchy structure [19].

4. Conclusion

The modified cassava flour-B selected from the organoleptic test to be used as the main ingredient of banana fritter flour has whiteness of flour (60.42%), WHC (41.17%), OHC (21.15%), moisture content (4.4 %), ash content (1.75%), crude fiber content (1.86%), and starch content (67.31%), which approached SNI quality of cassava flour (SNI No. 01-2997-1992).

Banana fritter flour from modified cassava flour-B with a concentration of 60%; 15% sago flour; 10% corn starch; 10% sugar; 1.2% salt; 0.7% baking soda; 0.9% gum arabic; and 2.2% vanilla, selected from the simple linear regression based on WHC, OHC, whiteness of flour, moisture content, ash content, crude fiber content, and starch content. The results of physicochemical analysis for banana fritter flour formulation are water holding capacity 31.20%; oil holding capacity 16.33%; whiteness of flour 76.10%; moisture content 7.98%; ash content 1.86%; crude fiber content 2.61%; and starch content 70.37%

References

[1] Carolina 2009 The innovation opportunity based on sampeu manggu (*Manihot esculenta* Linn.): agricultural ecosystem case in Tanjung Siang – Subang *Proceedings of the Grassroot Innovation Workshop* PPTTG – LIPI Subang D14 1-8
[2] The National Standarization Agency 1998 *SNI 01-4476-1998 for Seasoned Flour* (Jakarta: The

National Standarization Agency)
[3] Saguilan A A, Huicochea E F, Tovar J, Suarez F G, Meraz F G and Perez L A B 2005 Resistant starch-rich powders prepared by autoclaving of native and lintnerized banana starch: partial characterization *Journal Starch* **57** 405-412
[4] Faridah D N 2011 Changes of characteristics for garut starch crystalline (*Maranta arundinaceae* L.) in the development of resistant starch type III *Dissertation* School of Post Graduate IPB
[5] Association of Official Analytical Chemists 2005 *Official Methods of Analysis of The Association of Official Analytical Chemists* (Washington DC: AOAC)
[6] Soekarto S T 2006 *Organoleptic Assessment* (Jakarta: Bhatara Karya Aksara Press)
[7] BeMiller J N and Huber K C 2008 *Carbohydrates* In Amudaran S, Parkin K L, and Fennema O R (ed.) *Food Chemistry* (New York: CRC Press New York)
[8] Sultanry, Rubianty and Kasenger B 1985 *Food Chemistry* (Bandung Technology University Press)
[9] Meilgaard M, Civille G V and Carr B T 2000 *Sensory Evaluation Techniques (*Boca Raton Florida: CRC Press)
[10] Risti 2013 Effect of egg addition to protein content, fiber content, level of elasticity, and acceptance of gluten-free wet noodles made from composite flour University of Diponegoro
[11] The National Standarization Agency 1992 *SNI 01-2997-1992 for Quality Requirements of Cassava Flour* (Jakarta: The National Standarization Agency)
[12] Niba L L, Bokanga M M, Jackson F L, Schlimme D S, Li B W 2002 Pshysochemical properties and starch granular characteristics of flour from various *Manihot esculanta* (cassava) genotypes *Journal of Food Science* **67** 1701-1705
[13] Kulp K and Joseph G P 2000 *Handbook of Cereal Science and Technology* (New York: Marcel Dekker)
[14] Ashwar B A, Gani A, Shah A, Wani I A, Masoodi F A and Saxena D C 2016 Production of resistant starch from rice by dual autoclaving-retrogadation treatment: invitro digestibility, thermal and structural characterization *Food Hydrocolloids* **56** 108-117
[15] Swinkels J J M 1985 *Sources of Starch, Its Chemistry and Physics* In Van Beynum G M A and Roels J A (eds) *Starch Conversion Technology* (New York: Marcel Dekker Inc.)
[16] Zayas J F 2012 *Functionality of Proteins in Food* (USA: Springer Science & Business Media)
[17] Adie 2007 Studying the chemical and physical characteristics of tapioca and modified cassava flour as coating of nuts in the salted peanut products Faculty of Agricultural Technology IPB
[18] Rauf R 2015 *Food Chemistry* (Jakarta: Andi Press)
[19] Magdalena 2010 Formulation and prediction of shelf life of fried chicken seasoning flour made from modified cassava flour Faculty of Agricultural Technology IPB
[20] Arimi, J.M., E. Duggan, M. Sullivan, J.G. Lyng, E.D. O'Riordan. 2010. Effect of water activity on the crispiness of biscuit (cracker bread): mechanical an acoustic evaluation *Food Research International* **43** 1650-1655
[21] Nielsen S S 2003 *Food Analysis* (New York: Kluwer Academic Plenum Press)
[22] Winarno F G 2004 *Food Chemistry and Nutrition* (Jakarta: Gramedia Press)
[23] Abera S and Rakshit S K 2003 Comparison of physicochemical and functional properties of cassava starch extracted from fresh root and dry chips *Journal Starch* **55** 287-296
[24] Huang D P and Rooney L W 2001 *Starches for Snack Foods* In Lusas R W and Rooney L W *Snack Foods Processing* (Washington DC: CRC Press) pp 115-136
[25] Hee-Joung A 2005 Effects of ozonation and addition of amino acids on properties of rice starches *Dissertation* Faculty of Louisiana State Univ and Agricultural &Mechanical College
[26] Kumar A J, Singh R R B, Patel A A and Patil G R 2006 Kinetics of colour and texture changes in *Gulabjamun* balls during deep-fat frying *LWT-Food Sience and Technology* **39** 827–833
[27] Ketaren S 1985 *Introduction to Essential Oil Technology* (Jakarta: Balai Pustaka Press)
[28] Hasta L 2013 The Effect of tapioca starch with salted egg and time of Steaming for crispiness crackers *J. Ilmu Peternakan* **1** 307-313
[29] Yuyun A 2007 *Making of Crispy Fish* (Jakarta: Agromedia Pustaka Press)

Characteristics of Chemical and Functional Properties of Modified Cassava Flour (*Manihot esculenta*) by Autoclaving-Cooling Cycles Method

Raden Cecep Erwan Andriansyah, Taufik Rahman, Ainia Herminiati, Nurhaidar Rahman, Rohmah Luthfiyanti

Development Center for Appropriate Technology – Indonesian Institute of Sciences
Jl. KS Tubun No 5 Subang, West Java, Indonesia

Email: cecep_erwan@yahoo.com

Abstract. The modified cassava flour can be made using the method of the autoclaving cooling cycle (AAC). The stability of the warming can be seen from the decreasing value of breakdown viscosity, while the stability of the stirring process can be seen by the decreasing value of setback viscosity. The stages of research include: (1) the making of cassava flour, (2) the making of modified cassava flour by the method of treatment of ACC with a variety of flour concentration and autoclaving time, (3) chemical analysis of the moisture, ash, fat, protein, carbohydrate; The functional properties of the pasting characteristics to the initial temperature of the pasting, peak viscosity, hot paste viscosity, breakdown viscosity, cold paste viscosity and setback viscosity. The result shows that cassava flour modified by treatment of flour concentration 16% and autoclaving time 41 minutes having pasting code and pasting viscosity which is resistant to high temperature. Flour with this character is flour that is expected to maintain the texture of processed products with a paste form that remains stable. Utilization of modified cassava flour by the ACC method can be applied to the pasting product such as noodle and spaghetti, hoping to support for food diversification program to reduce dependence on wheat flour in Indonesia.

1. Introduction

In Indonesia, tubers are very potential to be developed as raw materials and supporting material for food and nonfood industry. One of the tubers that have potential as a source of carbohydrates for food diversification and as a functional food ingredient is cassava (*Mani hot esculenta*). Part of cassava tuber contains about 25-28% starch and 10% fiber. In the processing of cassava into starch, obtained tapioca 20-25% and 11% cassava [1], whereas tapioca containing fat 0.08 to 1.54%, 0.03 to 0.06 % protein, and 0.02 to 0.33% ash [2]. One way to increase the efficiency of cassava is to create a modified cassava flour in order to obtain the properties of flour suitable for a particular application. Modified flour is given special treatment so resulted in better properties to correct previous properties and other properties, mainly physicochemical and functional properties. According to [3], the treatment can be classified physically and chemically. Chemically modified is a modification of starch or flour which the hydroxyl group has been converted through a chemical reaction (esterification or oxidation) or by disrupting its original structure. Modification of starch or flour physically involves several factors including: temperature, pressure, cutting, and the water content of the starch. In this study, cassava flour is done by physical modification using the method of autoclaving cooling cycles (ACC). According [4] heating treatment by using the method of autoclaving can increase the production of resistant starch to 9%.

The processing of starch modification can increase a value of functional food and has better quality. One of the food ingredients that are the result of starch modification and can be used as raw material for the manufacture of functional food is resistant starch [4]. The benefit of modified starch in the form of resistant starch can make the product more crispy, better in terms of mouthfeel, color and flavor if compared with traditional ingredients added products such as insoluble dietary fiber. Resistant starch also has low calory value, which amounted to 11.7 kJ / g RS or 1.9 kcal / g, so it can be used as an ingredient for low-calorie food [5].

The purpose of this study is to determine the characteristic of the chemical and functional modified cassava flour (*Manihot esculenta*) by using autoclaving-cooling cycles.

2. Research methods

2.1. *Materials and tools*

The study is conducted at the Laboratory of Post-Harvest Development, the center for Appropriate Technology Development, Indonesian Institute of Sciences, Subang. The raw material used is cassava Manggu variety with harvesting age of 9 months, obtained from cassava farmer in the village district of Gandasoli, Tanjung Siang, Subang district. Supporting material is a chemical used for chemical analysis and functional properties. Tools used include slicer; autoclave; cabinet dryer; vibrator screen; refrigerator; and tools for the analysis of moisture, ash, fat, protein; and Rapid Visco Analyzer (RVA) Newport RVA with Thermo Haake K20 for profile analysis of modified cassava flour paste.

2.2. *Making cassava flour*

The processing of making cassava flour refers to method Cassava [6] peeled and washed to remove sap and grime. Furthermore, the downsizing of the tubers sample use a slicer. Cassava that has been thinly sliced dried using a cabinet dryer at a temperature 50 °C to a moisture content up to 12%. The next phase, sliced tubers are dry powdered using a disk mill, and then sieved using a screen vibrator until the degree of fineness of 70 mesh. Flow diagram of the process of making cassava flour can be seen in figure 1.

2.3. *Making Modified Cassava Flour*

Manufacture of modified cassava flour refers to a method [6] arrowroot starch modified. Manufacture of cassava starch modified by the method of autoclaving cycles cooling is done by 3 cycles. Cassava flour is suspended with the addition of water (table 1) according to treatment. Then the suspension is gelatinized at a temperature of 70-85 ° C. Further heated by using an autoclave at 121 ° C 1 atm pressure and adjusted for the treatment time. The heating temperature used is 121 ° C referring to previous research [7]. The suspension is heated further cooled at room temperature until the temperature drops, followed by storage at 4 ° C for 12-16 hours. The next stage is drying using a cabinet dryer at temperature 50 °C to a maximum moisture content of 12%. Modified cassava flour which has been dried subsequently crushed and sieved use 70 mesh sieving. In figure 2 can be seen a flow diagram of the process of making cassava modified starch.

The resulting modified cassava flour then testing the characteristics of the chemical properties include: analysis of water content used gravimetry method; ash content used method gravimetry; protein content used Kjeldahl analysis method; fat content used sokhlet analysis methods [8], analysis of carbohydrates by the difference. Analysis of functional properties pasting characteristics includes initial pasting temperature; peak viscosity; hot paste viscosity; breakdown viscosity; cold paste viscosity; and setback viscosity, using the Rapid Visco Analyzer (RVA) Newport RVA with Thermo Haake K20.

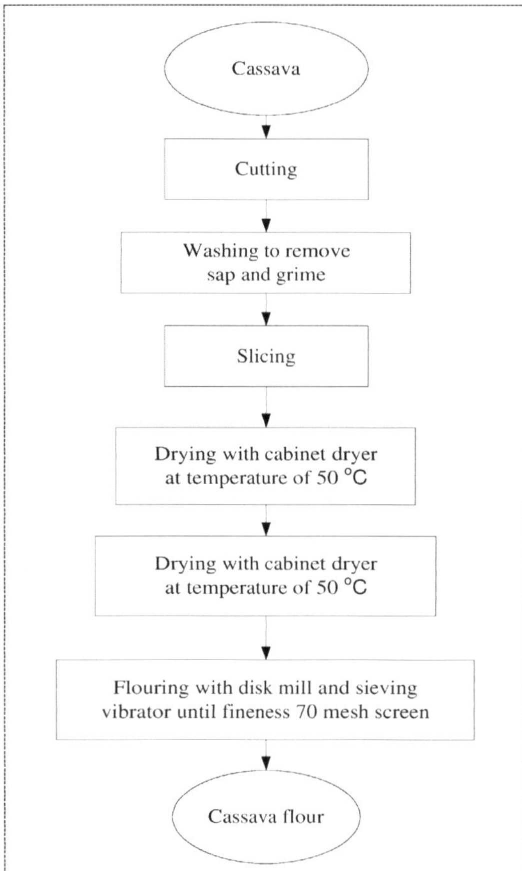

Figure 1. Flow diagram of the processing of making cassava flour [6].

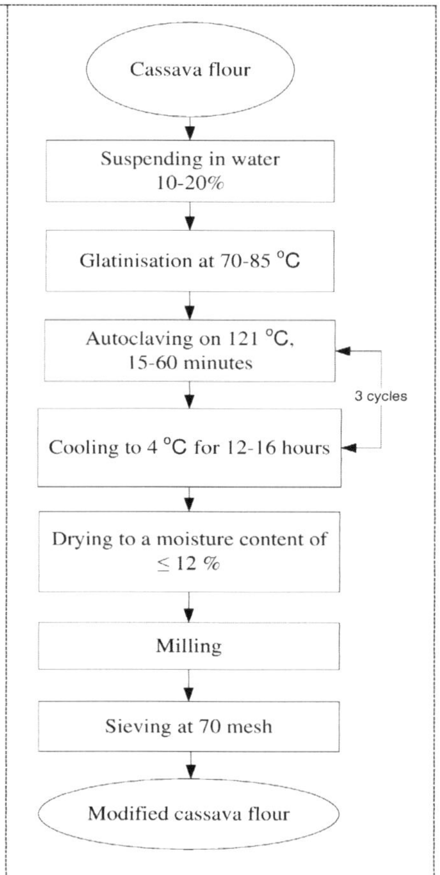

Figure 2. Flow diagram of the processing of making cassava flour [6].

2.4. Treatment and data analysis

The research design that used in this study uses software Design Expert 7. System (DX-7. DX 7 system obtained 16 experimental design. The system generates design data of the input concentration and time autoclave. In table 1 we can see the design of experiment using Design Expert 7.

Table 1. The design of experiment is using design expert 7.

Treatment	Concentration (%)	Autoclaving time (minute)
1	13	15
2	20	60
3	10	60
4	10	18

5	16	25
6	20	36
7	20	36
8	20	60
9	10	30
10	10	18
11	20	15
12	10	60
13	16	41
14	15	60
15	11	45
16	20	15

3. Research methods

3.1. Proximate analysis results of modified cassava flour

The proximate analysis result of modified cassava flour is presented in table 2. The modified cassava flour has a moisture content of 3.20-11.4%, according to standard for products processed from cassava edible (edible cassava flour) with a maximum moisture content of 13% [9].

The ash content in cassava modified starch is 1.54-2.3%. The ash content indicates total mineral in food. Mineral and inorganic salt in a small amount expressed as ash which is a food residue after the combustion process into carbon-free [10]. In the quantitative, value of ash content in flour and starch derived from mineral in the fresh tuber, fertilizer use, and can also come from the soil and air contamination during processing [11]. Maximum ash content is 3%, which is still allowed in edible cassava flour [9].

Table 2. Result of proximate analysis of modified cassava flour

No	Treatment	Parameter (%)				
		Water	Ash	Fat	Protein	Carbohydrate
1	1	6.21	1.70	1.17	0.20	90.72
2	2	7.13	1.65	0.44	0.96	89.82
3	3	11.4	1.64	1.23	0.98	84.75
4	4	6.81	1.70	1.10	0.20	90.19
5	5	7.60	1.54	1.05	0.40	89.41
6	6	8.10	1.60	1.14	0.40	88.76
7	7	7.90	1.80	1.15	0.60	88.55
8	8	5.70	1.60	0.66	0.60	91.44
9	9	6.20	1.90	0.51	0.98	90.41
10	10	5.50	2.20	0.60	0.60	91.10
11	11	3.70	2.30	0.16	0.40	93.44
12	12	5.50	1.90	1.13	0.98	90.49
13	13	5.40	1.80	1.76	1.76	89.28

14	14	3.30	1.60	1.51	2.45	91.14
15	15	3.60	1.90	0.94	1.51	92.05
16	16	3.20	1.80	1.17	1.14	92.69

Modified cassava flour is produced from edible cassava. Therefore, the quality requirement of modified cassava flour may refer to CODEX STAN 176-1989 (Rev. 1-1995) of edible cassava flour (can be seen in table 3). Moisture content and ash from modified cassava flour is according to the requirements of CODEX STAN 176-1989 (Rev.1-1995).

Table 3. Quality requirement of edible cassava flour in CODEX STAN 176-198 (REV.1-1995)

No	Parameter	Unit	Quality Requirements
1	Water	%	Max 13
2	Ash	%	Max 3
3	Crude fiber	%	Max 2
4	HCN	mg/kg	Max 10
5	Pesticide residues	-	Confirm
6	Heavy metal	-	Not detectable
7	Food additive	-	Not detectable

Source: [9]

Fat level obtained from the analysis of modified cassava flour is 0.16% -1.76%. According [12] The fat content is low and is at the boundary between 0.1-1.14%, except for treatment of 13 which has relatively high fat content. High-fat content is able to form a complex of amylose-fat that can inhibit the release of starch granules.

Levels of a protein produced in modified cassava flour are relatively low, 0.20 to 2.45%. Generally bulbs - not including the root source of protein, so the value of its protein content is relatively low inversely proportional to the value of carbohydrate level. The test result shows that carbohydrate content of modified cassava flour has a high value ranged between 84.75-93.44%, according to research results [7] that cassava is a crop that contains high carbohydrate with the low level of amylose and high amylopectin, so it can be used as food ingredient source of carbohydrate instead of rice. According to [13], in addition to carbohydrate, fat and protein in the flour are very important to the formation of pasta, which will affect the texture of the food produced.

3.2. Analysis results of pasting characteristics for modified cassava flour

Cassava starch pasting profile changes before and after the modification can be seen in Table 4, includes initial pasting temperature, peak viscosity, hot paste viscosity, breakdown viscosity during heating, cold paste viscosity and setback viscosity.

Early temperature formation of pasta provides information on the required temperature of the starch granules begin to swell in the presence of water by the starch binding; time and peak viscosity can provide information related to the water holding capacity of starch and trends disintegration; breakdown viscosity (BDV) to provide information regarding resistance to warming; setback viscosity (SBV) to provide information regarding the potential retro gradation and syneresis, and the final viscosity to provide information regarding the potential ability to form gels [14]. Change of profile manioc starch paste is described as follows:

3.2.1. Early temperature pasting. Table 4 shows the effect of concentration and time ACC modification processing to changed pasting temperature from cassava flour pasta. Pasting temperature decreases from natural cassava flour, from 61.20 °C into 50.30 °C after ACC processes that occur at 20% concentrations and 20 minutes autoclaving. Delay the process of pasting shows that modified starch granules ACC is more stable and resistant to the heating process. This increase in pasting temperature is thought to be due to increasingly rigid granule due to changes in the crystalline starch region. The combination of concentration and time of heating used provide Significant pasting temperature changes on 4 and 11 of treatment compared with natural cassava flour. The results of this study is different from [15], that conducts research on starch ammorphophalus, where the initial temperature pasting after modification HMT increases from 82.13 °C to be 86.54 - 88.56 °C. The same trend is happened also in sago starch [16], corn starch, peas, and lentils [17].

3.2.2. Peak viscosity. Peak viscosity is the ability of starch granules to inflate optimally when the flour pasta is heated. Cassava flour that undergoes ACC modification at various concentrations of flour and heating time decreased the peak viscosity value compared with natural cassava flour. The higher the concentration and the time of the modification process, the more decreased peak viscosity. The highest decrease in the viscosity of the peak occurs in modified cassava flour on treatment 3 (concentration of 10% flour, 60 minutes), and 14 (concentration of 15% flour, 60 minutes). According to [18], this peak viscosity indicates that HMT starch granules have inflated power and lower water absorption. This is influenced by the structure of the increasingly rigid starch granule as result of inter interaction and intramolecular granules that are getting stronger and tighter which will inhibit the penetration of water so that the swelling is limited.

Based on the research results [21], a presence of amylose-amylose chain interactions with the amylose-amylopectin chain occurring during the HMT process, causing the bond between the molecules are becoming tighter and water is increasingly difficult to penetrate into the granules. According [10] that the decrease in peak viscosity is influenced by the presence of fat amylose bond during the HMT process which can limit the interaction of starch molecules with other molecules outside the granule. This is thought to occur because cassava starch still contains small amounts of fat.

3.2.3. Hot pasta viscosity and breakdown viscosity. Measurement of hot paste viscosity and breakdown viscosity is to determine the effect of treatment ACC process to the stability of the starch paste during the heating process. Changes hot paste viscosity and breakdown viscosity of ACC cassava flour. The rate of the decrease occurred in various concentrations of cassava flour in variations of heating time. The result shows that the starch of ACC treatment has better thermal stability than natural starch. Breakdown viscosity value is getting lower with increasing the heating time. [19] suggested that the warming effect of HMT on white sorghum flour has a breakdown value which decreased significantly. The breakdown viscosity is one of the important factors when starch is applied to a product that demonstrates the stability of starch paste on heating. The smaller the value of the breakdown viscosity then starch will be more stable on heating condition. The Result of research [20] states that all of the parameters on the characterization of commodity starch paste on Indian water chestnut, which includes peak viscosity, hot paste viscosity, viscosity breakdown, and setback viscosity shows a declin.

Table 4. Result of pasting profile analysis of modified cassava flour

Treatment	**Parameter of Cassava *pasting profile***					
	Pasting Temp. (°C)	**Peak Viscosity (cP)**	**Hot Paste Viscosity (cP)**	**Breakdown Viscosity (cP)**	**Cold Paste Viscosity (cP)**	**Setback Viscosity (cP)**
Native cassava	61.20	2810	1138	1672	1741	603

flour :						
Cassava Flour						
ACC :						
1	63.20	1822	1314	508	1759	445
2	59.60	2543	1068	1475	1560	492
3	-	87	64	23	127	63
4	51.10	3746	1325	2421	1983	658
5	57.95	2029	1182	847	1669	487
6	50.20	2273	1130	1143	1720	590
7	61.60	1922	1282	640	1710	428
8	61.95	30.82	1254	1828	1934	680
9	-	582	329	253	631	302
10	59.60	2130	1089	1041	1621	532
11	50.30	4590	1282	3308	2033	751
12	57.55	3425	1296	2129	2055	759
13	62.75	1863	1061	802	1586	525
14	-	89	57	32	102	45
15	58.00	2829	1250	1579	1993	743
16	77.25	748	470	278	740	270

3.3. Response surface method (RSM)

3.3.1. RSM for treatment autoclaving cooling cycle and concentration of flour on parameter ash content. Three-dimensional contour plot graph in Figure 3 illustrates the shape of the surface of the combination of the interaction between the component influences each other to test response value ash content. The difference in surface height demonstrates the value of different responses on any combinations of components. The low area shows low ash content test grade while the high area indicates that the ash content test value is high.

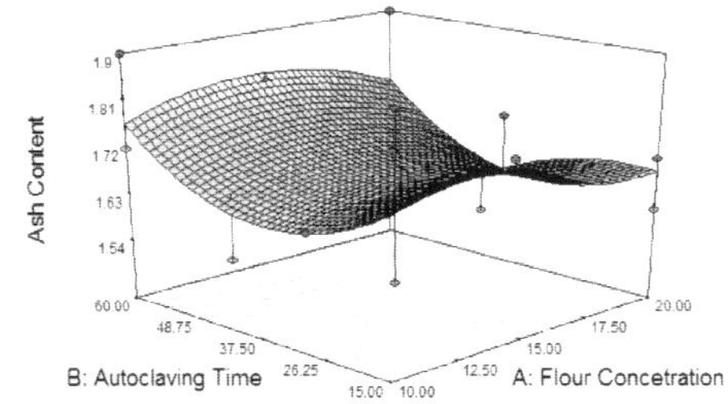

Figure 3. Three-dimensional graph of test results on the ash content of the response time of autoclaving processing optimization and concentration of cassava flour.

*3.3.2. RSM for treatment autoclaving cooling cycle and concentration of flour on parameter ash content.*Three-dimensional contour plot graph in Figure 4 illustrates the shape of the surface of the combination of the interaction between the components influence each other to test response value carbohydrate content. The difference in surface height demonstrates different response value for each combination of components. Low area indicates low carbohydrate test response rate while the high area indicates high carbohydrate test response rate.

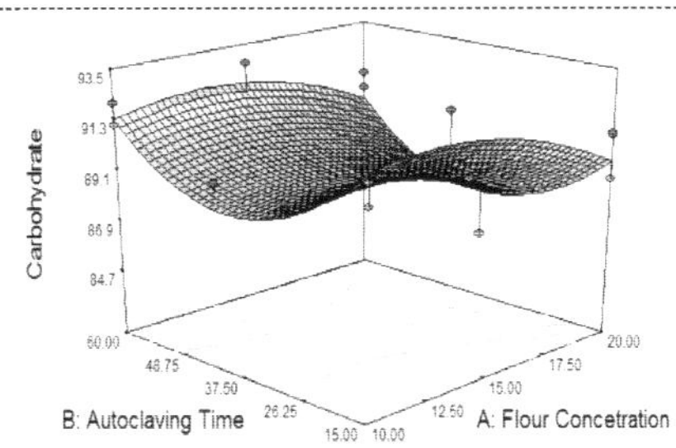

Figure 4. Three-dimensional graph of test results on the carbohydrate content response time autoclaving process optimization and concentration of cassava flour.

3.3.3. RSM for treatment autoclaving cooling cycle and concentration of flour on the breakdown viscosity (BV). Three-dimensional graph in figure 5 illustrates the results of a combination of surface response form two factors that influence the response of breakdown viscosity modified cassava flour. The difference in surface height indicates different response value for each combination of components The low area shows a low breakdown viscosity value, whereas the high area shows a high viscosity breakdown value. The different colors also show the value of the response of breakdown viscosity. Blue color indicates the lowest breakdown viscosity, and green color indicates the highest viscosity breakdown response.

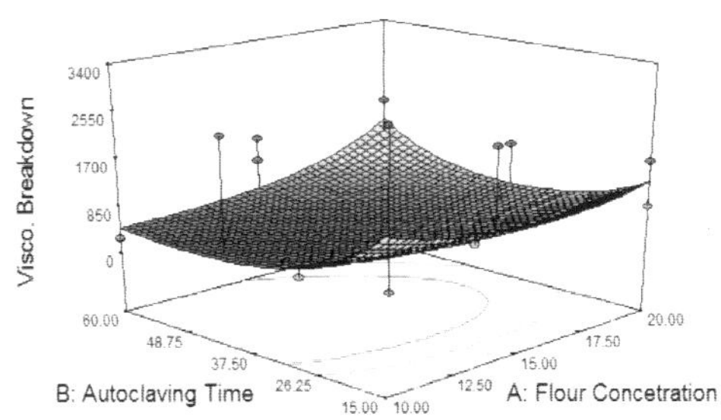

Figure 5. The three-dimensional graph of test results Setback viscosity response at the time ofautoclaving process optimization and concentration of cassava flour.

*3.3.4. RSM for treatment autoclaving cooling cycle and concentration of flour on the setback viscosity.*Contour three-dimensional graph plots can be seen in Figure. 6. The lines consisting of dots on the graph plots contour show a combination of the two components with different amounts of certain desirability produce the same value. Point prediction in Figure 6 shows the combined concentration of cassava flour and autoclaving time that produce certain desirability value. Three-dimensional graph shows the projection of contour plot. The low area on the three-dimensional graph shows the low desirability value, while the high area shows a high desirability value.

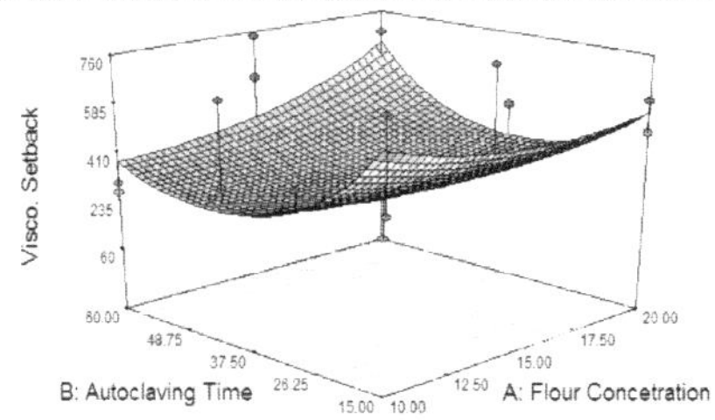

Figure 6. The three-dimensional graph of test results Setback viscosity response at the time ofautoclaving process optimization and concentration of cassava flour.

4. Conclusions and recommendations

Modification of cassava flour from the varieties Manggu can be done with the method of autoclaving cooling cycle, so can increase the stability of the flour to the heating and stirring. Characteristics of the functional properties of cassava modified flour are determined by (1) the initial temperature of the formation of pasta which provides information about the required temperature of the starch granules to begin expanding in the presence of water binding by starch; (2) the time and peak viscosity that provide information related to the starch water binding and disintegration trends (3) breakdown viscosity (BDV) is to provide information regarding resistance to heating; (4) setback viscosity (SBV) providing information regarding potential retro gradation and syneresis, and (5) the final viscosity to provide information regarding the potential ability to form gel.

The research result shows that modified cassava flour with a flour concentration treatment of 16% and autoclaving time 41 minutes has to paste the code and paste viscosity which is resistant to high temperature. Flour with this character is a flour that is expected to maintain the texture of processed products with a paste form that remains stable. Utilization of modified cassava flour by this ACC method can be applied to the pasta product such as noodle and spaghetti, hoping to support food diversification program to reduce dependence on wheat flour in Indonesia

Acknowledgment

The authors acknowledged financial and facility supports from Center for Appropriate Technology Development, Indonesian Institute of Sciences and the authors would like to thank to the contribution of Siti Khudaifanny, Neneng Komalasari, and Antonius Sukarwanto for his valuable assistances.

References

[1] Dziedzic SZ and Kearsley MW 1995 *The technology of starch production* In: SZ Dziedzic and MW Kearsley (Eds) *Handbook of starch hydrolysisproducts and their derivatives* London : (Blackie Academic and Professional)

[2] Rickard JE, Blanshard JMV and Asaoka M 1992 Effects of cultivar and growth season on the gelatinization properties of cassava (*Manihot esculenta*) starch *J. Food Science Agri* **59** 53–58

[3] Beynum GM AV and Roels JA 1985 *Starch conversion technology* Marcel Dekker Inc. New York and Basel.

[4] Sajilata MG, Singhal RS and Kulkarni PR 2006 Resistant starch–A review .Comprehensive reviews in foodscience and foodsafety **5 (1)** _1–17

[5] Taggart P 2004 Starch as an ingredients : manufacture and applications In Ann Charlotte

Eliasson (ed) Starch in food: structure, function, and application CRC Press Baco Raton Florida

[6] Sugiyono R, Pratiwi and Faridah DN 2009 Modifikasi pati garut (*Marantha arundinacea*) dengan perlakuan siklus pemanasan suhu tinggi - pendinginan (*Autoclaving-Cooling Cycling*) untuk menghasilkan pati resisten tipe III". *J. Teknologi Industri Pangan* **XX(1)** 17-61.

[7] Nazhrah EJ and Masniary L 2014 Pengaruh proses modifikasi fisik terhadap karakteristik pati dan produksi pati resisten dari empat varietas ubikayu (*Manihot esculenta*) *Jurnal Rekayasa Pangan Pertanian* **2(2)** 1-9.

[8] Association of Official Analytical Chemists 2005 *Official Methods of Analysis of The Association of Official Analytical Chemist.* Washington DC Association of Official Analitycal Chemists.

[9] Codex Alimentarius Commision 1995 Edible Cassava Flour (CODEX STAN 176-1989 (Rev. 1-1995I) Codex Alimentarius Commision USA

[10] Elliason AC 2004 *Starch in food: Structure, function, and applications* Boca Raton (US): CRC Press, Taylor & Francis Group

[11] Soebito S 1988 *Analisis Farmasi* Yogyakarta (ID) Gadjah Mada University Press

[12] Gunaratne A and Hoover V 2002 Effect of heat-moisture treatment on the structure and physicocemichal properties of tuber and root starches *Carbohydrate Polymers* **49(4)** 425-437

[13] Marco C Rosell CM 2008 Functional and rheological properties of protein enriched gluten free composite flours *J Food Eng* **88** 94-103.

[14] Adedokun MO and Itiola OA 2010 Material properties and compaction characteristics of natural and pregelatinized forms of four starches *Carbohydrate Polymers* **79** 818-824.

[15] Murdiati 2012 Pengaruh perlakuan*Heat-Moisture Treatment* (HMT) terhadap karakteristik fisiko kimia dan fungsional pati umbi walur (*Amorphophallus Campanulatus* VarSylvestris) Skripsi Fakultas Teknologi Pertanian Institut Pertanian Bogor.

[16] Pukkahuta C Suwannawat B Shobsngob S and Varavinit S 2008 Comparative study of pasting and thermal transition characteristics of osmotic pressure and heat-moisture treated corn starch *Carbohydrate Polymers* **72** 527-536.

[17] Chung HJ, Liu Q and Hoover R 2009. Effect of singel and dual hydrothermal on the crystalline structure, thermal properties, and nutritional fractions of pea, lentil, and navy bean starches *Food Research International* **43** 501-508 doi 10.1016/j.foodres.2009.07.030

[18] Hormdok R, Noomhorn A 2007 Hydrothermal treatments of rice starch for improvement of rice noodle quality LWT – *Food Science and Technology* **40** 1723-1731.

[19] Olayinka OO, Adebowale K and Olu Owolabi OBI 2008 Effect of heat moisture treatment on physicochemical properties of white sorgum starch. *Food hydrocolloids* **22** 225-230.

[20] Yadav BS, Prixit G and Ritika BY 2013 Hydrothermal modification of Indian water chestnut: Influence of heat-moisture treament and annealing on physicochemical, gelatinization and pasting characteristics *LWT-Food Science and Technology* **53** 211-217 doi 10.1016/j.lwt.2013.02.007

[21] Gunaratne A and Corke H 2007 Effect of hydroxypropylation and alkaline treatments in hydroxypropylation on some structural and physicochemical properties of heat-moisture treated wheat, potato, and waxy maize starch. *J. Carbohydrate Polymers* **68** 305-313 doi 10.1016/j.carbpol.2006.12.004

Characteristic of Fermented Drink from Whey Cheese with Addition of Mango (*Mangifera* x *odorata*) Juice

D Desnilasari[1] and R Kumalasari[1]

[1]Development Center for Appropriate Technology-Indonesian Institute of Sciences

Email: dewi.desnilasari@gmail.com, dewi019@lipi.go.id

Abstract. Whey cheese could be utilized become product such as fermented drink which is added by mango kweni juice to improve their acceptance. The aim of this research was to characterized physicochemical, sensory, and microbiology of fermented drink based on whey cheese with addition different concentration mango kweni juice of (0%, 5%, 10%, and 15%) by *Lactobacillus casei*. Color scale, viscosity, pH, total soluble solid, total free acid, fat, protein, total *L. casei* and sensory evaluation from panelist were examined after 24 hour of fermentation. Result showed that addition mango juice significantly affects the color scale, viscosity, pH, protein and number of *L. casei* of the product. The color of the product becomes more dark, red, and yellow. The product becomes more viscous. pH of the product become more acid and reduces protein content. Respectively total number of *L. casei* of the product increased 1 log. But addition of mango juice significantly did not affect sensory acceptance, total soluble solid, total free acid, and fat of the product. Sensory acceptance of the product range in dislike slightly and slightly like score that means formulation of the product need to be improved again.

1. Introduction

Whey cheese was released as a by-product during cheese manufacturing when milk was coagulated by acid and or rennet or physico-chemical process which often disposed as a waste and caused high environmental contamination [1,2,3]. It is important for utilization whey cheese to reduce the environmental pollution. Moreover whey contains about 80-90% of the volume of the milk that was coagulated [2]. Furthermore, liquid whey is composed of lactose (5%), water (93%), proteins (0.85%), and minimum amount of fat (0.36%) [3]. Even though the protein content in whey cheese is small, it contains the essential amino acids of good quality protein [1]. No wonder that whey cheese could be utilized become product that have value added.

Fermented drink using lactic acid bacteria (LAB) is one of the most attractive products for human diet [2]. LAB could fermentation whey, decrease lactose content in whey, producing mainly lactic acid and other metabolites such as volatile compounds contributing to the flavor and texture and increasing carbohydrate solubility and sweetness of the end product [4]. But whey fermented drink product sometimes have undesirable odor of cooked milk and salty-sour flavor of fresh whey and to overcome this, the product could be added by fruits juice that has good natural odor and flavor [2].

One of the fruits that have good natural odor and flavor is Mango kweni (*Mangifera x odorata*). Kweni (*Mangifera odorata* Grift) has typical exciting strong flavor because of the volatile compound that potential for flavoring purpose. Based on gas chromatography -olfactometry analysis, ethyl butanoate, ethyl crotonoate, a-pinene, ß-myrcene and two unknowns were the character impact compounds of Kweni [5]. Per 100g Mango kweni contain of water 80g, protein 0.9g, fat 0.1g carbohydrates including fiber 18.5 g, ash 0.6 g, beta-carotene equivalent 0.36 mg, thiamine 0.04 mg, riboflavin 0.06 mg, niacin 0.7 mg and vitamin C 13 mg. The energy value is 290 kJ/100 g [6]. Research by Okereke *et al* [7] showed that mango fruits juice could be the good medium for *Lactobacillus casei* that indicated from the decrease pH value and sugar content in 24 hour fermentation [7]. Research about characteristic of fermented drink product based whey with addition mango kweni juice was limited and it is potential to be develop become new product that could be accepted by consumer. The aim of this research was to characterize physicochemical, sensory, and microbiology of fermented drink based on whey cheese with addition of mango kweni juice.

2. Material and Methods

2.1. Materials

Whey cheese collected from waste of mozzarella process made with rennet, *Lactobacillus casei* isolated from Yakult in Development Center for Appropriate Technology, Sucrose, Skim Milk and Kueni mango (*Mangifera* x *odorata*) provided from traditional Market in Subang, West Java, Indonesia.

2.2. Production Fermented Drink from Whey Cheese with Addition Mango Juice

Starter *L.casei* was made by tree step, first by cultured 2 ose of *L.casei* in 10 ml Skim milk in 37^0C for 24 hour, and then cultured 5% from the first step result to 100 ml whey cheese in 37^0C for 24 hour, the last 5% from the second step was cultured in 500 ml whey cheese in 37^0C for 24 hour. Media for Fermented Drink was made based Whey Cheese with addition Mango Juice (0, 5, 10, and 15%) and sucrose 5%. Culture media was pasteurized in 75^0C along 15 minute using water bath (GFL type 1086). Furthermore fermented media was inoculated 5% starter *L.casei* and incubated in 37^0C for 24 hour. Each formulation of fermented drink has 3 repetitions of samples

2.3. Physical Analysis for Fermented Drink from Whey Cheese

Fermented Drink that already made was evaluated for the viscosity using DV-E Viscometer (Brookfield) with spindle no.61. Color of fermented drink product was analysis using colorimetric (3nh) to evaluated score of L, a, and b.

2.4. Sensory Analysis for Fermented Drink from Whey Cheese

Fermented drink product was evaluated for the sensory analysis (color, odor, viscosity, taste and overall) by 30 non trained panelists using 6-points hedonic scale.

2.5. Chemical analysis of Fermented Drink from Whey Cheese

The pH values of fermented drink product determined using pH meter (SI Analytics). Total soluble solid obtain using refractometer (Atago). Total Free Acid was determined with NaOH Titration. Total Fat and Protein was measured using Milkoscan^{TM} FT120 (FOSS).

2.6. Microbiology Analysis of Fermented Drink from Whey Cheese

Fermented Drink product was evaluated total lactic acid bacteria (*L.casei*) by serial dilutions. 10 ml product was aseptically mixed with 90ml Buffer Peptone Water and homogenized by shaking. Subsequent decimal dilutions were prepared with the same diluents and in all case duplicate-counting plates were prepared of appropriate dilutions. Total viable count of *L.casei* was carried out using the pour plate method [2].

2.7. Statistical Analysis

Data analyzed statistically by Analysis of Variance (ANOVA) using completely randomized design with significance at $p<0.05$. The analysis was performed using SPSS 14.0.

3. Results and Discussion

3.1. Physical Characteristic of Fermented Drink from Whey Cheese

Table 1. Score of Colorimetric of Fermented Drink from Whey Cheese

Formulation	L	a	B
0% of mango juice	47.96^{a}	-1.95^{a}	4.96^{a}
5% of mango juice	46.82^{b}	-1.76^{b}	7.83^{b}
10% of mango juice	44.77^{c}	-1.83^{b}	8.64^{c}
15% of mango juice	42.38^{d}	-1.65^{c}	9.35^{d}

*Notation different letters in the same column indicate significant difference at a significance level of 5%

Colorimetric result showed that scale of L, a, and b of the product with different portion of addition mango juice was significantly different (Table 1). L scale was slightly decreased that indicated the addition mango juice make product become darker. A scale of the product was increased that indicated addition more mango juice make color of product become more red. Coordinate of the b scale of the product increased slightly with the increased the portion of the mango juice which was indicated that product become more yellow. This phenomenon same like research by Moreira [9] that product mango juice with addition probiotic *Lactobacillus rhamnosus* GG has scale color a and b higher than mango juice without addition probiotic. The results of the color scale maybe due to the color of the mature kweni mango that has color yellowish and reddish [6]. The yellow colored of mango due to higher carotenoid content, wherein β-carotene followed by violaxanthin was the major carotenoid compounds that increased during ripening. The red colored of mango were characterized with higher anthocyanins with cyanidin-3-O-monoglucosides and peonidin-3-O-glucosides as the major anthocyanins [8].

Table 2. Viscosity of Fermented Drink from Whey Cheese

Formulation	Viscosity (cP)
0% of mango Juice	2.17^{a}
5% of mango juice	3.7^{b}
10% of mango juice	3.33^{b}
15% of mango juice	6.5^{c}

*Notation different letters in the same column indicate significant difference at a significance level of 5%

Viscosity value of the fermented drink from whey cheese showed that addition 15% mango has the highest value. The addition of mango juice significantly can increase viscosity value of the product. It is due to the ability mango juice coagulate the product more consistent. Moreover mango juice has pectin that has an ability to be gelling agent of the product. Pectin in kweni mango somehow increases during ripening [10]. Pectin whey based mango beverage in Chavan research has 0.29% pectin [2].

3.2. Sensory Characteristic of Fermented Drink from Whey Cheese

Table 3. Sensory Characteristic of Fermented Drink from Whey Cheese

Formulation	Color	Odor	Viscosity	Taste	Overall
0% of mango juice	2.97[a]	3.6[a]	3.3[a]	3.1[a]	3.27[a]
5% of mango juice	3.77[b]	3.53[a]	3.7[a]	3.77[a]	3.87[a]
10% of mango juice	3.8[b]	2.97[a]	3.7[a]	3.5[a]	3.5[a]
15% of mango juice	4.33[b]	3.07[a]	3.7[a]	3.67[a]	3.83[a]

*Notation different letters in the same column indicate significant difference at a significance level of 5%

Score 1: extremely dislike, 2: dislike, 3: dislike slightly, 4: slightly like, 5: like, 6: extremely like

Color parameter sensory evaluation showed that addition 5-15% mango juice significantly increased the score from panelist become slightly like. It is due to the color of the mango that the more high concentration of mango juice the more yellow the product and the panelist more accepting the product . However other parameter sensory such odor, viscosity and taste showed that did not significantly difference with the addition of mango juice. Those affect panelist acceptance to the overall score of the product. The addition mango juice to the product did not make the panelist more accepting the product. This finding indicated that formulation of the fermented drink from whey cheese need to be improved.

3.3. Chemical Characteristic of Fermented Drink from Whey Cheese

Table 4. Chemical Characteristic of Fermented Drink from Whey Cheese

Formulation	pH	Total Soluble Solid (Brix)	Total Free Acid (%)	Fat (%)	Protein (%)
0% of mango Juice	3.88[c]	8.63[a]	0.63[a]	0.64[a]	1.08[b]
5% of mango juice	3.77[a]	9.2[a]	0.72[a]	0.67[a]	1.07[b]
10% of mango juice	3.77[a]	8.8[a]	0.68[a]	0.61[a]	1.03[a]
15% of mango juice	3.85[b]	8.67[a]	0.67[a]	0.63[a]	1.03[a]

*Notation different letters in the same column indicate significant difference at a significance level of 5%

Chemical characteristic of fermented drink from whey cheese showed that addition mango juice significantly decrease slightly pH value the product. It is due to pH of the mango juice that already acid. pH whey with addition mango juice during fermentation 24 hour by *Lactobacillus casei* becomes more acid because lactose and carbohydrate from media was degrading by enzyme from *L.casei*. The breakdown of lactose by β-galactosidase into lactic acid thus increasing acid levels that in line with the fermentation time. Increased lactic acid will be followed by increase of H+, which means decrease in pH [11]. Okereke et al research mango juice as fermented media by *L.casei* showed pH of media getting more acid than orange, watermelon, and grape. It is indicated that mango juice gives the best nutrient than others [12].

Total soluble solid of the product range was between 8.63-9.2 brix. Statistically this parameter did not significantly difference. Furthermore with total free acid and fat parameter of the product also statistically did not significantly difference although with difference concentration mango juice that has been added. Research Harjiyanti et.al also finding the same result that addition 1-5% mango pulp to yogurt drink did not significantly affect the total soluble solid and total free acid [13].

Protein content of the fermented whey decreased with the addition mango juice. It is due to the portion of the whey that has higher protein more less substitute with high concentration mango juice.

Statistically addition mango juice significantly decreases protein content of the product. Lowering protein content of the product not to high it is due to high number of *L.casei* growth up in that media. Number *L.casei* will affect protein content of the product. Protein content of the product range 1.03-1.08% higher than in Chavan et al research whey based mango beverage protein content only 0.70% [2].

3.4. Microbiology Characteristic of Fermented Drink from Whey Cheese

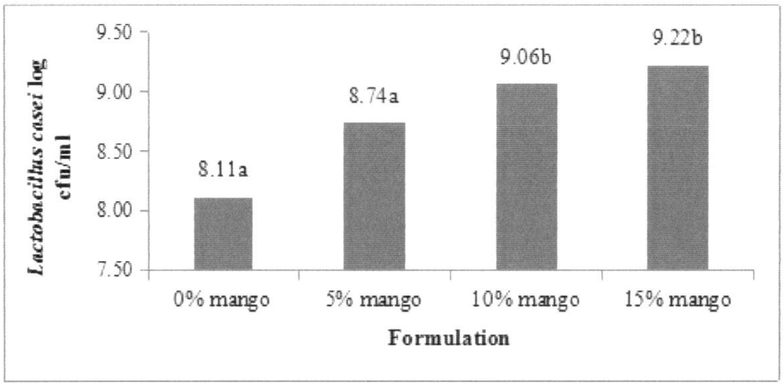

Figure 1. Total Lactic Acid Bacteria on Fermented Drink from Whey Cheese

Figure 1 showed that the more high concentration mango juice that has been add to the fermented drink increase total number probiotic *L.casei*. That is statistically significantly difference between the treatments . The highest total number L.casei was fermented drink whey cheese with addition 15% mango juice until 9.22 log cfu/ml or 1.7 x 10^9cfu/ml which was increase until 1 log than fermented drink 0% mango. This indicated that mango juice could give *L.casei* nutrient for growth up in numerous numbers. Mango juice based on Okereke et al research was the best media for growth up *L.casei* as probioticated product [12]. Probiotic beverage from whey and pineapple juice showed lower probiotic only 8.8 x 10^8 cfu/ml [14]. Total *L.casei* in fermented drink based whey with addition mango juice already full fill the criteria of the probiotic product that must has lactic acid bacteria content higher than 10^6 cfu/ml.

4. Conclusion

Addition variation concentration mango juice in fermented drink based whey cheese affected physic-chemical, sensory, and microbiology characteristic of the product. The color of the product becomes more dark, red, and yellow. The product becomes more viscose. pH of the product become more acid and reduces protein content. Respectively number *L.casei* of the product increase 1 log. However the treatments did not affect sensory acceptance, total soluble solid, total free acid, and fat of the product. Sensory acceptance of the product range in dislike slightly and slightly like score that means formulation of the product need to be improve

5. Acknowledgments

The authors acknowledge and thanks to Development Center for Appropriate Technology, Indonesian Institute of Sciences, for support funding and facilities to this research.

6. Reference

[1] Gupta C and Prakash D 2017 Therapeutic Potential of Milk Whey *Beverages* **3** 31

[2] Chavan RS Nalawade T Anit Kumar 2015 Studies of The Development of Whey Based Mango Beverage *Research and Review : Journal of Food and Dairy Technology* **3** 1-5

[3] Pescuma M Hébert EM Mozzi F and de Valdez GF 2008 Whey fermentation by thermophilic lactic acid bacteria: Evolution of carbohydrates and protein content *Food microbiology* **25** 442-451

[4] Mauriello G Moio L Moschetti G Piombino P Addeo F Coppola S 2001 Characterization of lactic acid bacteria strains on the basis of neutral volatile compounds produced in whey *Journal of applied microbiology* **90** 928-942

[5] Wijaya CH Apriyantono A May T Raharja H. And Ngakan TA 1999 Flavor of Kweni (Mangifera odorata Griff) an Exotic Tropical Fruit. *Flavor Chemistry of Ethnic Foods*(New York: Springer Science+Bussiness Media) ed F Shahidi *et al* Chapter 12 pp 119-125

[6] Bompard JM 1992 Mangifera odorata Griffith *Plant Resources of South-East Asia* **2** 218-220

[7] Okereke HC Anjunwa KV Owanta JO 2016 Physico-chemical Properties of Fresh and Probioticated Fruit Juices with Lactobacillus casei Produced in Nigeria *Asian Journal of Natural and Applied Sciences* **5** 2-6

[8] Karanjalker GR Ravishankar KV Shivashankara KS Dinesh MR Roy TK and Rao DS 2017 A Study on the Expression of Genes Involved in Carotenoids and Anthocyanins During Ripening in Fruit Peel of Green, Yellow, and Red Colored Mango Cultivars *Applied Biochemistry and Biotechnology* 1-15.

[9] Moreira RM Martins ML Júnior BRDCL Martins EMF Ramos AM Cristianini M da Rocha Campos AN Stringheta PC Silva VRO Canuto JW de Oliveira DC de Souza Pereira DC 2017 Development of a juçara and Ubá mango juice mixture with added Lactobacillus rhamnosus GG processed by high pressure *LWT-Food Science and Technology* **77** 259-268

[10] Bello-Pérez LA García-Suárez FJ & Agama-Acevedo E 2007 Mango Carbohydrates *Food* **1** 36-40

[11] Nursiwi A Nurhartadi E Utami R Sari AM Laksono PW & Aprilia EN 2017 Characteristic of Fermented Whey Beverage with Addition of Tomato Juice (Lycopersicum esculentum) *IOP Conference Series: Materials Science and Engineering* **193** p012009

[12] Okereke HC Ajunwa KV & Owanta JO 2016 Physico-chemical Properties of Fresh and Probioticated Fruit Juices with Lactobacillus casei produced in Nigeria *Asian Journal of Natural & Applied Sciences* **5** 2-6

[13] Harjiyanti MD Pramono YB & Mulyani S 2013 Total asam, viskositas, dan kesukaan pada yoghurt drink dengan sari buah mangga (Mangifera indica) sebagai perisa alami *Jurnal Aplikasi Teknologi Pangan 2* 104-107

[14] Shukla M Jha YK Admassu S 2013 Development of Probiotic Beverage from Whey and Pineapple Juice *J Food Process Technol* **4** 1-4

Quality Assessment of Physical and Organoleptic Instant Corn Rice on Scale-Up Process

R Kumalasari, R Ekafitri and N Indrianti
Development Center for Appropriate Technology-Indonesian Institute of Sciences

E-mail address : rima_kumalasari@yahoo.com/rima001@lipi.go.id

Abstract. Development of instant corn rice product has been successfully conducted on a laboratory scale. Corn has high carbohydrate content but low in fiber. The addition of fiber in instant corn rice, intended to improve the functioning of the product, and replace fiber loss during the process. Scale up process of Instant corn rice required to increase the production capacity. Scale up was the process to get identic output on a larger scale based on predetermined production scale. This study aimed to assess the changes and differences in the quality of instant corn rice during scale up. Instant corn rice scale up was done on production capacity 3 kg, 4 kg and 5 kg. Results showed that scale up of instant corn rice producing products with rehydration ratio ranges between 514% - 570%, the absorption rate ranged between 414% - 470%, swelling rate ranging between 119% - 134%, bulk density ranged from 0.3661 to 0.4745 (g/ml) and porosity ranging between 30-37%. The physical quality of instant corn rice on scale up were stable from the ones at laboratory scale on swelling rate, rehydration ratio, and absorption rate but not stable on bulk density and porosity. Organoleptic qualities were stable at increased scale compared on a laboratory scale. Bulk density was higher than those at laboratory scale, and the porosity was lower than those at laboratory scale.

1. Introduction

Indonesian staple food is rice, corn, sago and tubers. The shift in values, causing other commodities did not have a place as a staple food. So that people's reliance on rice is very high. In order to improve food security to comply the basic food needs, it is necessary to developing alternative staple food products.

Corn is one of the cereals that contain carbohydrates equivalent to rice (72-73%). Corn is generally processed by boiled, baked, or fermented. Corn can be processed into corn rice, as an alternative staple food. Furthermore, corn rice processed into instant products to save time in the presentation. Development of instant corn rice product has been successfully conducted on a laboratory scale [1]. Corn has high carbohydrate content but low in fiber. The addition of fiber in instant corn rice, intended to improve the functioning of the product, and replace fiber loss during the process. The novelty value of this study is the addition of dietary fiber to replace the missing corn fiber during the milling process.

Dietary fiber have many function such as increasing the number of beneficial bacteria in the digestive tract, improving immunity and resistance to pathogenic bacteria, facilitating defecation, and preventing cancer [2]. Fiber loss in corn grains during the process into grits is about 10%, from 2.2% in corn grains to 1.98% in corn grits [1]. Addition of agar fiber at 9.3% during the process of cooking corn rice, produce a product with the fiber content of 2.71% (36.86% increases) [3].

Laboratory scale is the basis of a production process. Scale up process required for industrial-scale production to fulfill consumer needs. Before entering the industrial scale, transition from laboratory scale to pilot plant scale to produce a prototype is needed. Pilot plant stage usually used to evaluate the idea of developing new products, new food supplies, or different operating conditions. The pilot plant stage is also used to evaluate the development of products, reduce costs, overcome technical problems, and evaluate new product ingredients, process variables, process production, optimization studies, and flavor profile.

Scale up is an activity to get the products that are identical (if possible) on a larger scale than the scale of production that has been previously defined. Product development (source and formulation), the operating unit testing, performance development of the tool, and the critical point determination process is required to be able to scale up. Scale up process requires strength analysis in determining the steps to be taken, including an analysis of operating conditions, design and optimum process [4], and expected to know the condition of the processing for certain, because the adjustment in cooking appliance used on a pilot scale will affect the quality of instant corn rice. Scale up is the increase of production capacity related equipment or technology that is larger than the previous production equipment [5].

Increasing capacity will affect the amount of raw materials, energy, and utilities. However, in scale up process, not necessarily to produce a similar quality to the laboratories scale results [6]. Large-scale processes will not produce a product that is identical to the original product, but will produce a product that resembles the original product. Based on the process and production levels, scale up process that is quite difficult to apply [7]. Therefore, research on scale up production of instant corn rice from lab scale to pilot scale, based on the best treatment baseline laboratories scale research is needed. This study aims to assess the changes and differences in the quality of instant corn rice during scale up.

2. Material and Methods

2.1. Date and place

This research was conducted from April 2011 to September 2011 at the Laboratory of Food Processing, Laboratory Testing of Food and Feed, Centre of Appropriate Technology Development Indonesian Institute of Sciences and Laboratory of Services Analysis, Bogor Agricultural University.

2.2. Materials and tools

The main materials used in this research are hybrid varieties of corn grits P 21 obtained from farmers in Bogor, West Java. The fiber material used is agar-agar powder. The support material is sodium citrate. The equipment used in the manufacture of instant corn rice is the sink, soaking tub, chopper, polisher, *Aron* Tools, steamers, stove, deep freezer, cabinet dryer, sealer, plastic containers for storage.

2.3. Instant Corn Rice Making Process

Corn grits washed and soaked in sodium citrate solution 1% for 2 hours. This soaking process aims to create a more porous structure of the material. The use of sodium citrate solution 1% on corn grits is to produce the fastest rehydration time [8]. Furthermore, corn grits washed and *"Aron"* processed (boiling corn grits in water at a ratio of 1: 4, while stirring continuously) for 25 minutes at a temperature of 85 - 93°C. *Aron* process aims to pre-gelatinization of starch in corn grits. The volume of water used is one part of corn grits and 4 parts of water [9].

2.4. Experimental design and Statistical analysis

This study design was a single factor, is the process of instant corn rice production on a scale of 3 kg, 4 kg and 5kg and a capacity of 2 kg (lab scale) as a control. Each treatment was repeated 3 times. Pilot-scale quantities per batch are determined based on the capacity of the available equipment; especially the dryer (cabinet dryer) with a maximum input capacity is 20kg. Base on a lab scale, from 2 kg of corn grits produce 6.5 kg of corn rice (yield 325%). Corn rice is the input material into the cabinet dryer. The results of the above analysis were compared by a one-way variance analysis (ANOVA). A Duncan's multiple range test with the option of homogeneous groups ($p < 0.05$), was carried out to determine significant differences among instant corn rice samples. SPSS 16.00 for Windows was used for this aim.

2.5. Analysis procedures

Samples were analysed physical properties include bulk density [10], porosity [11], rehydration ratio [12], water absorption, swelling rate. The stability of the organoleptic quality of instant corn rice qualitatively was analysed using different test. Organoleptic properties observed are the overall acceptance of the products, using 30 semi-trained panellists.

3. Results and Discussion

3.1. Bulk Density

Bulk density shows the ratio between the weights of a substance to its volume. Bulk density of the product ranged from 0.3661 to 0.4745 (g / ml) (Table 1). The process of scaling up production produce products with significantly different densities, which means that density, was unstable at scale up process.

Table 1. Comparison of bulk density of Instant Corn Rice on Lab Scale to Upscale

Mean of bulk density (gr/ml)			F-Test
Upscale		Lab Scale (2kg)	
3 kg	0.4745	0.3959	b
4 kg	0.4385		b
5 kg	0.3661		b

A-n: a = not significantly different ($p>0.05$)
b = significantly different ($p < 0.05$)

MPanual stirring on aron process causing agar gel is not well mixed, thus causing product sphericity decreases. The empty space between the particles increases causing product density decreases. A high sphericity of kernel causes a neat arrangement and the cavity of kernels getting smaller [13].

3.2. Porosity

Porosity is an important property on instant product. Pores formed during the process will facilitate and accelerate the process of rehydration. Loopholes or pores formed in the instant rice will facilitate the transfer of water and heat during cooking, producing a more tender cooked rice [14].

Table 2. Comparison of Porosity of Instant Corn Rice on Lab Scale to Upscale

Mean of Porosity (%)			F-Test
Upscale		Lab Scale (2kg)	
3 kg	36	37	b
4 kg	32		b
5 kg	30		b

A-n: a = not significantly different ($p>0.05$)
b = significantly different ($p < 0.05$)

The porosity of products ranges between 30-37%, the highest porosity (37%) was resulted from the production capacity of 2 kg (lab Scale) and the lowest porosity (30%) was resulted from the production capacity of 5 kg (Table 2). Different production capacity was resulted products with significantly different porosity, which means that porosity was unstable at scale up process. Manual stirring on aron process causing adhesion of agar gel on corn rice became uneven; causing porosity of the product is uneven.

3.3. *Rehydration ratio*

Rehydration ratio was the ability of a material to absorb water. Additionally rehydration aims to determine quality (appearance, colour, and aroma) of dry product after absorbing water. Rehydration was strongly influenced by the cell wall elasticity, loss of differential permeability in the membrane of protoplasm, loss of cell turgor pressure, protein denaturation, starch crystallinity and hydrogen bonding macromolecule [15].

Table 3. Comparison of Rehydration ratio of Instant Corn Rice on Lab Scale to Upscale

Mean of Rehydration ratio (%)			
Upscale		**Lab Scale (2kg)**	**F-Test**
3 kg	526	570	a
4 kg	514		a
5 kg	562		a

A-n: a = not significantly different (p>0.05)
b = significantly different (p < 0.05)

Rehydration ratio of instant corn rice ranged between 514% - 570%, the highest rehydration ratio (570%) was resulted from the production capacity of 2 kg (lab scale) and the lowest rehydration ratio (514%) was resulted from the production capacity of 4 kg (Table 3). Different production capacity was resulted products with not significantly different rehydration ratio, which means that rehydration ratio was stable at scale up process. Rehydration ratio of instant corn rice produced from a variety of production scale was higher than the rehydration ratio of instant porridge fortified with carrot flour (447.96%) [16].

3.4. *Absorption rate*

Absorption rate of instant corn rice ranged between 414% - 470%, the highest rehydration ratio (470%) was resulted from the production capacity of 2 kg (lab scale) and the lowest rehydration ratio (414%) was resulted from the production capacity of 4 kg (Table4). Different production capacity was resulted products with not significantly different absorption rate, which means that absorption rate was stable at scale up process. Absorption rate of instant corn rice produced from a production scale of 2-5 kg was higher than absorption rate of instant corn rice fortified with agar-agar powder and carrot pulp (306%) [3]. High water absorption ensures compactness in food products [17].

Table 4. Comparison of Absorption rate of Instant Corn Rice on Lab Scale to Upscale

Mean of Absorption rate (%)			
Upscale		**Lab Scale (2kg)**	**F-Test**
3 kg	426	470	a
4 kg	414		a
5 kg	462		a

A-n: a = not significantly different (p>0.05)
b = significantly different (p < 0.05)

3.5. *Swelling rate*

Swelling rate of instant corn rice ranged between 119% - 134%, the highest rehydration ratio (134%) was resulted from the production capacity of 3 kg and the lowest rehydration ratio (119%) was resulted from the production capacity of 4 kg (Table 5). Different production capacity was resulted products with not significantly different swelling rate, which means that swelling rate was stable at scale up process. Swelling rate of instant corn rice produced from a production scale of 2 -5 kg was higher than swelling rate of instant corn rice fortified with agar-agar powder (9.3%) and fortified with agar-agar powder mixed with carrot pulp (44-119%) [3]. during the process of water absorption, swelling of the particles of rice causing changes in the form of particles of rice [18].

Table 5. Comparison of swelling rate of Instant Corn Rice on Lab Scale to Upscale

Mean of Swelling rate (%)			F-Test
Upscale		Lab Scale (2kg)	
3 kg	134	122	a
4 kg	119		a
5 kg	126		a

A-n: a = not significantly different (p>0.05)
b = significantly different (p < 0.05)

3.6. *Quality stability In Organoleptic*

The stability of the organoleptic quality of instant corn rice was done qualitatively using different test. Table 6 show that samples at a capacity of 5 kg, 4 kg, and 3 kg overall quality has no significant difference with control (R). Organoleptic qualities (over all characteristic) of up scaled instant corn rice were stable from the ones at laboratory scale.

Table 6. Comparison of Over All Organoleptic Characteristic of Instant Corn Rice on Lab Scale to Upscale

Over All Organoleptic Characteristic		F-Test
Upscale	Control (R)	
3 kg	Laboratory Scale (2 kg)	a
4 kg		a
5 kg		a

A-n: a = not significantly different (p>0.05)
b = significantly different (p < 0.05)

4. Conclusion

Scale up of instant corn rice making process producing products with rehydration ratio ranges between 514% - 570%, the absorption rate ranged between 414% - 470%, swelling rate ranging between 119% - 134%, bulk density ranged from 0.3661 to 0.4745 (g/ml) and porosity ranging between 30- 37%. The physical quality of scale up instant corn rice making process were stable from the ones at laboratory scale on swelling rate, rehydration ratio, and absorption rate but not stable on bulk density and porosity. Organoleptic qualities (over all characteristic) of scale up instant corn rice making process were stable from the ones at laboratory scale. The density bulk was higher than those at laboratory scale; meanwhile the porosity was lower than those at laboratory scale.

On increasing the scale of production, it is advisable to make automatic "aron" equipment to obtain optimal results

Acknowledgment

We would like to thank to The Ministry of Research and Technology that has provided research funds and co- researchers, engineers, and analysts who have helped during the research process.

References

[1] Darmajana DA, Kumalasari R, Ekafitri R, Afifah N, Solichah E, Wijayanti F and Luthfiyanti R 2010 Perancangan Sistem Produksi Produk Olahan Berbasis Jagung : Jagung Instan Berserat *Laporan Kegiatan Program Insentif Peneliti dan Perekayasa LIPI.*

[2] Gong J and Yang C Advances in the Methods for Studying Gut Microbiota and Their Relevance to the Research of Dietary Fiber Function *Food Research International Journal* **48** 916–929

[3] Kumalasari R, Setiyoningrum S and Ekafitri R 2015 Karakteristik Fisik dan Sifat Fungsional Beras Jagung Instan Akibat Penambahan Jenis Serat dan Lama Pembekuan *Jurnal Pangan* **24** 37-48

[4] Hulbert G 1998 Design and Construction of Food Processing Operations. http://cpa.utk.edu/pdffiles/adc18.pdf. [8 Februari 2014].

[5] Valentas K J, Levine L and Clark J P 1991 Food Processing Operations and Scale Up *(New York: Marcel Dekker)*

[6] Dewi I A, Mulyadi A F and Ikawati N Q F Penggandaan Skala Mi Kering Dari Ubi Jalar *(Ipomea Batatas L.) Jurnal Teknologi Pertanian* **16** 41-50.

[7] Scott D D, Timothy J, Bowser P E and McGlynn W G 2007 Scaling Up your Food Process http://www.fapc.okstate.edu/factsheets. [14 Januari 2014]

[8] Husain H, Muchtadi T R, Sugiyono and Haryanto B 2006 Pengaruh Metode Pembekuan Dan Pengeringan Terhadap Karakteristik Grits Jagung Instan *Jurnal Teknologi dan Industri Pangan* **17** 189 – 196

[9] Lalitya N 2009 Optimasi Teknologi Pengolahan Dan Penyusunan Standard Operating Procedures (SOP) Penanakan Beras Jagung Dengan Alat Penanak Nasi Otomatis (Rice Cooker) *Skripsi. Fakultas Teknoloi Pertanian* Institut Pertanian Bogor

[10] Muchtadi TR and Sugiyono 1992 Ilmu Pengetahuan Bahan Pangan Pusat Antar universitas Pangan dan Gizi IPB Bogor

[11] Suliantri 1988 Pengaruh Penambahan Lipid Terhadap Sifat Fisiko Kimia Beras Instan.*Tesis. Program Pasca Sarjana* IPB Bogor

[12] Oktavia RY 2002 Pengaruh Larutan Na2HPO4 dan Na Sitrat dan Suhu Pengeringan Pada Pembuatan Nasi Instan Skripsi Fakultas Teknologi Pertanian IPB Bogor

[13] Sadeghi M H, Araghi A and Hemmat A 2010 Physico-mechanical Propertiesof Rough Rice (*Oryza sativa* L.) Grain as Affected by Variety and Moisture Content *Agriculture Enggineering International CIGR Journal* **12** 129 – 136.

[14] Prasert W and Suwannaporn P 2009 Optimization of Instant Jasmine Rice Process and Its Physicochemical Properties *Journal of Food Engineering* **95** 54–61.

[15] Neuma H J 1972 Dehydrated Celery Effect of Predrying Treatment and Rehydration Procedure Are Reconstitution *Journal of Food Science* **93** 437-441.

[16] Slamet A 2011 Fortifikasi Bubur Wortel Dalam Pembuatan Bubur Instan Untuk Peningkatan Provitamin A *Jurnal Agrointek.* **2** 1-8.

[17] Housson P and Ayenor G S 2002 Appropriate Processing and Food Functional Properties of Maize Flour *African J Sci Technol* **3** 126-121.

[18] Muramatsu Y, Tagawa A, Sakaguchi E and Kasai T 2006 Water Absorption Characteristic and Volume Changes of Milled and Brown Rice During Soaking *Cereal Chemistry Journal* **83** 624-631.

Development of modified starch technology (*maltodextrin*) from commercial tapioca on semi production scale using oil heater dextrinator

Agus Triyono, Raden Cecep Erwan Andriansyah, Rohmah Luthfiyanti, Taufik Rahman

Development Center for Appropriate Technology – Indonesian Institute of Sciences
Jl. KS Tubun No 5 Subang, West Java, Indonesia

E-mail: ateluyono@gmail.com

Abstract. One way to improve functional starch is by modification of starch into dextrin or maltodextrin. Maltodextrin is used in the food industry as a food substitution. Development of enzymatically modified starch technology has been performed with the use of α-amylase at optimum pH of 5.5, temperature 75-85 °C, with enzyme activity of 135 KNU/g. The maltodextrin produced from commercial tapioca has the quality requirements for food according to SNI 1992. The yield of maltodextrin obtained is about 80%. The use of the optimum amount of the α-amylase enzyme is 0.07 % v/w and the substrate amount of tapioca starch is 35%. Analysis of the feasibility of modified starch with the assumption of production scale of 300 kg per day, the economic value of 10 years business, the price of starch is IDR 8,350/kg, from tapioca starch (tapioca) IDR 4,000 – IDR 4,500/kg.

1. Introduction

Starch utilization is widely used as an ingredient for the food and pharmaceutical industry, especially for the food industry, as substitutes, mixtures, thickening agents, and fillers. The addition of modified starch to food products can enhance the superiority and good quality of physical appearance, taste, consistency, color, and nutrients. The modified starch is the starch whose hydroxyl group has undergone a change. Modified starch can also be hydrolysis (acid and enzyme) reactions, depolymerisation, crosslinking, dextrinization and typical chemical reactions (cationization, carboxymethylation and grafting). The modified starch will experience changes in properties that can be adapted for specific purposes [1]. Desirable properties generally are starch which has stable viscosity at high and low temperature, good mechanical shearing resistance and thickness resistance to acidic and sterilization temperatures [2].

Starch is a carbohydrate group of a glucose compound polymer composed of two main fractions, those are amylose and amylopectin. Amylose is highly hydrophilic, because it contains many hydroxyl groups. Collection of amylose in water is difficult to form gel, although high concentration. Starch molecules are not readily soluble in water at room temperature or cold water, in contrast to amylopectin that structure is branched; the starch will easily expand and form colloids in water. One example of starch modification product is dextrin/maltodextrin.

Dextrin is the intermediate product of acidic or enzymatic hydrolysis process. According to [3], dextrin is a product of starch hydrolysis, in the form of amorphous substances white to yellowish. [4] states there are several levels in the hydrolysis reaction of starch. The starch molecule initially breaks into shorter glucose chain units called oligosaccharides, and when further process is broken down into maltose and eventually maltose breaks into glucose.

Maltodextrin is defined as a product of starch hydrolysis containing a unit of α-D-glucose that is largely bonded through a 1.4 glycosidic bond with a DE less than 20 [5]. The general formula of maltodextrin is [(C6H10O5) nH2O)]. Maltodextrin with water will form a gel that can melt or dissolve and resemble a fatty structure. Consistency, appearance and organoleptic properties are acceptable. The use of maltodextrin in food production can also reduce calories by more than 70 percent [6]. The process of enzymatic starch hydrolysis showed in Figure 1.

$$\underset{\text{starch}}{(C_6H_{10}O_5)_n} + nH_2O \xrightarrow[\text{heat}]{(\alpha\text{-amylase enzyme})} \underset{\text{oligosaccharide}}{(C_6H_{10}O_5)_m.H_2O} + \underset{\text{maltose}}{(C_6H_{10}O_5)_2.H_2O} + \underset{\text{glucose}}{C_6H_{12}O_6}$$

Figure 1. Process of enzymatic starch hydrolysis

[7] The use of enzyme in the hydrolysis process is widespread due to several advantages compared with the use of acid solution. Enzyme in small quantities can decompose starch in large quantities, so the cost is relatively cheaper. [8] The enzyme works specifically, so it is easy to control, otherwise production equipment is easier than production using acid that requires acid-resistant equipment. The enzymatic hydrolysis process in starch to make maltodextrin more popular is done because it will produce a more controllable starch polymer fraction as compared to chemical processes with acid addition.

In 2002 Indonesia imports starch modification products by 80,000 tons, while import of dextrin product is estimated at 55-65% of total imports of starch modification. The price range of dextrin is US $ 425 per ton for white dextrin and US $ 450 per ton for yellow dextrin. Every year Indonesia's import value of dextrin and other starch modification products increases, reflecting the increasingly levels of modified starch requirements for food, pharmaceutical and cosmetics industries. Besides, the availability of sources of starch in Indonesia not only type and amount, became the background of assessment and development activities of dextrin products from sweet potato starch and starch from sources of other tubers (cassava, Garut bulbs). To produce semi-scale production dextrin done several stages of activity, namely the stage of techno-economic feasibility analysis of the planning stage of the equipment and lay out the location of the tool, And the development stage of other starch modification products such as the development of glucose and fructose manufacture, and its utilization as functional food substitution material [12].

The purpose of this study is to develop starch modification product from tuber source to semi-production scale and test technological ability of dextrinator or converter equipment shocked with heater oil heater. In order to realize the integrated technology package of modified starch of cassava starch (tapioca) at semi-production scale with quality requirement meets the requirement of dextrin quality for SNI food 1992 (Indonesian National Standard).

2. Research methods

2.1. Materials and equipment

2.1.1. Material. Fresh cassava material obtained from around the Subang regency and commercial bulk tapioca starch is obtained in the general market. Commercial α-Amylase enzyme, from thermal enzyme type from *Bacillus licheniformis* [14] is a solution (liquozyme supra) with activity of 135 KNU/g from Halim Sakti Pratama Company. In experimentation of the process of making modified starch (maltodextrin) by enzymatic method, to determine the amount of enzyme and the amount of optimum starch substrate with variation of concentration and process temperature, at pH 5.5 .0 (optimum). The main experiment is the trial of maltodextrin production from commercial cassava starch (tapioca) on the market and using a dextrinator (converter) with heater oil heater.

2.1.2. Equipment. Solvent / shredder, hydraulic press, dryer, shifter, scale laboratory dextrinator/ converter and semi-production scale dextrinator (converter) in PPTTG-LIPI Subang

2.2. Methodology

Development of modified starch technology (maltodextrin), carried out by the following method stages (Figure 2); Determination of temperature and length of dextrination by Lugol method and spectrometry. The determination of the amount of starch substrate concentration and enzyme concentration. Furthermore, the development of modified starch technology is done by trial production at semi-production scale from commercial tapioca starch material. Physico-chemical characterization resulting from the development of modified starch technology (maltodextrin) is based on the quality requirement of dextrin for food, which is SNI (Indonesian National Standard). Semi-modified starch technology development activity in addition to physico-chemical tested, also conducted and analyzed techno economy and feasibility.

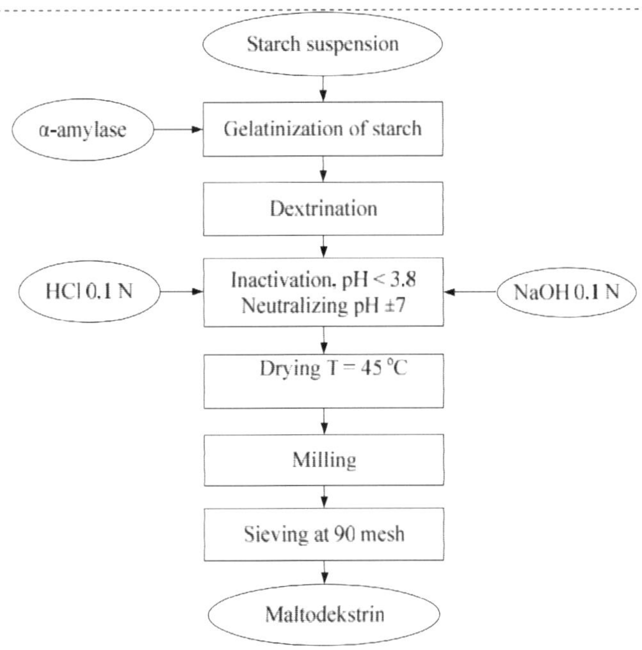

Figure 2. Diagram of modified starch processing (maltodextrin)

IOP Conf. Series: Earth and Environmental Science **101** (2017) 012026 doi:10.1088/1755-1315/101/1/012026

3. Results and discussion

3.1. Making starch
The stages of the starch-making process begin with cleaning, dissolving, and extraction Separation of starch is done by precipitation for 12 hours. The upper water is removed and the precipitate is dried. The starch is still in the form of dried lumps, then milled and sifted.

3.2. Physico-chemical analysis of cassava starch
The material of cassava starch is obtained from fresh cassava around Subang regency. The result of physico-chemical characteristic starch analysis of fresh cassava from Subang regency and commercial cassava starch (tapioca) can be seen in Table 1.

Table 1. The result of physico-chemical characteristic starch analysis of fresh cassava from Subang regency and commercial cassava starch (tapioca)

No	Test criteria	Unit	Value	
			Fresh cassava starch	Cassava starch commercial
1	Water content	%	8.16	12
2	Ash content	%	0.24	0.23
3	Starch content	%	88.82	84.60
4	Fat content	%	0.21	0.64
5	Protein content	%	0.31	0.30
6	Crude fiber content	%	-	0.45
7	White degree (percent BaSO4)	(%)	126.22	84.1

The results of water content of cassava starch analysis is smaller than commercial starch is 8.16%, according to SNI 3451-2011 maximum water content is 15%. The results of ash content showed almost the same value of 0.23% for commercial starch and 0.24%, fresh cassava starch still included in the maximum limit of SNI 3451-2011 is 0.6%. The starch content of fresh cassava starch is greater than commercial starch which is about 88.82%. White degree for fresh cassava starch is whiter and higher value is 126.22%, where the minimum SNI value is 94.5%. Overall, the quality of fresh cassava starch and commercial starch (tapioca) produced still fulfil the quality limit requirement by SNI 3451-2011.

3.3. Experimental result of making modified starch (maltodextrin) from cassava starch

3.3.1. Making starch modified in cassava starch

3.3.1.1. Determination of temperature and length of the dextrination process. Starch will form a blue color complex when it is spilled with an Iodine reagent. According to [9], dextrin will produce a brownish red complexion if Iodine reacted. In a preliminary experiment to determine the optimum temperature and length of the dextrination process, using the thermal type α-amylase enzyme is performed on variations in the amount of starch substrate; Starting at 15%, 25%, 35%, at pH 5.5 and the amount of 0.05% *(v/w)* α-amylase enzyme concentration [13].

Table 2. Determination of temperature and length of dextrination based on absorbance method spectrometer from residual starch on starch substrate amount 15%

Time (minutes)	Absorbance value		
	80 °C	**90 °C**	**100 °C**
15	0.570	0.436	0.312
30	**0.329**	**0.249**	**0.530**
45	0.371	0.284	0.778
60	0.374	0.312	
75	0.413		
90	0.545		
105	0.657		
120	0.756		

The length of the dextrination process at the concentration of starch substrate (15%, 25%, and 35%) is more than 30 minutes. This means that the concentration of starch solution of 15% requires about 30 minutes, at a concentration of 25% starch solution takes more than 50 minutes, and at a concentration of 35% starch solution takes time About 70 minutes is enough to produce dextrin (maltodextrin) (Table 2).

Table 3. The determination of the length of dextrination based on the color change of the Iodine reagent at different starch substrate concentrations 15%, 25%, 35% and 0.05% enzyme concentration at 85°C

Time (minutes)	Substrate concentration of cassava starch (%)		
	15	**25**	**35**
0	Dark blue	Dark blue	Dark blue
20	Blue	Blue	Blue
30	Purple	Blue	Blue
40	Red chocolate	Purple	Blue purple
50		Red chocolate	Blue purple
60			Purple
70			Red chocolate

For experiment the manufacture of maltodextrin is carried out at 85 °C, and the treatment of substrate concentration (15%, 25%, 35%), and the addition of enzyme concentration varied 0.05% 0.07% 0.09%. For experiments, the manufacture of maltodextrin was carried out at 85 °C, and the treatment of substrate concentration (15%, 25%, 35%), and the addition of enzyme concentration amounted to 0.05% 0.07% 0.09% (Table 3). From the preliminary experiment results, the concentration of starch substrate and the optimum concentration of enzyme α-amylase were selected. Then the results obtained will be used in the main experiment of semi-production of maltodextrin. Equipment and production facilities such as dextrinators jacketed with an oil heater.

3.3.1.2. Physico-chemical analysis of modified starch (maltodextrin). The experimental result of maltodextrin preparation from the combined treatment of the amount of starch substrate (15%, 25%, 35%), with enzyme amount (0.05%, 0.07%, 0.09%), Optimum results are from the treatment of 35% starch substrate, 0.07-0.09% enzyme concentration. The characteristics of maltodextrin are as follow; water content 4.70 – 6.42% (<11%); ash content 0.16 - 0.18% (<0.5%); and white degrees 83.30 - 93.66%, and dextrose content <5.0%, cold water solubility 97.14 - 99.35% (> 97%) (Table 4).

Table 4. Result of physico-chemical analysis of maltodextrin from cassava starch

Treatment		Parameter				
		Water content (%)	**Ash Content (%)**	**Dextrose Content (%)**	**Solubility in water (%)**	**White Degrees (% $BaSO_4$)**
a_1 (15%)	b_1 (0.05 %)	5.65	0.18	3.21	98.69	85.63
	b_2 (0.07 %)	6.42	0.16	2.94	99.35	83.30
	b_3 (0.09 %)	5.40	0.18	2.98	97.14	85.80
a_2 (25%)	b_1 (0.05 %)	5.05	0.18	3.85	97.17	84.40
	b_2 (0.07 %)	5.11	0.18	3.83	97.12	87.30
	b_3 (0.09 %)	5.38	0.18	3.24	98.29	93.66
a_3 (35%)	b_1 (0.05 %)	6.11	0.17	2.79	99.34	85.26
	b_2 (0.07 %)	5.33	0.17	3.56	99.06	87.81
	b_3 (0.09 %)	4.70	0.16	3.69	97.50	89.55
Quality Requirement For SNI Food (1992)		Max 11	Max 0.5	Max 5	Min 97	White-yellowish

3.3.2. Production experiment of modified starch (maltodextrin) from tapioca commercial on semi scale production

In the preliminary trial of maltodextrin manufacture, the highest starch substrate concentration is 35% with a time of about 40 minutes. The optimal concentration of enzyme α-amylase (*Bacillus licheniformis*) 0.07% (v/b). Experiment on semi-scale production using a system of dextrinator equipment (converter) at various levels of tapioca substrate concentration. Aims to obtain optimization of the capability of producing maltodextrin, and is expected to be done on small and medium enterprises (SMEs).

The development of modified starch making technology by experimenting the production of maltodextrin from starch sources of bulb (cassava), is one of the attempts to substitute modified starch from import with domestic production using starch source from local source of tapioca (tapioca) overflow.

From the experimental results of production (Table 5) the results of maltodextrin trial experiments from various variations of the concentration of tapioca starch substrate (25%, 50%, 55%, 60%) in the amount of 0.07% (v / b) α-amylase enzyme concentration to 75%, and yield of maltodextrin rendemen is increasing, and the value of DE (dextrose equivalent) is still relatively below 20. [10] The value of DE (dextrose equivalent) is the percent of reducing sugar in sugar calculated as dextrose in a dry basis. [11] The range of DE values is from 0-100. The value of DE 3-20 is owned by maltodextrin; 20-75 is owned by glucose and above 75 is referred to as hydrolyzate.

Table 5. Maltodextrin result from various concentrations of tapioca substrate using dexrinator (convert) with heater oil heater

No	Substrate strach (%)	Dextrination time (minutes)	Rendemen (%)	Water content (%)	Dextrose equivalent (D.E)
1	25	50	66	7.36	13.95
2	50	70	74	5.91	11.80
3	55	90	80	4.41	11.40
4	60	120	85	4.50	11.25
	Concentration < 75%	Time <120	Rendemen <85%	Water content <11	D.E <20

The maximum limited capacity of dextrinators, is processing the maximum amount of substrate starch concentration is 60%. In the amount of starch substrate concentration more than 60%, the stirring system is experiencing difficulty because it is too thick, especially when gelatinization conditions are achieved. Subsequently, the experimental production of semi-scale maltodextrins with the use of 0.07% α-amylase enzyme concentration (v/w) in the amount of tapioca starch concentration is 35% and the result fulfilled the quality requirements of dextrin (maltodextrin) for food, that is SNI (1992) (Table 6).

Table 6. Maltodextrin analysis results from tapioca commercial

No	Parameter	Value	
		Maltodextrin from Tapioca	Requirement of Quality of Dextrin For Food, SNI (1992)
1	Color	White	White-yellowish
2	White degree (%$BaSO_4$)	93.66	
3	Water content (% b/b)	5.40	Max 11
4	Ash Content (% b/b)	0.18	Max 0.5
5	Dextrose Content (%),	3.24	Max 5
6	D.E	18.40<20	
7	The part that Soluble in Cold water (%)	98.25	Min 97
8	Viscosity (cP)	-	3-4
9	Degree of acid (0,01 N NaOH/100 g)	0.80	Max 5

3.4. Economic analysis and feasibility economic business

Economic analysis and business feasibility of product or business plan from modified starch production (maltodextrin) from cassava starch (tapioca). In general, the economic viability of the business is strongly influenced by raw materials (quantity, price, and starch yield). The yield of cassava starch (tapioca) can reach more than 25% due to harvest time of more than 9 months, so that the price of cassava starch (tapioca) is around IDR 3,000 to IDR 4,500 per kg.

The heating capacity of dextrinators with oil heater to produce maltodextrin is 250-300 kg per day. The input of starch substrate concentration is a maximum of 60% and will result in a maltodextrin rendemen of about 84%. Business economic feasibility result for 10 years, selling price of maltodextrin from cassava starch (tapioca) IDR 8,360/kg with price of tapioca starch IDR 3,000-IDR 4,500/kg.

4. Conclusions and recommendations

Maltodextrin (modified starch), obtained from partial process of starch hydrolysis by using commercial α-amylase enzyme, from a number of starch concentrations (about> 35%) of commercial cassava starch (tapioca) is widely available. The experimental production of maltodextrin (enzyme starch) enzymatically with commercial α-amylase enzyme, from *Bacillus licheniformis* (activity 135 KNU/g) is optimum at pH 5.5 and temperature 85 °C.

The results obtained in the development of technology of making modified starch (maltodextrin) from cassava (tapioca) enzymatically fulfilling SNI quality requirements for food (SNI, 1992), and for non-food (SNI, 1989). The yield of maltodextrin was obtained about 80%, from the use of enzyme enzyme α-amylase (*B. licheniformis*) 0.07% v/w on 35% tapioca starch substrate. Modified starch feasibility analysis is production capacity 300 kg per day and 10 years economic value for maltodextrin from tapioca starch, with the selling price IDR 8,360/kg.

Acknowledgment

The authors would like to thank to the contribution of Oktiana Palupi from Padjajaran University, Siti Kudhaifanny from PPTTG-LIPI for assistance in the development of modified starch not only the laboratory scale but also the semi-production scale.

References

[1] Fleche G 1985 Chemical modification and degradation of starch In: G. M. A. Van Beynum and J A Roles (eds.) *Starch Conversion Technology* Marcel Dekker Inc. New York and Bassel.
[2] Wurzburg O B 1989 *Modified starches:properties and uses* CRC Press Inc Boca Ranton Florida
[3] Dewan Standarisasi Nasional 1992 *Dekstrin untuk industri non pangan* Jakarta.
[4] Gaman P M and Sherrington K B 1992 *Pengantar ilmu pangan nutrisi dan mikrobiologi*. Translation Gadjah Mada University Press Yogyakarta.
[5] Griffin V K and Brooks J R 1989 Production and size distribution of rice maltodextrin hydrolized from milled rice flour using heat stable alpha amylase *Journal of Food Science* **54** 190-193.
[6] Roper H 1996 Starch: present use and future utilization in: Van Bekkum, H. H. Ropper and A. G J Voragen (eds.) *Carbohydrates as organic raw materials* **III** VCH Publisher Weinheim.
[7] Pomeranz Y 1991 *functional properties of food components* **2nd** edition Academic press.
[8] Tjokroadikoesoemo P S 1986 *HFS dari industri ubi kayu dan umbi lainnya* Gramedia Jakarta.
[9] Winarno F G 2002 *Kimia pangan dan gizi* PT Gramedia Pustaka Utama Jakarta.
[10] Kennedy J F, Knill C J and Taylor D W 1995 Maltodextrins in : Kearsley, M.W.J. dan Dziedzic, S.Z., (eds.) *Handbook of starch hydrolysis product and their derivatives* Blackie Academic & Professional London Chapter **3** 65-83.
[11] Kearsley M W and Dziedzic S Z 1995 Physical and chemical properties of glucose syrup. *Di dalam* : Kearsley, M.W.J. dan S.Z.Dziedzic (eds.) *Handbook of Starch Hydrolysis Product and Their Derivatives*. Blackie Academic & Professional, London. Chapter **5**. 129-154.
[12] Triyono A 2008 Karakteristik hasil optimalisasi usaha produksi pati termodifikasi secara enzimatik dari umbi-umbian dengan konverter sistim pemanas berjaked oli. *Prosiding Seminar Nasional Teknoin* **B1-B6**.
[13] Triyono A 2007 usaha peningkatan pati ubi kayu (manihot utilissima) dengan modifikasi secara enzimatik sebagai bahan untuk industri pangan *Prosiding seminar nasional* **D7-1-D7-8**.
[14] Sutanto A I 2001 *Pemanfaatan Pati sagu sebagai bahan baku pembuatan dekstrin secara enzimatis* FATETA IPB Indonesia.
[15] SNI 2011 SNI **3451-2011** *Tapioka*.

IOP Conf. Series: Earth and Environmental Science **101** (2017) 012027 doi:10.1088/1755-1315/101/1/012027

Comparative Study of Powdered Ginger Drink Processed by Different Method:Traditional and using Evaporation Machine

Wuri Apriyana[1], Vita Taufika Rosyida[1], Septi Nur Hayati[1], Cici Darsih[1], Crescentiana Dewi Poeloengasih[1]

[1]Research Unit for Natural Product Technology, Indonesian Institute of Science, Yogyakarta 55861, Indonesia

Email : wuri001@lipi.go.id

Abstract. Ginger drink is one of the traditional beverage that became one of the products of interest by consumers in Indonesia. This drink is believed to have excellent properties for the health of the body. In this study, we have compared the moisture content, ash content, metal content and the identified compound of product which processed with traditional technique and using an evaporator machine. The results show that both of products fulfilled some parameters of the Indonesian National Standard for the traditional powdered drink. GC-MS analysis data showed the identified compound of both product. The major of hydrocarbon groups that influenced the flavor such as zingiberene, camphene, beta-phelladrine, beta-sesquepelladrine, curcumene, and beta-bisabolene were found higher in ginger drink powder treated with a machine than those processed traditionally.

1. Introduction

Ginger (Zingiber officinale) is a member of the Zingiberaceae family of plants. The plant is native to Asia but now cultivated in the West Indies, Africa, India, and other tropical regions [1]. This rhizome has become popular spice and used widely in Indonesian cuisine as well as in other countries. Traditionally, it also used as a carminative and stimulant, to increase appetite, treat the digestive tract disorder such as nausea and vomit, to treat common cold, cough, FDVdiarrhea, malaria, fever, and arthritis [2]. Another study also described the benefits of ginger on the treatment of catarrh, rheumatism, nervous diseases, gingivitis, toothache, asthma, stroke, constipation and diabetes [3]. Some scientific research has proven that it has some pharmacological activity such as antiemetic, antidiabetic, antioxidant, antinausea, antihistaminic activity on duodenum immunomodulator, antimicrobial in vitro antihelminth, antifungal and anticancer [4].

Some literature has reported the major compound that obtained from ginger rhizomes such as 4-hydroxy-3-methoxyphenyl moieties, gingerols, acetyl gingerol derivatives mono-and diacetylgingerdiol derivatives, shogaols, paradols, dihydroparadol, zingerones. 1-dehydrogingerdione, diarylheptanoide and metal ether derivatives [1-5]. Among these compounds, 6-gingerol was found to possess anti-oxidative activity through the inhibition of phospholipid peroxidation [5].

In Indonesia, ginger can be processed into various ready to consume products such as essential oil, ginger powder, ginger syrup, ginger sweet, ginger candy and ginger crystal [6]. The supporting trend of ginger demand is the current shift in lifestyle in which more people are opting to get jamu, or traditional or herbal medicine, to keep them healthy or to remedy some illnesses despite the major development in the chemical based pharmaceuticals industry. Nowadays, people can obtain jamu in powder form. They can deserve their own beverage more practice. With a sweet instant powder, the quality of the product can be maintained, not easily contaminated, not easily infected with the disease, and the product without preservatives [7].

According to Phoungchandang et al. [8], one of the important keys to making beverages in the form of powder is the evaporation process. Evaporation is an operating unit used to remove some of the water contained in liquid food by boiling. Evaporation aims to increase the solids and still in the form of puree. In another case, Menon et al. [9] had studied the effect of heat processing on the flavor compounds of fresh Indian ginger (Zingiber officinale Roscoe). They reported that gerental (24.2%) and zingerone (14.2%) were decreased during the thermal processing. Purnomo et al. [10] and Okyne et al. [11] also reported a significant active compound decomposition during thermal processing of ginger.

Almost all of ginger drink small industry in Indonesia still used the traditional technique to produced the ginger drink powder. They used traditional stove and pan, boil the ginger solution with the high temperature, and stirred manually until the powder obtained. In this study, we have been comparing the quality of ginger drink product that produced with traditional technique and using evaporation machine. Some parameter such as moisture content, heavy metal content and compound content was analyzed.

2. Materials and Methods

2.1. Materials

The fresh samples of ginger rhizome of 9-month-old, sugar, and spice such as cinnamon (Cinnamomum zaylanicum), cardamom (Amomum compactum), and cloves (Syzygium aromaticum) were obtained from Beringharjo Market, Yogyakarta Province, Indonesia.

2.2. Methods

Ginger drink production :

Traditional process: 2.5 kg of ginger rhizomes are washed, ground with a machine and extracted with water. A total of 10 kg of refined sugar put into the pan, added 1 Liter of water, stirred on the stove until all the sugar melted, added ginger extract and spice solution, stirred continuously until it became powder.

With evaporation machine: 2.5 kg of ginger rhizomes are washed, grounded and extracted with water. The machine has operated by heating the oil in the bath vessel for preparing the temperature conditions on the evaporative tube. After a heat transfer process, the temperature at the evaporative tube monitored reaches 100 °C, 10 kg of refined sugar introduced until completely melted, then the ginger extract and spice solution added. The evaporation process begins, followed by a continuous stirring process at a temperature of 100 °C which is the boiling point of water solvent. At that stage the temperature inside the vessel is kept constant by the Thermo control, the flame of the stove can be reduced and enlarged automatically to maintain the desired temperature. The inside material stirred continuously until granules are formed and become dry.

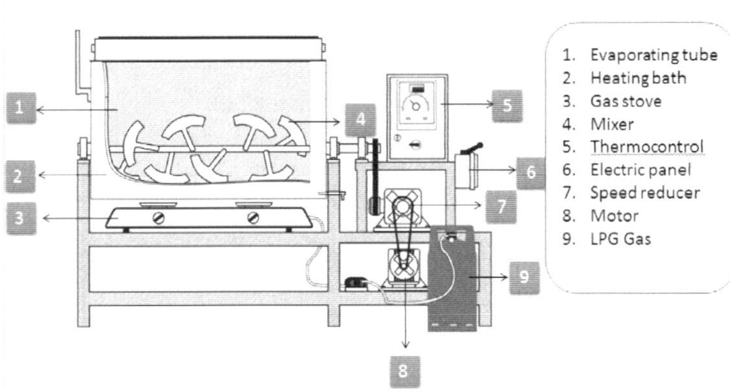

Figure 1. Scheme of evaporation machine

Sample Analysis:

Moisture content :
The moisture content of ginger powder samples was determined gravimetrically by drying the samples at 105 °C for 24 h. The moisture content was calculated as the percentage of water removed from the initial mass sample.

Ash content :
The ash content of samples was determined gravimetrically by heating the samples at 550 °C for 5 hours. The ash content was calculated as the percentage of inorganic residue remaining after the calcination process from the initial mass sample.

Chemical element content :
Chemical element content of ginger power samples was analyzed using Atomic Absorption Spectrometer (AAS) Perkin Elmer 5100PC for Cd(326.1 nm), Pb(283.3 nm), Cu(324.8 nm), Sn (284.0 nm) and Zn(213.9 nm).

GC-MS Analysis :
The ginger powder extracted with petroleum ether, then the filtrate was analyzed with Shimadzu GCMS-QP2010 S with the following condition: column used was Rti-5MS, 30 m length, and the inner diameter of 0.25 mm and the initial column temperature was 280 °C (5 °C/minute), while the injector temperature was 300 °C, the flow rate was 40 mL/minute.

3. Results and Discussion

The instant ginger powder produced with two different techniques did not visually differ significantly. The results of the water content analysis, as well as the metal content of the products, are presented in Table 1. The moisture content for the manually processed samples has a moisture content of 0.39%, while the sample processed using a machine has a moisture content of 0.27%. It is lower than the moisture content of ginger powder produced by Phoungcangdang [8] which ranged from 1-1.25%. Both of water content values have qualified the SNI criteria for traditional powder drink (SNI 01-4320-1996). Based on the standards set by Indonesian National Standard (SNI 01-4320-1996), the value of water content for the traditional powdered drink is at a maximum of 3%. Low moisture content has a closely related to the quality and durability of these products. Ash content for both products did not differ significantly, 0.13% for traditional technique and 0.12% for the product processed by a machine. Indonesian National Standard (SNI 01-4320-1996) requires that the maximum ash content for traditional powder drink is equal to 1.5%.

The metal content of the product is identified using Atomic Absorption Spectroscopy and presented in Table 1. Both products have similar profiles of cadmium (Cd), lead(Pb), copper(Co), and tin(Sn), while Zn metals found with levels of 3.65 mg/kg for products which traditional technique, and 2.90 mg/kg for machine-treated products. The maximum Zn content required by SNI is 50 mg/kg. Thus the content of metal Cd, Pb, Cu, Zn and Sn already qualified the SNI for traditional powder drink.

Table 1. Analysis results of ginger drink powder produced by different technique

	Manual	**Machine**
Moisture content	0.39%	0.27%
Ash content	0.13%	0.12%
Cadmium (Cd)	undetected	undetected
Lead (Pb)	undetected	undetected
Copper (Cu)	undetected	undetected
Zinc (Zn)	3.65 mg/kg	2.90 mg/kg
Tin (Sn)	undetected	undetected

Limit of detection : Cd=0.01 ppm; Pb= 0.01 ppm; Pb= 0.01 ppm; Sn= 0.25 ppm

Identification compound of both products appeared in Table 2. The major compound identified in the ginger drink with the traditional technique were alpha-pinene (11.02%), zingiberene (10.82%), decanoic acid ethyl ester (9.69%), hexadecane (7.31%) and citral (5%). Other compounds detected in low quantities such as beta-bisabolene (2.29%), beta-sesquepelladrine (2.26%) and beta-pelladrine (2.06%). Different results showed by beverage powder manufactured by machine. The major detected compounds were zingiberene (17.46%), citral (17.05%), camphene (6.93%), curcumene (6.18%), beta-sesquepelladrine (5.88%), 1.8- Cineole (5.61%), and Borneol (4.3%).

Based on these data, it showed that different processing techniques would cause the different content of the products. The content of alpha pinene, camphene, beta-phellandrene and curcumene have influenced the flavor of ginger [10]. The taste of ginger is primarily affected by monoterpenes (Camphene, limonene, myrcene, beta-phellandren, alpha-pinene, borneol, 1,8-cineol, citronellol, geranial, geraniol, geranyl acetate, linalool, neral and others) [3]. Some of these compounds could arise from dehydration of oxygenated compounds [10]. In both samples, these compounds were detected, so it can be seen that the ginger flavor that appears on the ginger drink is a representation of the content of the compounds.

The [6]- gingerol compounds did not find in the GC-MS data; it is possible that these compounds suffered damage due to heating treatment during the process of making ginger powder. The molecular structure of gingerol consisted of hydroxyl keto functional group which was thermally labile [12]. Therefore, in a ginger-based drink, it was found that the content [6] -gingerol 35-80 times lower when compared in fresh ginger [13].

The major of hydrocarbon groups that influenced the flavor such as zingiberene, alpha-pinene, camphene, beta-phelladrine, beta-sesquepelladrine, curcumene and beta-bisabolene had been compared (Figure 2). It showed that the amount of these compound in the ginger drink produced by machine was higher than those produced by a traditional method, except alpha-pinene. It may be due to a closed heating system, heating at a relatively lower temperature and a relatively faster processing so that the compound content still maintained. However, sensory evaluation should be done to obtain a complete data on the flavor of the ginger drinks.

Table 2. Identified compounds of ginger drink powder extract produced by traditional technique and by evaporation machine.

Peak	Traditional				Machine			
	Compounds	Molecule structure	Retention time (min)	Amount (%)	Compounds	Molecule structure	Retention time (min)	Amount (%)
1	Alpha-Pinene	$C_{10}H_{16}$	13.517	11.02	Alpha-pinene	$C_{10}H_{16}$	13.586	7.45
2	1-Decene	$C_{10}H_{20}$	15.057	1.98	Camphene	$C_{10}H_{16}$	14.325	6.93
3	Beta-myrcene	$C_{10}H_{16}$	15.205	1.31	Beta-myrcene	$C_{10}H_{16}$	15.402	1.27
4	Methyl-heptenone	$C_8H_{14}O$	16.252	1.25	Methyl-heptenone	$C_8H_{14}O$	16.457	1.01
5	1-Limonene	$C_{10}H_{16}$	17.303	1.65	1-Limonene	$C_{10}H_{16}$	17.015	1.2
6	Beta-pheladrine	$C_{10}H_{16}$	18.533	2.06	Beta-pheladrine	$C_{10}H_{16}$	17.270	1.98
7	1-Undecene	$C_{11}H_{22}$	18.669	1.29	1,8-Cineole	$C_{10}H_{18}O$	17.486	5.61
8	Undecane	$C_{11}H_{24}$	18.791	0.85	Linalool	$C_{10}H_{18}O$	20.294	1.96
9	1,8-Cineole	$C_{10}H_{18}O$	19.485	0.84	Phellandral	$C_{10}H_{16}O$	20.917	0.46
10	Linalool	$C_{10}H_{18}O$	21.877	1.31	Limonene oxide	$C_{10}H_{16}O$	22.467	1.24
11	Phellandral	$C_{10}H_{16}O$	21.997	0.93	Camphor	$C_{10}H_{16}O$	22.850	1.24
12	Citronelol	$C_{10}H_{20}O$	22.631	0.69	Limonene oxide	$C_{10}H_{16}O$	23.040	0.52
13	Citral	$C_{10}H_{16}O$	22.767	2.85	Terpineol	$C_{10}H_{18}O$	23.217	0.4
14	Citral	$C_{10}H_{16}O$	24.652	2.15	Borneol	$C_{10}H_{18}O$	23.546	4.3
15	Undecenoic acid	$C_{11}H_{20}O_2$	25.044	0.93	Alpha-terpineol	$C_{10}H_{18}O$	23.867	1.21
16	Ethyl octanoate	$C_{10}H_{20}O_2$	25.146	1.00	Citronelol	$C_{10}H_{20}O$	24.762	1.35
17	Nonanoic acid, ethyl ester	$C_{11}H_{22}O_2$	25.880	2.92	Citral	$C_{10}H_{16}O$	25.745	7.52
18	Tetradecane	$C_{14}H_{30}$	26.040	0.44	Citral	$C_{10}H_{16}O$	26.630	9.53
19	Methyldecanoate	$C_{11}H_{22}O_2$	26.787	1.20	Geranyl Acetate	$C_{12}H_{20}O_2$	28.748	2.93
20	Tetradecane	$C_{14}H_{30}$	26.967	0.72	Hexadecane	$C_{16}H_{34}$	29.042	5.84
21	1-Pentadecene	$C_{15}H_{30}$	28.037	1.63	Heptadecane	$C_{17}H_{36}$	29.251	2.89
22	Geranyl Acetate	$C_{12}H_{20}O_2$	28.124	1.44	Curcumene	$C_{15}H_{22}$	32.310	6.18
23	Ethyl cis-4-decenoate	$C_{12}H_{22}O_2$	28.502	0.61	Zingiberene	$C_{15}H_{24}$	32.543	17.46
24	Decanoic acid, ethyl ester	$C_{12}H_{24}O_2$	28.848	9.69	Beta-bisabolene	$C_{15}H_{24}$	33.033	3.64
25	Allo-aromadendrene	$C_{15}H_{24}$	30.674	1.22	Beta-sesquepelladrine	$C_{15}H_{24}$	34.039	5.88
26	Hexadecane	$C_{16}H_{34}$	31.397	7.31				
27	Heptadecane	$C_{17}H_{36}$	32.284	2.39				
28	Curcumene	$C_{15}H_{22}$	33.183	1.40				
29	Zingiberene	$C_{15}H_{24}$	33.607	10.82				
30	Beta-bisabolene	$C_{15}H_{24}$	35.568	2.29				
31	Beta-sesquepelladrine	$C_{15}H_{24}$	36.807	2.26				
32	Hexadecane	$C_{16}H_{34}$	41.252	2.24				
33	3-Undecen-5-yne, (E)-	$C_{11}H_{18}$	42.274	7.23				
34	Tetradecanoic acid, methyl ester	$C_{15}H_{30}O_2$	45.444	6.79				
	Total Area			94.71				100

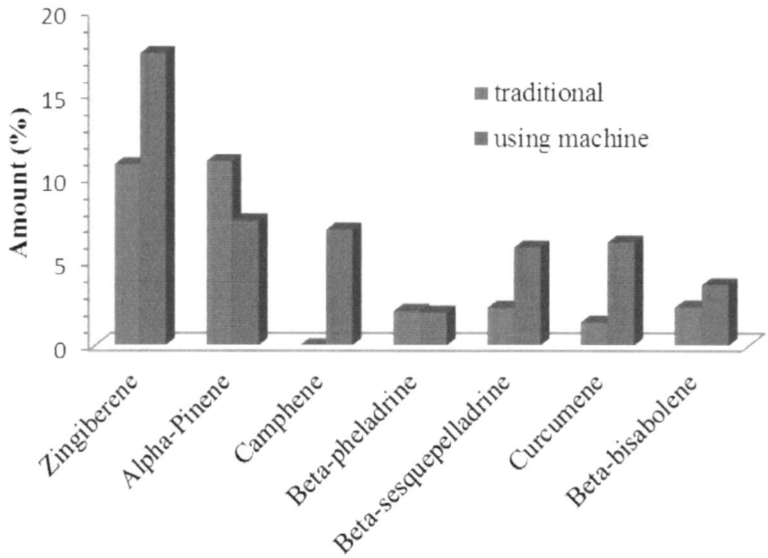

Figure 2. Comparison of hydrocarbon compounds contribute to the flavor of ginger in both product

4. Conclusion

Based on the study, the production of instant ginger beverages produced with two different processes fulfilled the criteria of traditional powder drink based on water content criteria, ash content and heavy metal content. The results of identification using GC-MS show the content of some typical compounds contained in ginger. However, some compounds detected in low amounts may be due to the stage processing that involving heating at high temperatures.

Acknowledgement

The authors acknowledge the Indonesian Institute of Sciences for providing financial support through Kegiatan IPTEKDA 2016, CV. Anugrah Sukses Mandiri for the collaboration and UD. Rekayasa Wangdi for manufacturing the evaporation machine.

References

[1] Singletary K 2010 *Nutrition Today* **45** 171-183.
[2] Solikhah E.N. 2016 *J Med Sci* **48** 226-239.
[3] Chrubasik S Pittlerc M.H. and Roufogalis B.D. 2005 *Phytomedicine* **12** 684-701.
[4] Al-Awwadi N.A.J 2017 *J. Pharmacognosy Phytother.* **9** 111-116.
[5] Aeschbach R, Loliger J, Scott BC, Murcja A, Butler J, Halliweii B 1994 *Food Chem Toxicol* **32** 31–6.
[6] Wiedayati 2016 Ginger: Superior, Hot Export Commodity for Your Health *Export News Indonesia*
[7] Mailoa M.N., Setha B. and Febe, F.G. 2015 *Indian J. Sci. Tech* **8** 154–157.
[8] Phoungchandang S, Sertwasana A, Sanchai P and Pasuwan P 2009 *World Appl Sci J* **6** 488-493.
[9] Menon A.N., Padmakumari K.P., Kutty B. S., Sumathikutty M.A., Sreekumar and Arumugham, C. 2007. *J. Essent. Oil Res.* **19** 105-110.
[10] Purnomo H, Jaya F and Widjanarko SB 2010 *Int. Food Res. J.* **17** 335-347.

[11] Okyne R.O., Patterson J., Walker L.T. and Verghese M. 2015 *Food and Nutrition Sciences* **6** 445-451.
[12] Girhepunje N.S., Dumore, N.G., Taksande S.U., Chaudari D.V and Shende S.N. 2017 *Indian J. Med. Res. Pharm. Sci.* **4** 24-28.
[13] Cho S., Lee D.G, Lee S., Chae S. and Lee S. 2015 *J Appl Biol Chem* **58** 377−381.

IOP Conf. Series: Earth and Environmental Science **101** (2017) 012028 doi:10.1088/1755-1315/101/1/012028

Evaluation of temperature and relative humidity on two types of zero energy cool chamber (ZECC) in South Sulawesi, Indonesia

Andi Dirpan[1], Muhammad Tahir Sapsal[1], Abdul Kadir Muhammad[2], Mulyati M Tahir[1], Rahimuddin[3]

[1] Department of Agricultural Technology, Hasanuddin University Makassar 90245, Indonesia
[2] Center for Mechatronics and Control System, Mechanical Engineering Department, State Polytechnic of Ujung Pandang, Jl. Perintis Kemerdekaan KM 10 Makassar, 90245, Indonesia
[3] Department of Marine Engineering, Hasanuddin University Makassar 90245, Indonesia

E-mail:dirpan@unhas.ac.id

Abstract. Zero Energy Cool Chamber (ZECC) is a cooling chamber for storing fruits and vegetables from the viewpoints of low cost and energy savings. The aim of the present study is to evaluate temperature and relative humidity (RH) on two types of zero energy cool chamber (ZECC) in South Sulawesi, Indonesia. The first category was placed underground while the second category was on the surface. Then, the performance of the ZECC was measured by calculating temperature and relative humidity. The results show that the ZECC was constructed on the surface produce lower temperature and higher RH compare to ZECC which placed underground. In average, the temperature in the outside (28.0^{0}C) is greater than in inside (26.2^{0}C) of the ZECC. On the other hand, the relative humidity in the outside (72.9%) is less than in inside (87.2%) of the ZECC. It was concluded that the ZECC where was constructed on the surface is more suitable than ZECC in the underground for decreasing temperature and increasing relative humidity.

1. Introduction

Recently, in developing countries, the postharvest handling practices regarding vegetables throughout the marketing channels remain inadequate. The farmers have no storage facilities, while the transport and commercialization channels also lack storage facilities. As a result, most harvested fruits and vegetables are usually stored in the open, exposed to high temperatures and low relative humidity conditions, until an intermediary, retailer purchases those commodities [1]. The combination of higher humidity and lower temperature facilitates extended shelf life. These conditions can be achieved by using a semi underground storage system with moistened walls. The use of a shading curtain is also an effective means of reducing the temperature. Higher humidity and lower temperature inside the zero energy cool chamber (ZECC) offers a unique advantage for maintaining the firmness of fruits and vegetables by lowering the physiological loss in weight (PLW) and other metabolic processes[1,2].

Some studies on constructing and applying ZECC have been conducted. Islam and Morimoto[3–5] have developed a ZECC consisting of a brick wall cooler and a storage container coated with antimicrobials. In this study, the ZECC can maintain relatively low inside temperature and high relative humidity as compared with outside temperature and relative humidity. Ganesan et al. [6] standardized the quantity of water applied in zero energy storage. In his study, eggplants were stored in zero energy storage and at room temperature. Three different levels of watering operation (100, 75 and 50 l per day) were added to drip irrigation system to moisten the sand in the cool chamber. The physiological loss in weight and rotting percentage increased at room temperature but decreased with increasing water level up to 100 l per day. As a result, the shelf-life of Brinjal enhanced to 9 days with the addition of 100 l of water per day[7]. Narayana et al. [8] investigated the quality of banana packed with low-density polyethene bag with or without ventilation under different storage conditions, namely, ambient temperature, zero energy cool chamber (brick-sand evaporative cooler) and 13.5^{0}C low temperature. Shelf-life of the fruit was highest at 13.5^0 C and minimum under ambient conditions in both vented and unvented poly bags. Non-significant variation was observed for total soluble solids of fruit under all three storage conditions. The acidity increased gradually up to the end of green life and then either decreased (in unvented poly bags) or remained unchanged (in ventilated poly- bags and unpacked control) until the end of life (yellow). Reducing sugars and total sugars increased gradually throughout the storage period and were highest in the control fruit and fruit packed in ventilated poly bags. Spoilage was maximum in unvented poly bags under ambient and zero energy cool chamber conditions. Storing banana fruit in unvented poly bags at low temperature could extend the shelf-life of the fruit up to 19.33 days[7]. From the above studies, it can be seen that all of their ZECC was constructed on the surface. Therefore, in this study focusing on evaluating ZECC which create underground.

To the best of the author's knowledge, this is the first study to evaluate ZECC which construct underground, particularly in South Sulawesi, Indonesia. The objective of this study is to assess temperature and relative humidity inside two different of zero energy cool chambers (ZECC) in South Sulawesi, Indonesia.

2. Material and Method

This experiment was conducted at Makassar, South Sulawesi, Indonesia from June to July 2017. Two ZECCs were set up located at the complex residential for lecturer of Hasanuddin University.

2.1. *Structure of the ZECC*

The design was based on the evaporative cooling of a porous body [1,7,9,10]. Two different of ZECC were constructed. The first ZECC was placed underground while the second ZECC was on the surface as can be seen in Figure 1. Figure 2 shows the complete randomized block design of the ZECC. The dimensions of the outer and inner brick walls were 100 (L) × 100 (W) × 50 (H) cm and 80 (L) × 80 (W) × 50 (H) cm, respectively. The 7.5-cm gap left between the outer and inner wall was filled with sand as the use of a porous material for a particular type of evaporative cooler could reduce ambient temperature by as much as 15°C. Tap water was supplied to the sand area through drip irrigation using a valve. The cooling chamber's storage area was 45 (L) × 45 (W) × 50 cm (H) in size. A foam mattress made frame measuring 55 (L) × 55 (W) cm was used to cover the chamber.

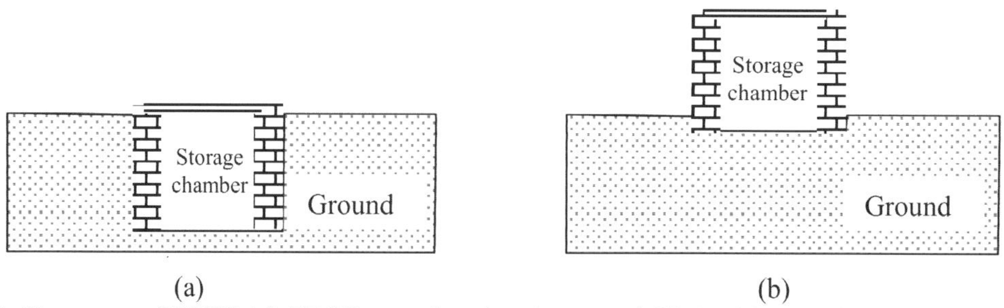

Figure 1. Two type of ZECC (a) ZECC was placed underground (b) ZECC was placed on the surface

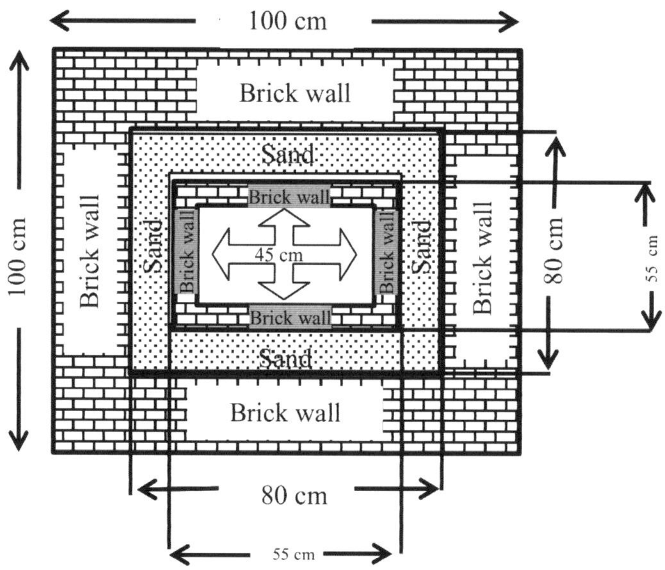

Figure 2. Pictorial (top) view of the ZECC

Shading curtain was used to cover ZECC which made from Palm's roof. The use of a shading curtain prevents an increase in ZECC temperature effectively [1].

2.2. Watering

The level of water (20 litres per day) was added to drip irrigation system to moisten the sand in the cooling chamber In the ZECC. Water is distributed from a water supply to low-pressure drip nozzles through a programmable flow valve for wetting the layer of sand.

2.3. Measurements of temperature and relative humidity

The temperatures at all places were simultaneously measured by using a digital thermometer (DS18B20 Waterproof Temperature Sensor Probe with an accuracy of ± (0.4%+0.9°C) with two thermocouples (0.3 mm d.). One thermocouple was placed inside storage chamber in the centre; another one has been put outside the ZECC for measuring the outside temperature. The relative humidity (RH) of the ZECC was measured simultaneously using a thermohydrometer (SHT11 humidity sensor module SHT 11). All instruments have data logger function. The data were recorded at ten-minute intervals for 24 hours. Thus, about 270 points of data for two days were obtained.

3. Results and discussion

3.1. Temperature

Figure 3 explains the typical diurnal changes temperature in inside and outside of the ZECC under the watering condition where the ZECC placed in the underground and surface. The highest temperature (34.0°C) and lowest temperature (23.5°C) were recorded in outside of the ZECC. In average, the temperature in the outside is higher than in inside of the ZECC, particularly in daylight. For instance, while the temperature in T1 (inside + underground + watering) and T2 (inside + surface + watering) of the ZECC at 5 hours is about 27.1^{0}C and 26.3^{0}C, respectively, the corresponding values for T0(outside) is about 33.3^{0}C. This result was consistent with a study by Islam and Morimoto[1], in which the temperature of the ZECC investigated; the temperature inside of the ZECC is lower than the outside one. As a result, the temperature inside of the ZECC can be potentially reduced using watering for the sand. This result happens as an effect of the evaporative cooling mechanism by water and the reduction of solar radiation by the shading curtain. Here, the vapour pressure of the moist sand and surrounding air and the outside temperature around the ZECC attempt to equalize. Liquid water molecules of the moist sand-zeolite mixture gasify under the influence of outside air through a process that uses energy to change the physical state. Convection and conduction were heat transitions that happened from the higher temperature of the air and brick walls to the lower temperature of water[1]. There is a drop in ambient temperature during this conversion process. This cooling temperature caused by the effect of evaporation cools down the inside temperature of the ZECC below the dry-bulb temperature. This is the result of the combined effects of underground temperature, the moist inside walls, and watering. Consequently, the inside air temperature of the ZECC became cooler. On the other hand, the sun's infrared rays were blocked 60% by the shading curtain, thereby lowering both the temperature inside the ZECC and that surrounding it[1,9,10].

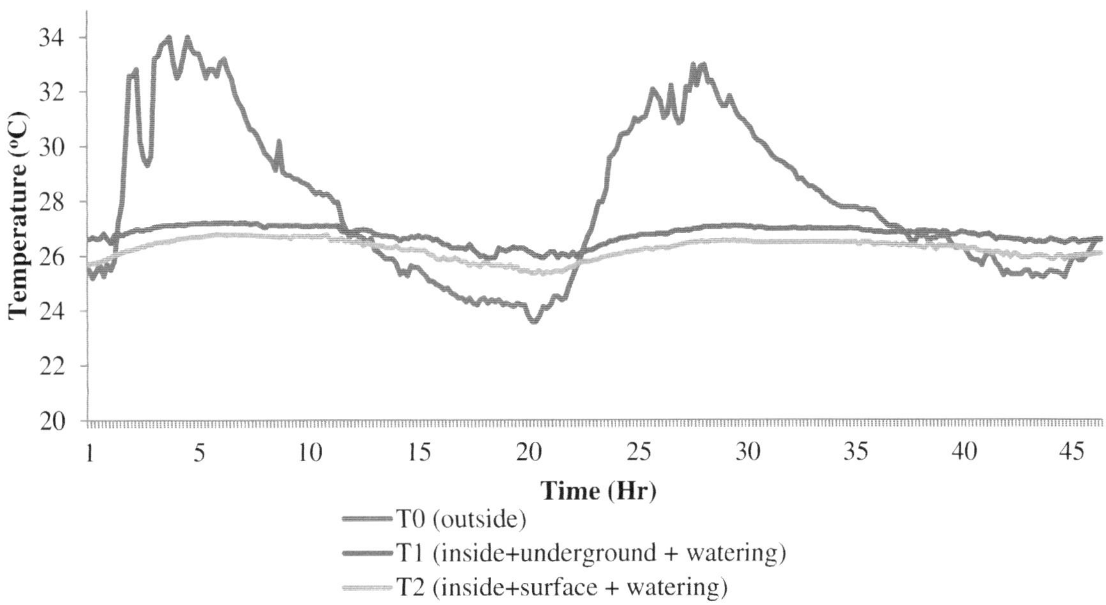

Fig. 3. Diurnal temperature in inside and outside of the ZECC under watering condition

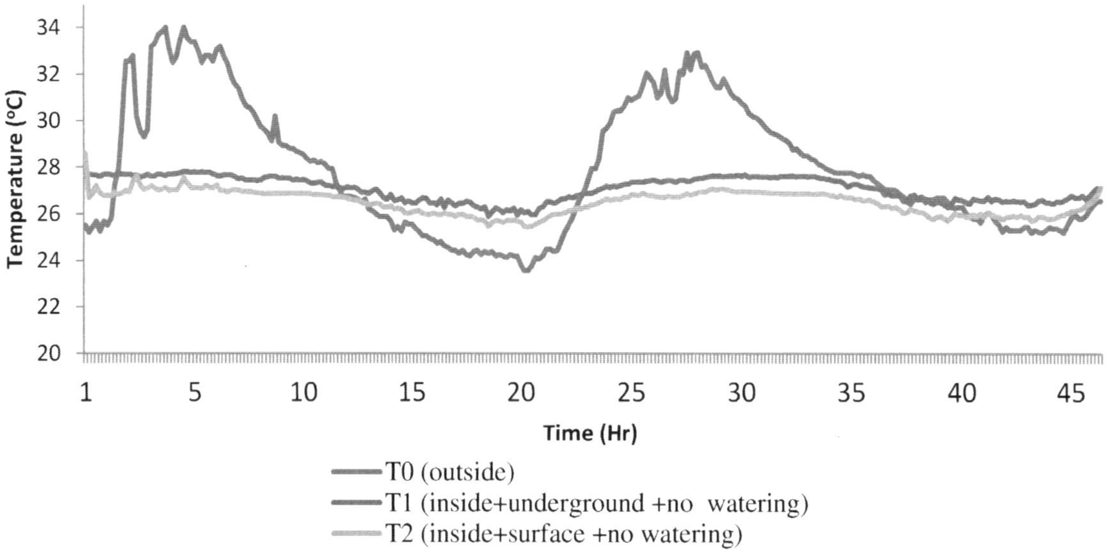

Fig. 4. Diurnal temperature in inside and outside of the ZECC without watering

Figure 4 shows the typical daily changes temperature in inside and outside of the ZECC without watering condition where the ZECC placed in the underground and on the surface. The highest temperature (34.0°C) and the lowest one (23.5°C) was recorded in outside of the ZECC. In average, the temperature in the outside is higher than in inside of the ZECC, particularly in daylight. The temperature in T1 and T2 of the ZECC for 5 hours was about 27.8°C and 27.1°C. The corresponding values for T (outside) is about 33.8°C. This result was also consistent with a study by Islam and Morimoto[1], in which the temperature of the ZECC investigated; the temperature inside of the ZECC is lower than the outside one.

3.2. Relative Humidity

Figure 5 shows the daily changes in inside and outside relative humidity of the ZECC over 45 hours. Overall, it can be seen that relative humidity in outside is lower than inside of the ZECC. In average relative humidity of RH0, RH1, RH2, RH3 and R4 were 72.9%, 92.9%, 94.1%, 83.1% and 79.2% respectively. Also, relative humidity of ZECC with watering condition (RH1, RH2) is always higher than ZECC without watering (RH3, RH4). Differences this relative humidity could be due to moisture in the sand caused by evaporative cooling method[6]. Lower temperature and higher relative humidity in cooling chambers contributed to the improved shelf life of fruit and vegetables. For instance, increasing humidity of the storage container for mushrooms from 76% to 96% decreased transpiration rate by 87% at 4^{0}C while decreasing the temperature from 16 to 4^{0}C decreased transpiration rate by 61%at 96% RH[11].

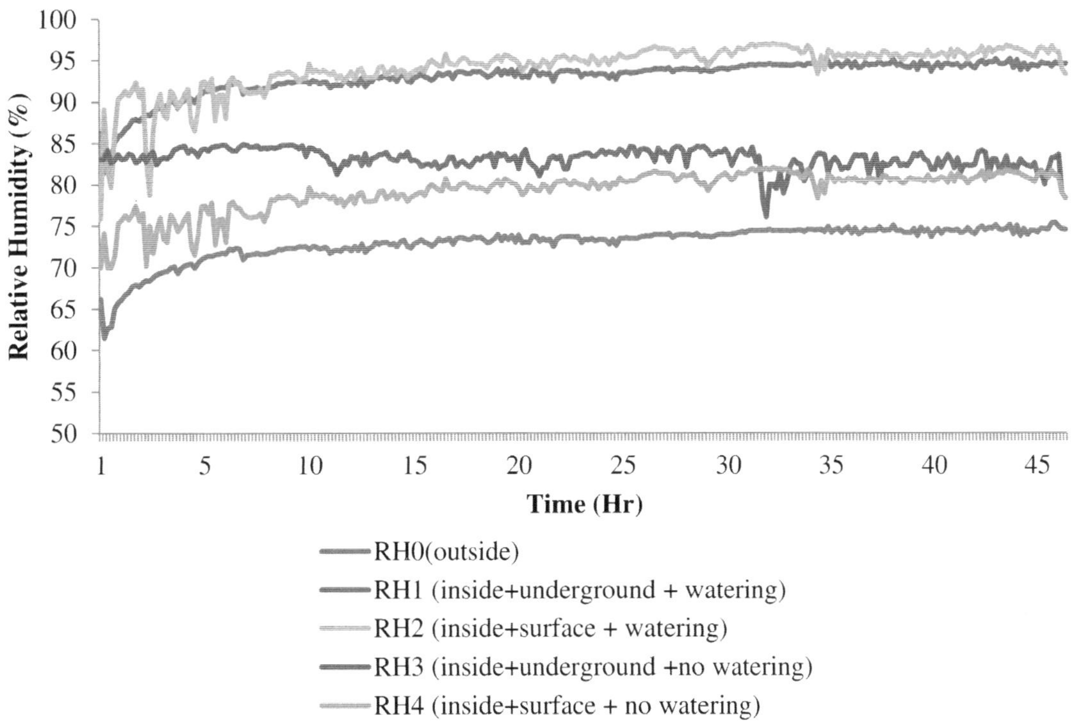

Fig. 5. Diurnal changes in inside and outside relative humidity (RH) of the ZECC under watering and no watering condition

4. Conclusion

Two types of ZECC were constructed; underground and on the surface, to evaluate temperature and relative humidity inside the ZECC compared to the outside one. A level of water (20 litres per day) was added to drip irrigation system to moisten the sand in the cooling chamber In the ZECC. The results show that the ZECC was constructed on the surface produce lower temperature and higher RH compare to ZECC which placed underground. In addition, the average of the temperature outside of the ZECC (28.0^{0}C) is higher than the inside one (26.2^{0}C). On the other hand, the relative humidity outside of the ZECC (72.9%) is less than the inside one (87.2%). It was concluded that the ZECC where was constructed on the surface is more suitable for decreasing temperature and increasing relative humidity inside the ZECC compared to the underground one. The use of a watering can potentially reduce the temperature inside of the ZECC.

References

[1] Islam M P and Morimoto T 2012 Zero energy cool chamber for extending the shelf-life of tomato and eggplant *Japan Agric. Res. Q.* **46** 257–67
[2] Morimoto T, Purwanto W, Suzuki J and Hashimoto Y 1997 Optimization of heat treatment for fruit during storage using neural networks and genetic algorithms *Comput. Electron. Agric.* **19** 87–101
[3] Islam M P and Morimoto T 2014 A new zero energy cool chamber with a solar-driven adsorption refrigerator *Renew. Energy* **72** 367–76
[4] Islam M P, Morimoto T and Hatou K 2013 Dynamic optimization of inside temperature of Zero Energy Cool Chamber for storing fruits and vegetables using neural networks and genetic algorithms *Comput. Electron. Agric.* **95** 98–107
[5] Islam M P and Morimoto T 2015 Evaluation of a new heat transfer and evaporative design for a zero energy storage structure *Sol. Energy* **118** 469–84
[6] Ganesan M, Balasubramanian K and Bhavani R . 2004 Studies on the application of different levels of water on Zero energy cool chamber with reference to the shelf-life of brinjal *J. Indian Inst. Sci* **84** 107–11
[7] Islam M P and Morimoto T 2015 Progress and development in brick wall cooler storage system *Renew. Sustain. Energy Rev.* **50** 277–303
[8] Narayana C, Mustafa M and Sathiamoorthy S 2002 Effect of packaging and storage on shelf-life and quality of banana cv. Karpuravalli *J. Dairying, Foods Home Sci.* **59** 113–7
[9] Islam M P, Morimoto T and Hatou K 2013 Optimization of Watering for Minimizing the Inside Temperature of Zero Energy Cool Chamber for Storing Fruits and Vegetables *IFAC Proc. Vol.* **46** 17–22
[10] Islam M P, Morimoto T and Hatou K 2012 Storage behavior of tomato inside a zero energy cool chamber *Agric. Eng. Int. CIGR J.* **14** 209–17
[11] Mahajan P V., Oliveira F A R and Macedo I 2008 Effect of temperature and humidity on the transpiration rate of the whole mushrooms *J. Food Eng.* **84** 281–8

IOP Conf. Series: Earth and Environmental Science **101** (2017) 012029 doi:10.1088/1755-1315/101/1/012029

Shelf life prediction of canned fried-rice using accelerated shelf life testing (ASLT) arrhenius method

M Kurniadi[1], R Bintang[2], A Kusumaningrum[1], A Nursiwi[2], A Nurhikmat[1], A Susanto[1], M Angwar[1], Triwiyono[1] and A Frediansyah[1]

[1]Research Unit for Natural Products Technology, Indonesian Institute of Sciences, Indonesia

[2]Science and Food Technology, Sebelas Maret University (UNS), Surakarta, Indonesia

Email address: hm_kur@yahoo.com

Abstract. Research on shelf-life prediction of canned fried rice using Accelerated Shelf-life Test (ASLT) of Arrhenius model has been conducted. The aim of this research to predict shelf life of canned-fried rice products. Lethality value of 121°C for 15 and 20 minutes and Total Plate count methods are used to determine time and temperatures of sterilization process.Various storage temperatures of ASLT Arrhenius method were 35, 45 and 55°C during 35days. Rancidity is one of the derivation quality of canned fried rice. In this research, sample of canned fried rice is tested using rancidity value (TBA). TBA value was used as parameter which be measured once a week periodically. The use of can for fried rice without any chemical preservative is one of the advantage of the product, additionaly the use of physicalproperties such as temperature and pressure during its process can extend the shelf life and reduce the microbial contamination. The same research has never done before for fried rice as ready to eat meal. The result showed that the optimum conditions of sterilization process were 121°C,15 minutes with total plate count number of 9,3 x 10^1 CFU/ml. Lethality value of canned fried rice at 121°C,15 minutes was 3.63 minutes. The calculated Shelf-life of canned fried rice using Accelerated Shelf-life Test (ASLT) of Arrhenius method was 10.3 months.

1. Introduction

Canned food products has been developed since several decades ago. In some countries, military ransom used special food of their own country. The United States of America (USA) has its pizza and the Phillipines has its canned rice with sausage from pork [1]. In Indonesia, rice is the main food that is consumed by many people there. More than 90% of Indonesian people consume rice as their main food [2],[3]. As the high consumption of rice in Indonesia, Indonesian people tend to process rice into other food products to avoid boredom and increase rice"s nutrition value. One of favourite processed rice products in Indonesia is fried rice. In its serving, fried rice is processed food that must be immediately consumed, because fried rice has short shelf-life relatively. The decrease that occur on fried rice is rancidity. Beside rancidity, there are microorganisms which cause decrease and harm on a human"s health. This is caused by chemical reactions taking place naturally, so that it will affect the taste, color, texture and nutrients in the food[4]. In the oil there are two types of main damage, which are rancidity and hydrolysis [5]. Rice-based food products are vulnerable to get contaminated by *B. cereus* that is growing and producing emetic toxins in a relatively short time on the rice storaged in room temperature [6]. Therefore, preservation is needed to extend shelf-life of canned fried rice.

In canning, sterilization process is the primary process in commercial sterile food production, especially to guarantee the security of commercial sterile food. The sterilization process in canning is heating process of the can as well as its content in many factors causing decay on food, without causing over cooking on the food [7]. In general, the canning process of food ingredients consisting of some stages, some of them are ingredient preparation, filling process, exhausting, sterilization, cooling, and storing [8]. After sterilization process, cooling process must be done immediately to prevent food over cooking and thermophilic bacteria growing. The temperature to reduce *Clostridium botulinum* number in a canned food is 121.1°C for 3 minutes [9]. The indicators of optimum sterilization process is generally done by ensuring that *Clostridium botulinum* has died. Thus, other microbes which are vulnerable to heat will be automatically dead if *Clostridium botulinum* can be killed successfully.

The use of can for fried rice without any chemical preservative is one of the advantage of the product, additionaly the use of physical properties such as temperature and pressure during its process can extend the shelf life and reduce the microbial contamination. The same research has never done before for fried rice as ready to eat meal. In Indonesia, rule regarding shelf-life of food ingredients is stated in Law of Food No. 18 year 2012 regarding food and Government Regulation No.28 year 2004 regarding security, quality and nutrient of food. To determine shelf-life of canne fried rice, it needs to conduct a research regarding shelf-life prediction. One of the methods to predict shelf-life is by using *Accelerated Shelf Life Testing* (ASLT) method with Arrhenius model and food storage with extreme temperature, in which the damage on food products occur faster then the shelf life is determined based on extrapolation to storage temperature.

Based on product identification that is done, it is found that quality factor that becomes parameter of the product"s quality decrease reaction kinetics. To find out quality decrease level, quality factor data are transformed in a plot kinetic and an appropriate kinetic parameter model will be obtained. Quality decrease process of food ingredients is stated by the following common equation [10]-[11]:

$$\frac{dQ}{dt} = k.Q^n \quad (1)$$

Q : Quality factor
t : time
k : constant rate that depends on temperature and water activity
n : degree factor or reaction order
dQ/dt : change of Q factor per time unit (positive sign if the decrease is stated in Q addition and negative if the measure one is Q decrease)

Most of all, decrease of food ingredient quality includes zero order (order 0) reaction and one order (order 1). Withevaluation of *constant rate* (k) at different three temperatures or more, Arrhenius correlation chart can be made, i.e. extrapolation with straight line of correlation between ln k and 1/T to predict kinetic reaction (k) from reactions at other temperatures [11]. Shelf-life is determined based on the most influential factor toward the product. One of the factors that is able to influence a product"s shelf-life is temperature. Determination of shelf-life and temperature limit factor can be done by Arrhenius kinetic approach. Zero order quality decrease reaction can be stated with the following equation [12]:

$$t = \frac{(A_1 - A_0)}{K} \quad (2)$$

And for order 1 decrease reaction can be stated by the following equation:

$$t = \frac{ln\frac{A_1}{A_0}}{K} \quad (3)$$

In which:
A_1 : final score
A_0 : initial score (day-0)

K : reaction constant rate
t : shelf-life (in day, month or year)

Here we report, for the first time, the using of ASLT method for predicting shelf life of canned fried rice product

2. Materials and Methods

2.1 Sample preparation
Ingredients which are used to make fried rice are rice type IR64, garlic, onion, chilly, margarine, soy sauce, fried beef sausage, salt, fried local chicken meat and food grade cylindric can (anamel material) in size of 301 x 205 mm (Ø x h) [13]. Chemical reagents used for analysis are TBA reagent, aquades, HCl 4N, PCA media, and NaCl 0.85%.

2.2 Instruments
"Tomy"Autoclave vertical series SS-325, "Ellab" Fo meter, Thermocouple CTF 9004, "varin" seamer machine and incubator at operation temperatures.

2.3 Accelerated Shelf-life Testing (ASLT) Arrhenius Method
Kinetics observation on canned fried rice quality decrease uses *Accelerated Shelf Life Test* (ASLT) method and *Arrhenius* model. The sample is stored at 3 different temperatures, i.e. 35°C, 45°C, and 55°C. The sample parameters are observed for 35 days to get decrease rate based on rancidity testing of the product during storage as the critical parameter of the quality of canned fried rice. Next, the data which were obtained from rancidity testing were plotted in chart of correlation of time (x axis) and average sensory score on each storage temperature (y axis). The determination of shelf-life of canned fried rice is determined by the biggest R^2 (determination coefficient) value among critical parameters (TBA score) [14].

2.4 Sensory Test
Sensory test are used for knowing organoleptic level of canned fried rice to 25 untrained panelists. Organoleptic level including of colour, smell, taste, texture.

2.5 Analysis Methods
2.5.1. Total plate count
The materials used are PCA media and NaCl of 0,85%. The equipment and the materials are sterilized.About 25 g of sample was taken and mixed up using mixer machine, followed by addition of 225 ml NaCl 0,85% and homogenized. Homogenous liquor was brought to Laminar Air Flow for dilution and platting. Fresh rice sample was diluted up to 10^{-5} and canned rice up to 10^{-3}. The dilution was done by taking liquid of 1 ml and poured in to the tube containing 9 ml NaCl 0,85%. The diluted sample was platted into the petridish about of 1 ml poured on to the homogenizeed media and let stand until clotting. After clotting, petridish is put into a closed container and filled up with CO (anaerobic condition). The container is stored in incubator at 37°C for 72 hours. At the end, the number of colony was calculated manually using colony counter.

2.5.2. Target lethality (Fo)
The analysis of heat lethality value (Fo) using Thermocouple CTF 9004 which its needle installed on the wall of the can. Afterwards, canned products is sterilized with the temperature and time required. During the sterilization process, thermocouple recorded the data to be re-written in Microsoft excel and analyzed. Condition of autoclave set up at 121°C and time of 15 and 20 minutes. Fo value can be estimated with equation [15]:

$$Fo = \int_{t=0}^{t} 10^{\left[\frac{Tautoclave-121}{z}\right]} dt \qquad (4)$$

2.5.3 TBA Analysis

TBA analysis was conducted once a week during 35 days. The sample was stored at 3 different temperatures of 35°C, 45°C, 55°C (based on actual temperatures in Indonesia from distribution to products sale) during 35 days. Sample of 10 grams and aquadest 50 ml were mixed up into blender then poured into destillation flask 1000 ml with addition of aquades 47,5 ml and HCL 4 M of 2,5 ml. Afterward destillation flask was installed on distillation apparatus and be heated until 50 ml of destilate was collected. Collected destilate then be filtered and poured into erlenmeyer followed by addition of TBA reagent of 5 ml (0,02 M thiobarbituric-acid solution in 90% of glacial acetate acid). Next step, the glass containing solution was heated on boiling water for 35 minutes, and cool down by using flowing water. Afterwards, absorbance was measured with wave length of 528 nm [16].

3. Result and Discussion

3.1. Optimum Temperature and time of Canned-fried rice Sterilization

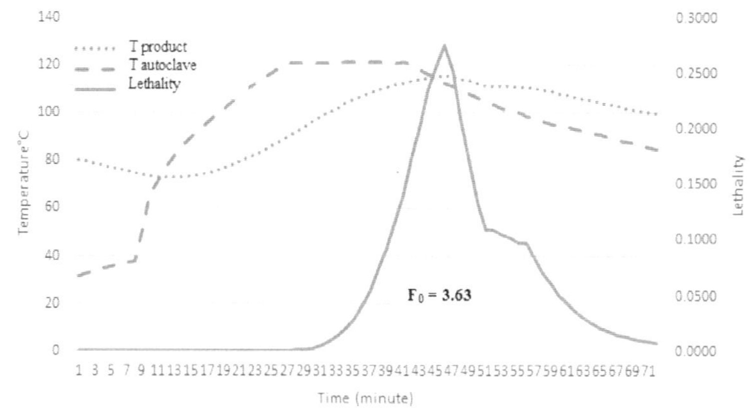

Figure 1. Heat sufficiency in the sterilization process of canned fried rice in temperature 121°C for 15 minutes

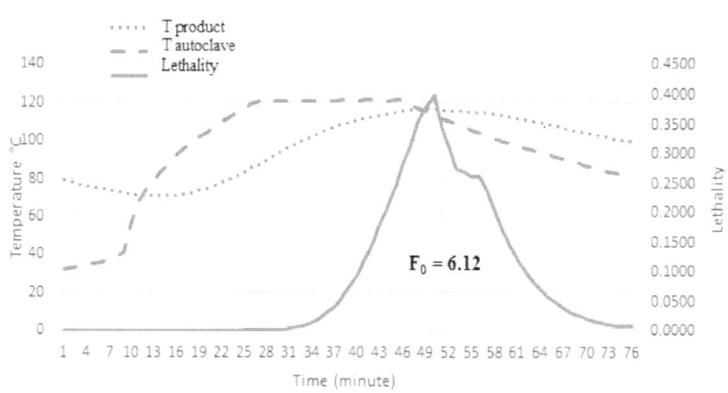

Figure 2. Heat sufficiency in the sterilization process of canned fried rice in temperature 121°C for 20 minutes

In determining optimum value of the sterilization of canned fried rice, we need to concern on Fo value in that process. Fo value is lethality equivalence of thermal processs with heating time at temperature of 250 F or 121°C. Fo value to reduce *Clostridium botulinum* by 12D is 3 minutes at minimum [17]. Microbes which can form spores has better heat resistant compared than those which cannot. Fo value that fulfill process standards is sterilization treatment at temperature of 121°C for 15 minutes that is Fo of 3.63 minutes.

Table 1. Total calculation of microbes number

Sample	Microbes Number (CFU/g)
Fresh Fried-Rice	2.3×10^4
Sterilized Fried Rice (121°C ,15 minutes)	9.3×10^1
Sterilized Fried Rice (121°C, 20 minutes)	9×10^1

Most of BPOM"s regulations state that canned products apply maximum limit by 1 x 10 CFU/g, the products are canned corned, canned sausage, canned fish, canned fruits and canned vegetables [18]. Fried rice is not acid food therefore it safe to interact with can materials. In canned food acid, acid can be interact with Pb so it is possible for food contaminated by metal ." Gudeg " canned that sterilized at 121°C during 20 minutes has been occured Fe (iron), S (Sulphur) and Al (alumunium) changes . No lead (Pb) has detected in the can before use as packaging . The study about metal migration during thermal process has been done in previous research[19]-[20]. This canned fried rice is safe to be consumed. It means that the process that is done during producing canned fried rice has processed well and appropriately. The can has met the requirement standard for canned food industry.

Table 2. Sensory test of canned fried rice

Sample	Colors	Favor	Taste	Texture	Overall
121°C, 20 minutes	4.32[a]	4.24[a]	4.36[a]	4.48[a]	4.40[a]
121°C,15 minutes	4.40[a]	4.40[a]	4.44[a]	4.52[a]	4.48[a]
Remarks:	Small alphabet letter for vertical comparison, and number that is followed by the same alphabet in each sample show real indifference in sig. 5%				

Sensory Score:
1 = dislike much
2 = dislike
3 = a bit like
4 = like
5 = like very much

Table 2 shows the result of organoleptic on all sensory parameters , i.e. colour , smell , taste , texture , overall are in the same *subset* . It shows that different treatments of temperature and sterilization time of canned fried rice in each sample do not show real difference among samples . It can be seen that the higher the temperature is, the longer time needed in the sterilization process, the assessment from 25 panelists will be getting decreased . It means that the panelists consider that the quality of the product decrease.

Table 3. TBA value change of canned fried rice during storage at temperature of 35, 45 and 55°C

Time (Days)	Average TBA value (mg malonaldehid/kg)		
	35°C	45°C	55°C
0	0.026	0.026	0.026
7	0.027	0.029	0.032
14	0.027	0.029	0.032
21	0.029	0.031	0.033
28	0.031	0.032	0.035
35	0.039	0.043	0.048

Rancidity limit of canned foods is 0.5 mg malonaldehid /kg. The selection of reaction order of rancidity increase on canned fried rice is done by comparing determination coefficient (R^2) in each linear regression equation at the same temperature of zero order with reaction ofone order. Reaction order with the biggest determination coefficient is the used reaction order.

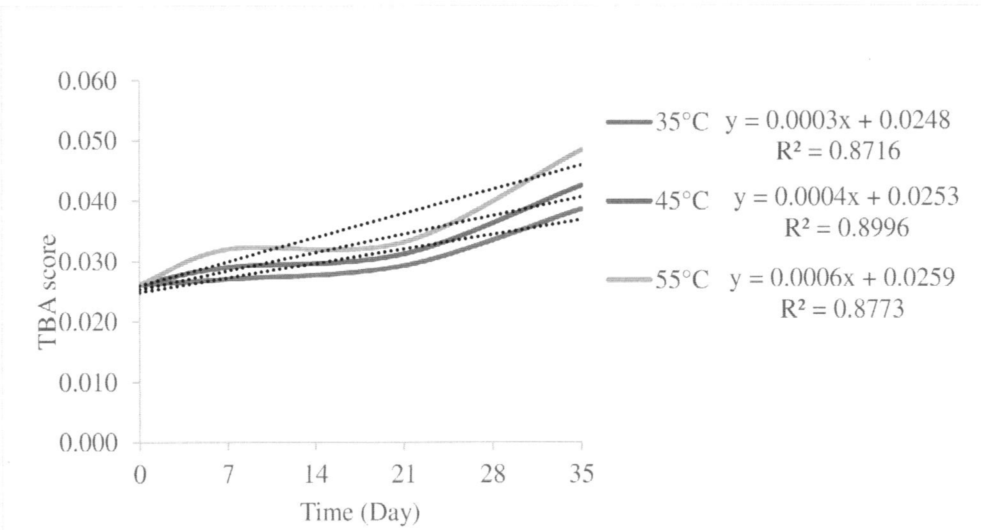

Figure 3. Chart of correlation between TBA score and time at Order 0

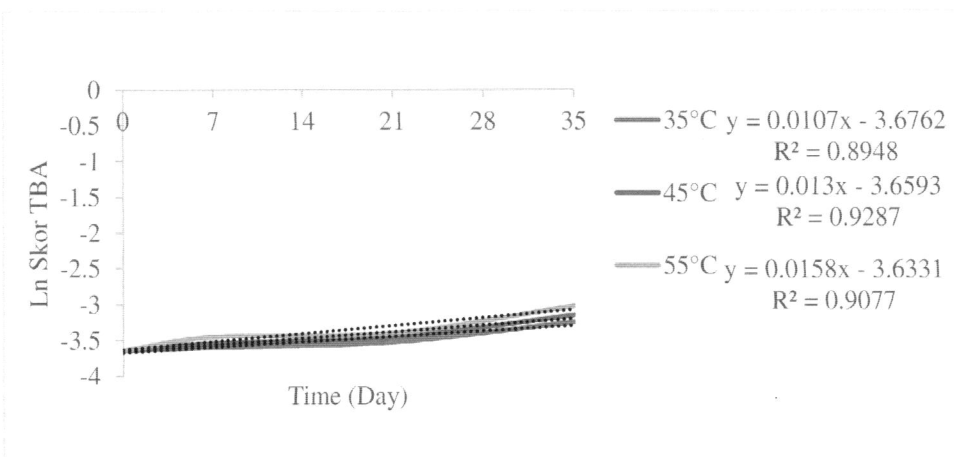

Figure 4. Chart of correlation between TBA score and time at Order 1

Table 4. K value in Order 1 of Linear regression equation

Temperature (K)	K	ln K	1/T
308	0.0107	-4.53751	0.003247
318	0.0130	-4.34281	0.003145
328	0.0158	-4.14775	0.003049

Figure 3 and Figure 4 show that R^2 value order 1 at temperature of 35, 45 and 55°C is bigger than order 0, so that order one is selected to determine Arrhenius equation. The determination of Arrhenius equation is done by making plot of ln K and 1/T (Table. 4) values in the reaction of TBA value change of canned fried rice.

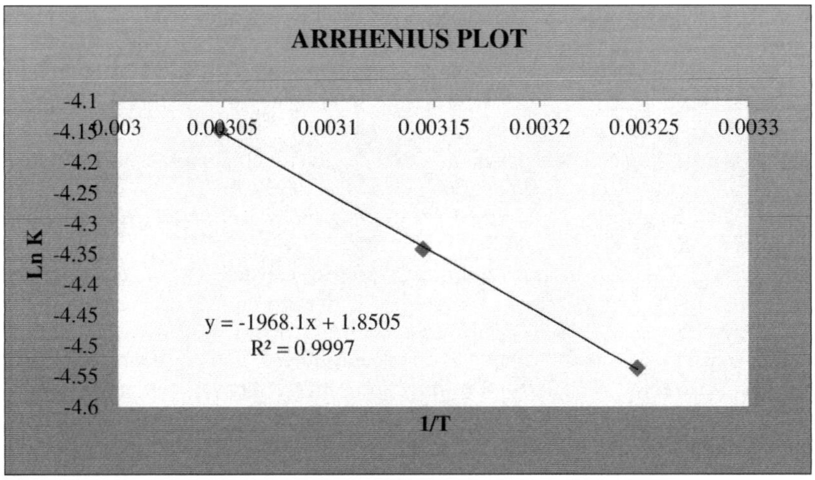

Figure 5. Arrhenius plot of TBA value of canned fried rice during storage

Linear equation from Arrhenius plot :
$y = -1968.1x + 1.8505$
For extrapolation, the storage temperature in normal condition 30 °C (303 K), Resulted :
$\ln K = -1968.1\left(\frac{1}{T}\right) + 1.8505$
$\ln K_{303K} = -4.6449$
$K_{303K} = 0.0096$

TBA critical point (A_I), resulted from previous research was 0.5 mg/kg; initial TBA (Ao) was 0.026 mg/kg, so then from equation (3) in order 1:

$$t\,(shelf-life) = \frac{\left(\ln\frac{0.5}{0.026}\right)}{0.0096}$$

$$shelf-life\ prediction = 307.6\ days = 10.3\ months$$

4. Conclusion

Optimum temperature and time of sterilization of canned fried rice is at 121°C for 15 minutes. The shelf-life prediction of canned fried rice by using ASLT Methods Arrhenius Model at storage temperature of 30°C is 10.3 months. We recommend to consumer to follow the guideline that has been published especially on heating step of the canned food before consumed to improve the texture and increase the appetite.

Acknowledgment
The authors would like to thank the project of Superior Research LIPI 2016 for the financial support

References

[1] Daniel 2014 *Thermal Process of canned nasi rendang: microbiology, chemical and organoleptic evaluation* Thesis (Yogyakarta: Gadjah Mada University)
[2] Biro Pusat Statistik 2008 *Statistik Indonesia: Statistical Yearbook of Indonesia* (Jakarta: Biro Pusat Statistik)
[3] Yuliawan T and Handoko I 2016 The effect of temperature rise to rice crop yield in Indonesia uses *Shierary Rice* model with geographical information system (GIS) feature *Procedia* **33** 214–220
[4] Bintoro V P, Legowo A M and Purnomoadi A 2015 Garlic antioxidant (*Allium sativum L.*) to prevent meat rancidity *Ital. Oral Surg.* **3** 137–141
[5] Macaire H, Tonfack F, Anjaneyulu B, Sri M, Karuna L, Badari R, Prasad M and Linder N 2016 Oxidative stabilization of RBD palm olein under forced storage conditions by old cameroonian green tea leaves methanolic extract *NFS* **3** 33–40
[6] Agata N, Ohta, M and Yokohama K. 2002 Production of bacillus cereus emetic toxin (*cereulide*) in various foods *Int. J. Food Microbiol.* **73** 23–27
[7] Kannan A and Sandaka P 2008 Heat transfer analysis of canned food sterilization in a still retort *J. Food Eng.* **88** 213–228
[8] Utami R 2012 Karakteristik pemanasan pada proses pengalengan gel cincau hitam (*Mesona Palustris*) Thesis (Bogor: Bogor Agricultural University)
[9] Ogbulie T E, Uzomah A and Agbugba M N 2014 Assessment of the safety of some on the shelf canned food products using PCR-based molecular technique. *Niger. Food J.* **32** 81–91
[10] Fu B and Labuza T 1997 *Shelf life testing : procedures and prediction methods for frozen foods.* (Boston: Springer) doi:https://doi.org/10.1007/978-1-4615-5975-7_19
[11] Labuza T and Riboh D 1982 Theory and application of arrhenius kinetics to the prediction of nutrient losses in food *J. food Technol.* **36** 66–74
[12] Brandão T R S and Silvaa C L M 2011 Dynamic approach for assessing food quality and safety characteristics : the case of processed foods *Ital. Oral Surg.* **1** 1015–1025
[13] Kurniadi M, Angwar M, Nurhikmat A, Susanto A, Triwiyono and Praharasti A 2015 The development of canned traditional food based military ration packaged in cans 1st *International Conference On Appropriate Technology Development* (Bandung) October 5-7, 2015 (Center for Appropriate Technology Development) pp 70-75
[14] Córdova A, Quezada C and Saavedra J 2011 A MALST method comparison over univariate kinetic modeling for determination of shelf-life in cereal snack of dried apples. *Ital. Oral Surg.* **1** 1045–1050
[15] Gonçalves E C, Minim L A, Coimbra J S R and Minim V P R 2005 Modeling sterilization process of canned foods using artificial neural networks *Chem. Eng. Process. Process Intensif.* **44** 1269–1276
[16] Devasagayam T P A, Boloor K K and Ramasarma T 2003 Methods for estimating lipid peroxidation : an analysis of merits and demerits *Indian J. Biochem. Biophys.* **40** 300–308
[17] Office of the Federal Register National Archives and Records Administration 2010 *Food and drugs: chapter 1-food and drug administration subchapter b-food for human consumption* (U.S: Government Printing Office)
[18] Badan Pengawas Obat dan Makanan Republik Indonesia 2016 Kriteria mikrobiologi dalam pangan olahan *Report Peraturan kepala badan pengawas obat dan makanan republik Indonesia* (Jakarta : Badan Pengawas Obat dan Makanan Republik Indonesia) pp 4–54
[19] Nurhikmat A, Suratmo B, Bintoro N and Sentana S 2015 The quality changes on canned gudeg 'Bu Tjitro' during storage *AGRITECH* **35** 353–357
[20] Dewi D C 2012 Determinasi kadar logam timbal (Pb) dalam makanan kaleng menggunakan destruksi basah dan destruksi kering *J. ALCHEMY* **2** 12-25

Characterization of Lactic Acid Bacteria as Poultry Probiotic Candidates with Aflatoxin B1 Binding Activities

E Damayanti[1], L Istiqomah[1], J E Saragih[2], T Purwoko[2], Sardjono[3]

[1]Research Unit for Natural Products Technology, Indonesian Institute of Sciences, (BPTBA-LIPI). Jl. Jogja-Wonosari Km. 31.5, Gading, Playen, Gunungkidul, 55861 D.I. Yogyakarta, Indonesia

[2]Faculty of Mathematics and Natural Sciences, Sebelas Maret University, Surakarta, Indonesia

[3]Faculty of Agricultural Technology, Universitas Gadjah Mada, Yogyakarta, Indonesia

E-mail: emadamayanti80@gmail.com

Abstract. Our previous studies have selected lactic acid bacteria (LAB) with antifungal activities from traditional fermented foods made from cassava (G7) and silage feed palm leaf (PDS5 and PDS3). In this study we evaluated their ability to bind aflatoxin B1 (AFB1) and probiotic characteristic. The probiotic characteristic assays of LAB consisted of resistance to acidic conditions (pH 3), gastric juice and bile salts 0.3%. We also carried out an in vitro evaluation of LAB aflatoxin binding ability in viable and non-viable cell for 24 and 48 hours of incubation. The measurement of aflatoxin content was performed by ELISA method using AgraQuant Total Aflatoxin Assay kit. The results showed that all isolates were potential as probiotics and the G7 isolate had the highest viability among other isolates in pH 3 (92.61 %) and the bile salts assay (97.71 %). The percentage of aflatoxin reduction between viable and non-viable cell from each LAB isolate were different. The highest aflatoxin reduction in viable cell assay was performed by G7 isolate (69.11 %) whereas in non-viable cell assay was performed by PDS3 isolate (73.75 %) during incubation time 48 hours. In this study, G7 isolate performed the best probiotic characteristics with the highest viability in acid pH assay, bile salt 0.3% assay and percentage of aflatoxin B1 reduction in viable cell condition. Molecular identification using 16S rRNA sequence analysis showed that G7 isolate had homology with *Lactobacillus plantarum* (99.9%). It was concluded that *Lactobacillus plantarum* G7 was potential as probiotic with aflatoxin binding activities.

Keywords: aflatoxin B1, lactic acid bacteria, probiotic

1. Introduction

Local feed still have many limitations in quantity and quality. One of its limitation is high contamination of mold and mycotoxins in feed, especially feed-based grains which still dominates between 60-70% in the composition of the feed formula. Mycotoxin contamination in feed in Indonesia shows that more than 80% of corn contaminated with aflatoxin B1 (AFB1) and the average is above the threshold of tolerance (> 50 ppb) [1]. The other report also stated that loss of materials and fodder due to fungal contamination worldwide is estimated at 5-10 % [2], while the The Food and Agriculture Organization (FAO) estimates that around 25% of foodstuffs and animal feed contaminated with mycotoxins [3]. The accumulation of mycotoxin contaminants in the diet have a very fatal effect for livestock because it causes the immunosuppresive effect, unproductive livestock and even death. Mycotoxins also cause health problems in both humans and animals such as carcinogenic, mutagenic and teratogenic [4].

Physical handling for mycotoxin contamination such as heating and chemical handling such as the addition of certain toxin binders still causes the contamination of both fungi and mycotoxins [3]. The addition of toxin binders in feed need high costs and could not be guaranteed free mycotoxin. In addition, the use of chemicals in the feed could not be guaranteed safety because the residue in has an effect on livestock products. The use of microorganisms with mycotoxin binding activitiy and mycotoxin detoxification become a promising alternative. The use of microorganisms and natural materials is expected to be safe and ensures consumer protection with the production of healthy and safe-to-eat livestock products.

Several previous studies showed that some *Lactobacillus* strains have inhibitory activities on growth fungi and aflatoxin production [5]. *Lactobacillus casei* has ability as AFB1 binder and aflatoxin detoxification in foodstuffs [6]. The administration *Lactobacillus rhamnosus* GG in mice was also known to improve the excretion of aflatoxin in feces [7]. *Lactobacillus* strain has higher ability to bind AFB1 than the *Pediococcus* and *Leuconostoc* strains [8]. And the other study revealed that some isolates of *Lactobacillus* isolated from human and dairy product have ability to bind AFB1 [9].

A preliminary study of lactic acid bacteria (LAB) strains isolated from fermented food and silage feed resulted the selected isolates which have antifungal activities. In vitro assay showed that lactic acid bacteria had inhibitory effect on the growth of mycotoxin-producing fungal biomass. *Lactobacillus* sp. G7 inhibited growth of *Aspergillus flavus* FNCC 6002, *A. parasiticus* FNCC 6033, and *Penicillium citrinum* FNCC 6111 with biomass growth inhibition around 63.17% [10]. The other result showed that *Lactobacillus* sp. PDS 3 and *Lactobacillus* sp. PDS5 isolated from silage feed have inhibition activities on growth of *A. flavus* FNCC 6002 [11]. Based on that previous studies it was important to determine the probiotic characteristic of selected LAB with aflatoxin binding activities.

2. Material and Method

2.1. Lactic acid bacteria isolates and culture conditions.

Three LAB strains (*Lactobacillus* sp. G7, *Lactobacillus* sp. PDS3 and *Lactobacillus* sp. PDS5) were used for this research for AFB1 binding activities. *Lactobacillus* sp. G7 was isolated from cassava fermentation whereas *Lactobacillus* PDS3 and *Lactobacillus* PDS5 were isolated from silage feed. LAB strains were maintained and collected in Microbiology Laboratory of Research Unit for Natural Product Technology, Indonesian Institute of Sciences (BPTBA LIPI) Yogyakarta, Indonesia. The isolates were identified by Gram staining, catalase tests, motility assay, morphological characteristic and grown on de Man Ragosa Sharpe Agar (MRSA) as a selective medium. LAB isolates were preserved on lyophilized cultures with skim milk as filler material.

2.2. Acid tolerant assay

Acid tolerance assay referred to modified method [12]. LAB cultures on MRS Broth were centrifuged at 4137 x g for 10 min at 4 °C. Pellets were washed two times by sterile phosphate buffered saline (PBS) and diluted in sterile PBS before being inoculated on MRS Broth (pH 2, with 1 M HCl addition). Cell viability was calculated by the total plate count (TPC) method on MRS Agar media on 0 and 90 min incubation times.

2.3. Gastric juice tolerant assay

Gastric juice tolerance was observed according to modified gastric juice simulation [12]. The selected LAB isolates were incubated on MRS Broth at 37 °C for 18 h. A total of 1 mL culture was centrifuged at 5000 g, 10 min, 4 °C then it was washed two times by using sterile PBS and diluted on 0.3 mL sterile PBS. A total of 0.2 mL dilution was taken and then mixed with 1 mL of artificial gastric juice. The mixture liquid was homogenized and incubated at 37 °C for 2 h and then sampled after 0, 1, and 2 h. Serial dilutions of samples was made on sterile PBS and then inoculated on MRS Agar media for cell viability observation. The artificial gastric juice was made from pepsin (Sigma) (3 g/L) dilution at pH 2.

2.4. Bile salt tolerant assay

Bile salt tolerance was determined by modified method [12]. A total of 1 mL LAB culture was centrifuged at 4137 *x g* for 10 min at 4 °C and washed twice by using sterile PBS. The cells were diluted in 0.3 mL of PBS then mixed with 0.2 mL of dilution and 1 mL PBS containing 0.3% (w/v) bile salt (Merck). The mixture was incubated at 37 °C for 3 h and then was sampled after 0 and 2 h. The cell viability was calculated by using serial dilution and plated on MRS agar media. Three replications were used for each treatment.

2.5. Aflatoxin binding assay

Aflatoxin binding assay was determined by using method with modification [4]. Solid AFB1 was suspended in benzene–acetonitrile (97/3, vol/vol) to obtain an AFB1 concentration of 1000 mg/L. The benzene–acetonitrile was evaporated by nitrogen stream and AFB1 was resuspended in methanol. The working solution of 5 mg/L AFB1 was prepared in PBS at pH 7.0 by using a stock solution of AFB1 in methanol. The cultures of each strain were divided in two aliquots. In one aliquot, a volume of the culture broth corresponding to approximately 1×10^{10} bacteria was transferred into tubes and centrifuged (viable cells). The second aliquot (approx. 1×10^{10} bacteria in tubes) was autoclaved at 121 °C for 20 min (nonviable cells) and centrifuged. Supernatants were removed and bacterial pellets (viable or nonviable cells) were washed with deionized water and centrifuged again. Aflatoxin solutions without cells served as controls.

The experiments were driven as following conditions. To study the effect of incubation time on the reduction of mycotoxin concentration, samples were vortexed shortly (5 s) and incubated for 48 h with soft agitation. The binding assay was performed at pH 7.0 for AFB1 after incubation at 48 h. Bacterial pellets with bound AFB1 were washed by suspending in 2 mL PBS (pH 7.0) containing no AFB1, and incubated for 10 min at 37 °C. The bacteria were pelleted, and a sample of the supernatant was collected for the quantification of AFB1 from bacteria. This washing procedure was repeated another two times. Supernatant samples (500 µL) were collected by centrifugation after incubation for 0, 24 and 48 h, and analyzed by Elisa Reader. The measurement of aflatoxin content was performed by ELISA method using AgraQuant Total Aflatoxin Assay procedure. They were compared against the control solution which were free of any bacteria but contained AFB1. The reductions in the quantities of AFB1 present in the test solutions was an indication of the capability of the bacterial strains in binding the AFB1 [13].

2.6. 16S rRNA sequence analysis

Genomic DNA were isolated from fresh culture on MRS Agar according to Genomic DNA Purification Kit (K0512) (Fermentas) protocol. DNA concentrations were measured using spectrophotometer (BioSpec-DNA/protein/Enzyme analyzer-Shimadzu) at λ260 nm and λ280 nm that was making a reaction formula for polymerase chain reaction (PCR). Primer used for PCR reaction was a universal primer 27f (5′-AGAGTTT GATCCTGGCTCAG-3′) and 1492r (5′-GGTTACCTTGTTACGACTT-3′) [14, 15]. Reaction volume of PCR was 50 µl which consist of 20 ng/µl of DNA concentration, 25 µl of GoTaq Green Master Mix, 1 µl of 1492 r primer (10 pm) and 1 µl of 27f primer (10 pm). PCR reaction was performed on 35 cycles for 3 h at 25 min using PCR machine (Takara-Thermal Cycler). PCR condition was set as follows : 4 °C with 94 °C for 2 min of pre-denaturation, 94 °C for 2 min of denaturation, 48,5 °C for 1 min of annealing, 72 °C for 1 min of elongation and 72 °C for 1 min of final extention [14]. PCR product was analyzed using gel electrophoresis agarose 1% [16] using electrophoresis apparatus (BioCRAFT BE 520) and 1 kb DNA ladder (Gene Rule-Fermentas) as ladder. Sequencing of 16S rRNA sequence was conducted by First BASE Laboratories (Singapore). Sequence data was showed in *abi format and was edited by FinchTV program and was contiq analyzed by BioEdit program. Contiq sequence was then analyzed using basic local alignment search tool (BLAST) program in Gene Bank of National Center for Biotechnology Information (NCBI) (website: www.ncbi.nlm.nih.gov).

2.7 Data analysis
The quantitative data were analyzed by using One-way analysis of variance (ANOVA) with post hoc test (Duncan multiple range test (P<0.05)) to distinguish the treatments means using CoSTAT statistical software [17]. The total of bacteria cell (cfu/mL) from viability test was converted to the logarithmic value before statistical analysis. Viability percentages was calculated by dividing a total of colonies in final incubation (log10 cfu/mL) with a total of colonies in initial incubation (log10 cfu/mL) and multiplied by 100%. Whereas the percentage of AFB1 binding for each bacteria was calculated by dividing the difference level AFB1 on certain and 0 minutes incubation time with the level AFB1 on 0 minutes incubation time and multiplied by 100%.

3. Result and Discussion

An important characteristic that must be possessed by lactic acid bacteria with probiotic ability were viability and survival ability on stress condition in digestive tract such as acid pH condition, gastric juice and bile salt exposure. Percentage viability of lactic acid bacteria on digestive tract simulation was shown in Table 1. The assay results on ~~acid~~ pH 3 conditions during 90 minutes incubation time showed that the viability of all isolates did not significant (P >0.05). The viability in this result was lower than the previous results. *Pediococcus acidilactici* R01 and *P. acidilactici* R02 isolated from broiler proventriculus had 103.4 - 120.74 % of viability on acid pH [18]. Different result was reported by Istiqomah [19] that the viability of *L. plantarum* isolated from digestive tract of chicken on pH 3 condition were 41. 9 – 54.6%. The variation result of viability of each LAB strain showed that species of LAB probiotics was different from LAB which is commonly found in fermented foods [20].

Table 1. The viability of lactic acid bacteria isolates on digestive tract simulation condition (acid pH, gastric juice and bile salt)

	Viability in simulation condition (%)		
LAB isolates	acid pH 3 (90 min)	gastric juice (45 min)	bile salt 0.3% (120 min)
G7	92.61 ± 7.87	99.60 ± 0.35	97.71 ± 1.64
PDS3	88.55 ± 2.84	100.23 ± 2.99	91.95 ± 6.83
PDS5	79.89 ± 10.91	98.97 ± 3.46	87.52 ± 2.14

Note: Means not significant (P<0.05)

In the gastric juice simulation test, the viability range were 98.97 - 100.23%. The viability of these isolates were higher than *P. pentosaceus* DB9 isolated from broiler chicken (75.66%) and *Lactobacillus* I72 isolated from the chicken intestine (81.76%) [21]. However, the viability of this study were lower than the isolates of *P. acidilactici* R01 and R02 isolated from broiler proventriculus of 100.35% - 102.71% and 100.02% - 102.65% respectively. While on the bile salt simulation assay for 1 - 2 hours incubation showed that G7, PDS3 and PDS5 isolates had higher viability than *P. acidilactici* R01 (59.69% - 76.53%) and *P. acidilactici* R02 (43.57% - 40.69%) [18].

The ability to survive on bile salt would be possible for LAB to deconjugate bile salt and would be effective to reduce serum cholesterol in broiler chicken. The high activity of bile salt hydrolyse in lumen of intestine could reduce bile salt conjugation ability to break down the lipid [12]. One of the essential characteristic of probiotic in order to give beneficial health for individual host was resistance to the gastrointestinal environment effects such as acid and bile salt in digestive tract [22]. This study showed that all LAB isolates had ability to survive on pH 3 after 90 minutes incubation. In the gastric juice tolerance assay, all LAB isolates also showed high viability after 45 minutes of incubation. Similar to the result in gastric juice tolerance assay, all LAB had high cell viability on 0.3% (b/v) bile salt at two hours of incubation. Based on the ability to survive on bile salt, gastric juice and low pH conditions, all LAB isolates had the same adaptation mechanism toward gastrointestinal stress condition.

Further assay for LAB isolates characterization was aflatoxin binding activity. The aflatoxin binding assay was performed using living bacterial cell (viable) and dead bacterial cell (non-viable). Based on previous research, it was known that alfatoxin binding of lactic acid bacteria was not based on detoxification mechanism by extracellular (enzymatic) metabolite. It was based on binding process by LAB cell wall. The results of aflatoxin binding ability assay were shown in Figure 1 for viable cell and Figure 2 for non-viable cell.

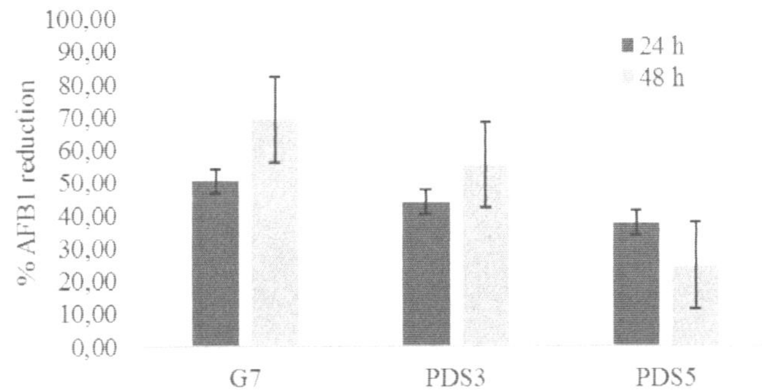

Figure 1. The AFB1 percentage reduction by viable cell of lactic acid bacteria isolates.

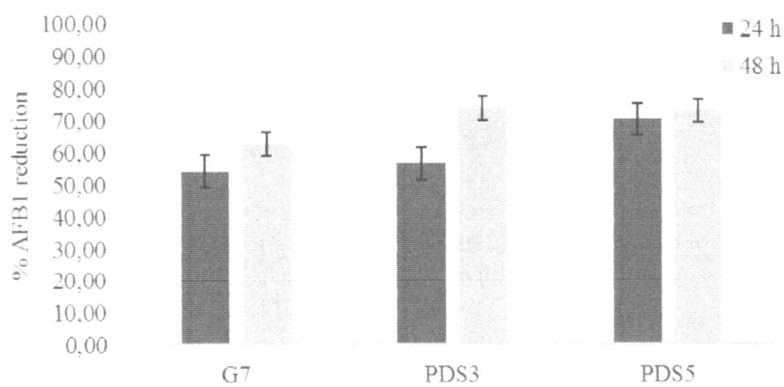

Figure 2. The AFB1 percentage reduction by non-viable cell of lactic acid bacteria isolates.

The results of aflatoxin binding assay occurred in strain-specific lactic acid bacteria and the longer incubation time had tendency to increase binding ability. The aflatoxin binding ability was performed by viable and non-viable lactic acid bacteria. Each bacteria had different ability to bind aflatoxin. Based on percentage of aflatoxin reduction, it was known that G7 isolate resulted the highest percentage to reduce AFB1 in viable cell condition (Figure 1) whereas PDS3 resulted the highest percentage to reduce AFB1 in non-viable cell conditions during incubation time 48 hours. As a mentioned in previous study, mycotoxin reduction by lactic acid bacteria was depend on specific strains [8]. The other study stated that the stability of the bacteria toxin complex was strain dependent and *L. casei* was a stronger binder of AFB1 compared with the other bacteria [6]. The AFB1 binding activity of G7 isolate at 48 hours incubation time was found to be higher (57.87%) than aflatoxin reduction of *L. acidophilus* P022 and *L. fermentum* subsp. *cellobiosus* 408 (15.5% - 42.8%) at 60 minutes of incubation [23]. Similar results were stated by Fazeli et al. [6], that *L.* casei, *L. plantarum*, and *L. fermentum* isolated from traditional Iranian sourdough and dairy products are known to have AFB1 binding ability between 25 – 61 % during incubation time 24, 48 and 72 hours.

Lactic acid bacteria had the ability of AFB1 binding both from viable and non-viable cells, this was indicated that AFB1 removal were not depend on bacterial viability and that the metabolic conversion of the toxin did not occur in bacteria. It was known that the removal of toxins occurs via adhesion to cell wall components rather than covalent binding or metabolic degradation, since dead cells still display binding ability [23]. Even though the mechanism of AFB1 removal by lactic acid bacteria is still unknown, it has been suggested that aflatoxins molecules are bound to the bacterial cell wall components of bacteria. AFB1 is bound to the bacteria by weak noncovalent interactions, such as associating with hydrophobic pockets on the bacterial surface [24]. The binding process was reversible and AFB1 was released by repeated aqueous washes [8].

In this study, G7 LAB isolate performed the best characteristic as probiotic with highest viability in acid pH assay, bile salt 0.3% assay and aflatoxin reduction percentage in viable cell condition. One of important characteristic of probiotic was colonization ability of live cell probiotic in host digestive tract. The selected lactic acid bacteria (G7 isolate) which has probiotic characteristic and the highest aflatoxin binding ability was identified as *Lactobacillus plantarum*. The identification procedure was confirmed by molecular test with 16S rRNA. The PCR product of the 16S rRNA gene is about 1,500 bp. Using a 1,500 bp fragment from the LAB isolates, the 16S rRNA gene sequencing revealed 99.9% similarity with the species *Lactobacillus plantarum*. This result similar with biochemical characterization using API 50 CHL which have closed relation to *Lactobacillus plantarum* (99.3 %) [10]. Several studies have reported that *L. plantarum* strain has binding aflatoxin activities. *L. plantarum* strain completely inhibited aflatoxin B1 production by *A. parasiticus* NRRL2999 (98.9 – 99.99%) [5]. *Lactobacillus plantarum* isolated from dairy product had removal aflatoxin B1 ability approximately 56% [6]. *Lactobacillus plantarum* MON03 (LP) isolated from Tunisian butter was found to perform significant binding ability to AFB1 and AFM1 in PBS (i.e. 82 % and 89 %, respectively) within 24 h of incubation and able to tolerate gastric acidity [25]. Among all the investigated *L. plantarum*, strains *L. plantarum* C88 shown the strongest AFB1 binding capacity in vitro, and was orally administered to mice with liver oxidative damage induced by AFB1. The binding rates of different strains (viable and heat-killed bacteria) with AFB1 ranged from 20.88% to 59.44% [26].

4. Conclusion
In this study, G7 lactic acid bacteria isolate has the best characteristic as probiotic with high viability in acid pH assay (92.61 %), bile salt 0.3% assay (97.71 %) and aflatoxin reduction percentage in viable cell condition (69.11 %). The identification procedure using molecular test with 16S rRNA revealed that the G7 isolate was identified as *Lactobacillus plantarum* (99.9% of homology).

5. References
[1] Agus A and Nuryono 2007 Mycotoxins in Indonesian feedstuffs *Biotracer General Meeting* November 14–16, 2007 (Athens, Greece)
[2] Yang E J and H C Chang 2010 Purification of a new antifungal compound produced by *Lactobacillus plantarum* AF1 isolated from kimchi *Int. Food Microbiol.* **139** 56-63
[3] Fuchs S, Sontag G, Stidl R, Ehrlich V, Kundi M and Knasmuller S 2008 Detoxification of patulin and ochratoxin A, two abundant mycotoxins, by lactic acid bacteria *Food Chem. Toxicol.* **46** 1398-1407
[4] Topcu A, Bulat T, Wishah R and Boyac I H 2010 Short Communication : Detoxification of aflatoxin B1 and patulin by *Enterococcus faecium* strains *Int. J. Food Microbiol.* **139** 202-5

[5] Gomah N H, Ragab W S and Bullerman L B 2010 Inhibition of fungal growth and aflatoxin B1 production by some *Lactobacillus* strains *Assiut J. Agric. Sci.* **40** 27-36
[6] Fazeli M R, Hajimohammadali M, Moshkani A, Samadi N, Jamalifar O, Khoshayand M R, Vaghari E and Pouragahi S 2009 Aflatoxin B1 binding capacity of autochthonous strains of lactic acid bacteria *J. Food Prot.* **72** 189-92
[7] Gratz S, Ta¨ubel M, Juvonen R O, Viluksela M, Turner P C, Mykka¨nen H and El-Nezami H 2006 *Lactobacillus rhamnosus* strain GG modulates intestinal absorption, fecal excretion, and toxicity of aflatoxin B1 in rats *Appl. Environ. Microbiol.* **72** 7398-400
[8] Zinedine A, Faid M and Benlemlih M 2005 In vitro reduction of aflatoxin B1 by strains of lactic acid bacteria isolated from Moroccan sourdough bread *Int. J. Agric. Biol.* **7** 67-70
[9] Hamidi A, Mirnejad R, Yahaghi E, Behnod V, Mirhosseini A, Amani S, Sattari S, Darian E K 2013 The aflatoxin B1 isolating potential of two lactic acid bacteria *Asian Pac. J. Trop. Biomed.* **3** 732-6
[10] Damayanti E, Indriati R Y, Sembiring L, Julendra H and Sakti A A 2014 Antifungal activities of lactic acid bacteria against *Aspergillus flavus*, *A. parasiticus* and *Penicillium citrinum* as mycotoxin producing fungi. *Proceed. the 6th AAAP Animal Science Congress vol II* November 10-14, 2014 (Yogyakarta: Universitas Gadjah Mada) 1742-45
[11] Damayanti E, Suryani A E, Sofyan A, Karimy M F and Julendra H 2015 Isolation of lactic acid bacteria for antifungal activity isolated from silage and animal digestives tract *J. Agritech* **35** 164-9
[12] Torshizi M A K, Rahimi S, Mojgani N, Esmaeilkhanian S and Grimes J L 2008 Screening of indigenous strains of lactic acid bacteria for development of a probiotic for poultry *Asian-Aus. J. Anim. Sci.* **21** 1495-500
[13] Hamidi A, Mirnejad R, Yahaghi E, Behnod V, Mirhosseini A, Amani S, Sattari S and Darian E K 2013 The aflatoxin B1 isolating potential of two lactic acid bacteria *Asian Pac. J. Trop. Biomed.* **3** 732-6
[14] Al-Jassim R A M, Scoot P T, Trebbin A L, Trott D and Pollit C C 2005 The genetic diversity of lactic acid producing bacteria in the equine gastrointestinal tract *FEMS Microbiol. Lett.* **248** 75–81
[15] Gong, J, Si W, Foster R J, Huang R, Yu H, Yin Y, Yang C and Han Y 2007 16S rRNA gene-based analysis of mucosaassociated bacterial community and phylogeny in the chicken gastrointestinal tracts: from crops to ceca *FEMS Microbiol. Ecol.* **59** 147-57
[16] Harisha S 2007 *Biotechnology procedures and experiments handbooks* (Massachusetts: Infinity Science Press)
[17] Cohort 2008 *CoSTAT Version 6.400* Cohort Software 798 (Moneterey)
[18] E. Damayanti, H. Julendra, A. Sofyan and S. N. Hayati 2014 Bile salt and acid tolerant of lactic acid bacteria isolated from proventriculus of broiler chicken *Media Peternakan* **3** 80-6
[19] Istiqomah 2015 *Isolation and characterisation of lactic acid bacteria as phytase producer from digestive tract of poultry and the phytase characterisation* (Indonesia: Universitas Gadjah Mada, Thesis)
[20] Pfeiler E A and Klaenhammer T R 2007 The genomic of lactic acid bacteria *TRENDS Microbiol.* **15** 546-53
[21] Damayanti, E, Herdian H, Angwar M, Febrisiantosa A and Istiqomah L 2012 Lactic acid bacterial screening from gastrointestinal digestive tract of native and broiler chicken for probiotic candidate purposes *J. Indonesian Trop. Anim. Agric.* **37** 168-75
[22] Kosin B and Rakshit S K 2006 Microbial and pocessing criteria for production of probiotics: a review *Food Technol. Biotechnol.* **44** 371-9
[23] Bueno D J, Casale C H, Pizzolitto R P, Salvano M A and Oliver G 2007 Physical adsorption of aflatoxin B1 by lactic acid bacteria and *Saccharomyces cerevisiae* : a theoretical model *J. Food Prot.* **70** 2148-54

[24] Haskard Haskard C A, El-Nezami H S, Kankaanpää P E, Salminen S and Ahokas J T 2001 Surface binding of aflatoxin b_1 by lactic acid bacteria *Appl. Environ. Microbiol.* **67** 3086-91

[25] Jebali R, Abbès S, Salah-Abbès J B, Younes R B, Haous Z and Oueslati R 2015 Ability of *Lactobacillus plantarum* MON03 to mitigate aflatoxins (B1 and M1) immunotoxicities in mice *J. Immunotoxicol.* **12** 290-9

[26] Huang L, Duan C, Zhao Y, Gao L, Niu C, Xu J and Li S 2017 Reduction of aflatoxin B1 toxicity by *Lactobacillus plantarum* C88: a potential probiotic strain isolated from Chinese traditional fermented food "Tofu" *PLoS ONE* **12** e0170109

IOP Conf. Series: Earth and Environmental Science **101** (2017) 012031 doi:10.1088/1755-1315/101/1/012031

Effect of urea treatment of cocoa pod on rumen fermentation characteristics *in vitro*

A S Anggraeni, H Herdian, A A Sakti, A Sofyan and M Ekaningrum

Research Unit for Natural Products Technology (BPTBA), Indonesian Institute of Sciences (LIPI), Jl. Jogja-Wonosari Km. 31.5, Gading, Playen, Gunungkidul, 55861 D.I. Yogyakarta, Indonesia

E-mail: ayus001@lipi.go.id / ayu.anggraeni07@gmail.com

Abstract. Indonesia is a third largest country in the world for cocoa production. A cocoa pod could be utilized as alternative feeds due to their sufficient quantity and availability throughout the year. On the other hand, low nutritional quality such as highly fibrous materials and low protein content usually characterized in agricultural and plantation by-products as it appears on cocoa pod. Ammoniation treatment using urea improve the nutritional quality of feedstuff. The objective of this study was to evaluate the effect of ammoniation treatments on a cocoa pod on *in vitro* feed fermentation and gas production on ruminal fluid. KA treatment gave highest gas production than other treatment. Total gas production during 48 hours of the cocoa pod was significantly affected by treatments ($P<0.05$). Total volatile acid (VFA), acetate (C2), propionate (C3), butyrate (C4) total VFA and A/P ratio indicated no significant difference among treatments. KA and KD treatments have tendency effect on IVDMD than control (KS) treatment. In this study ammoniation treatment using 5% of urea had a positive and effect on the best degradability and ruminal fermentation of cocoa pod.

1. Introduction

Increasing in feed cost will lead a decrease in performance of farm business especially in the small farm business. Grasses and legumes are the main sources of feeds to satisfy ruminant animals nutritional requirements, for maintenance, production or reproduction. In farm business feed cost needs 60-70% of total production cost [1], based on this condition; it is necessary to explore alternative feed to reduce feed cost and also to avoid competition with human needed. Some agricultural and plantation by-products have the potential as alternative feeds due to their sufficient quantity and availability throughout the year. On the other hand, highly fibrous materials and low protein content that indicated low nutritional quality usually characterized in agricultural and plantation by-products. This condition often leads the by-products to be treated, either physically, chemically and biologically before feeding to animals [2-4].

Indonesia a third largest country in the world for cocoa production after Ivory Coast and Ghana with total planting area in 2008 was 1,473,258 ha, total production of 792,791 tons [5]. In 2013 the area of cocoa was recorded at 1,745,789 ha, with cocoa seed production of 938.8 thousand tons. Comparison between cocoa pod: cocoa beans: fresh placentas are 74%: 24%: 2% of cocoa [6] respectively. Based on the ratio of cocoa beans with the cocoa pod, the potential of drying cocoa pod is 872,3 thousand tons/year [5]. When it used as an alternative feed fiber source to replaces forage (50% dry matter in rations), it can meet the need for 635,305 animal units (1 AU = 250 kg) or equivalent to 5,310,873 small ruminants [7]. However, the Indonesian people are still little bit utilized cocoa pod as an alternative feed. It is usually just thrown away, without any use and treatment. Cocoa pod can be used as animal feed cause it contains crude protein 6.8-11,71%, crude fiber 20,79%, fat 11,80%, BETN 34,90% NDS 55,30-73 , 90% and ADS 38.31-58.98% [6]. Behind the potential of the cocoa pod; it contained anti nutrients such as theobromine, lignin, and tannins. Theobromine is a harmless alkaloid but it can be destroyed by heating or drying, however continuous feeding of theobromine can reduce growth of animal [8]. The high content of lignin (between 12% to 19% dry matter higher than rice straw content) and silica, also causes low digestibility in animal [9-10]. Lignin is composed of a network of phenolic polymers that attach cellulose and hemicellulosic fibers, so the structure becomes very strong which it makes hard to digest [11]. While the presence of tannins performs complex bonds with proteins and carbohydrates that cause a reduction in rumen microbial activity in degrading protein and carbohydrates that decreasing the digestibility [6]. The cocoa pod can not be stored for long periods due to their high water content, which easily decomposed and moldy, resulting in the cocoa pod not palatable for livestock [7,12].

Therefore, to minimize the effects of antinutrients and maximize the use of cocoa pod and extend the shelf life it is necessary to do one of treatment with nitrogen sources, chemical and physical treatment [13-14]. Urea treatment of rice straw could increase its nutritive value [15-16]. Alkali source like urea has been reported to be effective in improving fibrous feed quality [17-18]. The urea also supplies N for rumen microbial growth besides improves fiber utilization. Volatile fatty acid (VFA), gas production and microbial biomass are the main product of feed organic matter fermentation in ruminant [19]. The objective of this research was to evaluate the effect of ammoniation treatments using urea on a cocoa pod on *in vitro* feed fermentation and gas production on ruminal fluid.

2. Materials and methods

2.1. Sample preparation

The cocoa pod was obtained from Nglanggeran village, Gunungkidul regency, Yogyakarta, Indonesia. The fresh cocoa pod was chopped manually around 1-2 cm thickness as homogenous as possible. Three kg of chopped cocoa pod was used for each treatment. All treatment kept for one month at room temperature. Each treatment was conducted in three replicates. The treated cocoa pod was analyzed for dry matter (DM) by following the procedure from [20].

2.2. Treatment and experimental design

This research was conducted based on Completely Randomized Design (CRD) with three treatments and three replications. Treatments described as follows:

KS: Cocoa pod without added urea.
KA: Cocoa pod ammoniation treatment used 5% urea in simple open clip packed.
KD: Cocoa pod ammoniation treatment used 5% urea in aerated treatment.

KA, treated cocoa pod with adding 5% urea were placed in an unused plastic banner that packed in simple open clip method. KD treated cocoa pod that puts in unused plastic banner without packed process only aerated a treatment. KS treated cocoa pod without adding urea placed in an unused plastic banner that packed in simple open clip method. All treatment incubated for 30 days at room temperature.

2.3. Fermentability and in vitro degradability assessment

The sample of cocoa pod each treatment and rumen liquid were prepared before *in vitro* assessment. Two ruminally fistulated Ongole crossbreed cattle used as rumen liquor donor, those adapted by feeding consisted of forage (*P*. hybrid) and concentrate (80:20 in dry matter basis). Rumen fluid was taken using aspirator, and immediately transported in pre-warmed vacuum flask (39 °C water temperature) and filtered.

In vitro gas production technique according to [21] used to evaluate *in vitro* fermentability. The exponential equation according to [22] was used to calculate gas production kinetics. The exponential function is P = a+b (1-e-ct) with describing P is cumulative total gas production, a is shared gas production from soluble fraction, b is the gas production from insoluble fraction, c is the rate of gas production, t is the time of incubation and e is Euler's constant (2.7183…). Fitting curve method using Neway Software [23] use to estimated value of *a*, *b* and *c*. 100 mL syringe glass (Fortuna model, Poulten and Graft Gmbh Germany) used for fermentation. Three syringes are containing rumen-buffer without sample (blank) used in the experiment. All of the syringes consisted of samples and blank were randomly incubated for 48 hours in an incubator at 39°C [24].

Gas production cumulative recorded at 0, 2, 4, 8, 12, 24, 36 and 48 hours incubation. Gas was released after 48 h incubation and the fluid contained in the syringe taken for analysis of VFA, and *in vitro* dry matter degradability. Substrate from each syringe measured according to [20] method to get dry matter (DM) value. Percentage of DM differences between initial and after incubation and corrected with blank were calculated as *in vitro* degradability as followed the formula as previously describe by [25]:

$$IVDMD = \frac{[DMf-(DMr-D\quad)]}{DMf} \tag{1}$$

Where: IVDMD *in vitro* dry matter degradability. DMf: dry matter of feed, DMb: dry matter of blank, DMr: dry matter of residue.

Volatile fatty acid (VFA) product from fermentation measured according to [26]. Meta-phosphoric acid added at the sample and stored at -20°C before analysis. Gas chromatography (Shimadzu type 8A) was used to analyzed VFA content using GP10% SP-1200/1% H_3PO_4 column with 80/100 Chromosorb WAW (Supelco, Bellefonte, PA).

2.4. *Data Analysis*

Variables measured were *in vitro* degradability (IVDMD), fermentability (gas production kinetics, *a*, *b* and *c*), and individual volatile fatty acids (VFA). Analysis of variance (ANOVA) was used to evaluated data, and post hoc test of Duncan's Multiple Range Test was used to analyzed differences among mean treatments that performed by the CoSTAT statistical software [27].

3. Result and discussion

Cocoa pod ammoniation treatment used 5% urea in simple open clip packed was generated the highest gas production compared to the others (figure 1).

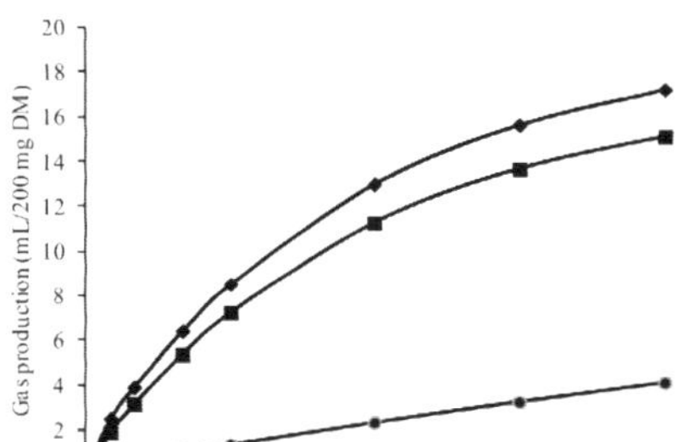

Figure 1. Cumulative gas production of cocoa pod ammoniation treatment used 5% urea in simple open clip packed (KA), cocoa pod ammoniation treatment used 5% urea in aerated treatment (KD), cocoa pod without added urea (KS) incubated during 48 hours.

Figure. 1 showed that KA treatment gives highest gas production than other treatment. Kinetic gas production parameters used as an indication of fermentability evaluation by *in vitro* gas production [26]. Based on a gas production kinetic curve, treatment KA and KD began to show the different than KS treatment up to 2 hours of incubation. But, after 12 hours of incubation gas production of the cocoa pod was significantly affected by treatment. The highest cumulative gas production found at KA treatment followed by KD and KS. It can say that ammoniation treatments using urea 5% (KA) produce the best cocoa pod degradability than other treatment. Urea treatment showed to be effective in improving the nutritive value of cocoa pod. The reason is that urea supplement may act as nonprotein nitrogen source that rumen microbes could utilize it and then passed as microbial protein for the host animal [4]. Five percent of urea using in this treatment is still safe and not produce toxic ammonia effect on microbes in the *in vitro* gas test. This result according to [28] that high level of urea used on *in vitro* gas production decrease in U3 treatments but increased from treatments U0 to U2. Added by [29] explained that urea would toxic to the rumen if present in large amount.

Table 1. Gas production kinetic, volatile fatty acid production and *in vitro* degradability of cocoa pod ammoniation treatments

Variable	Treatments		
	KA	KD	KS
Fermentability/ Gas production kinetic parameters			
a (mL)	0.95	0.52	0.29
b (mL)	18.50	16.70	14.72
c (mL/h)	0.04 ^{b}	0.04 ^{b}	0.01 ^{a}
a+b (mL)	19.45	17.22	15.01
Gas (48 h) (mL)	17.19^{b}	15.09^{b}	4.09^{a}
Volatile fatty acids			
Acetate (mM)	19.00	13.45	11.65
Propionate (mM)	5.76	4.22	3.28
Butyrate (mM)	3.13	2.07	1.64
Total VFA (mM)	27.89	19.74	16.57
A/P ratio	3.29	3.15	3.55
In vitro degradability			
IVDMD (%)	35.47^{b}	31.84^{ab}	24.24^{a}

KA = ammoniation treatment used 5% urea in simple open clip packed (KA) urea in simple open clip packed, KD = cocoa pod ammoniation treatment used 5% urea in aerated treatment, KS = cocoa pod without added urea (KS). Different superscript in the same row indicates significantly different ($p<0.05$)

Ruminal fermentability characteristic productions are shown in Table 1. Total gas production 48 hours of the cocoa pod was significantly affected by treatments ($P<0.05$) and gas production from soluble fraction (*a*) or total fraction (*a+b*) from cocoa pod had tendency affected by treatments. During the first 24 hours of incubation, fermentation is relatively intensive (lag phase), after which it reaches a stationary phase. The kinetics of gas production emerge to be determined by two distinct phases; the first one insoluble but potentially fermentable fraction and second suitable to the degradation of the soluble fraction of the tested mixtures. Positive results on gas production generated from the soluble fraction (*a*) indicated ruminal microbes already done adaptation time (*lag* phase) before degrading the soluble particle this results contrast with the previous study reported by [29], [26] that possibility there was negative "a" value arising from lag time of soluble fraction degradation activity by ruminal microbes and then to adhere to cellulosic fraction. Short-chain fatty acids (acetate, propionate, and butyrate), gases, and microbial cells were the product of carbohydrates fermentation when a feedstuff incubated with buffered rumen fluid *in vitro*. Gas production from protein fermentation is relatively small as compared to carbohydrate fermentation [25,30].

Productions of total volatile acid there is acetate (C2), propionate (C3), butyrate (C4) total VFA and A/P ratio showed no significant difference (P>0.05) among treatments (Table 1). Total VFA varied between 16-28 mM per mL of rumen fluid. VFA production was not sufficient condition for optimal rumen microbial protein synthesis because VFA range required for rumen microbial growth was 80-160 mM [31], added by [32] that total VFA production ranged from 106.67-165.81 mM. Other studies stated that the range of VFA in the rumen was 60 mM acetate, 20 mM propionate and ten mM butyrate [33]. Increased VFA concentration correlated with increasing microbial production [32]. An increase in the number of VFA shows quickly or not the feed is degraded by rumen microbes. The composition of VFA at rumen changes with the difference of physical form, the composition of feed, level and feeding frequency, and also feed processing. High VFA production is the energy sufficiency for ruminant. Production of VFA from KA 68.28% and KD 19.12 higher than KS. The low value of total VFA (Table 1) may be because sample measurement was done in another laboratory based on this condition less precise handling during sample preparation which caused volatile acid disappear before analysis. Although the low value of VFA, we assured all samples kept in similar condition.

A/P ratio seemed constant in all treatments with average value 3.33. In Table 1, cocoa pod treatment using urea (KA) and aerated (KD) resulted in the ratio of acetate to propionate (A/P) lower than the control (KS). A/P ratio of control was 3.52 while the A/P ratio KA and KD treatment were 3.29 and 3.15 respectively. It has been suggested that there is an increased proportion of propionate in the rumen compared to acetate. Urea treatment plays a role in glucose metabolism affected the production of propionate that was glucogenic. This result in line with [32] that supplementation of *H. tiliaceous* leaves led to a lower ratio of A/P. The proportion of C2: C3: C4 was constant about 69:21:10 for all treatments. Based on the ruminal fermentation stoichiometry, individual VFA proportion consisting of C2, C3, and C4 were 60-70%, 20-30% and 10-15% respectively [33]. Event among treatments indicated no significant difference, but in KA treatment show highest total VFA than other treatments. Urea treatment could improve the nutritive value of cocoa pod; it will reduce fiber content and increase the protein content of the cocoa pod. Production of VFA mostly depends on the fermentation of the carbohydrate feed, a fraction of the protein [35-36]. A lot of nutrient supply also provides more substrate for rumen microbes to grow and to yield more microbial protein from their biomass [37-38], which also observed in this study. When the microbes are abundant and active, an elevated VFA concentration expected as the end product of microbial fermentation in the rumen [39].

In vitro dry matter degradability (IVDMD) was present in table 1. KA and KD treatments have tendency effect on IVDMD than control (KS) treatment. Urea treatment emerged to be effective in improving the nutritive value of cocoa pod by reducing ADF content of cocoa pod [4]. Zain [40] observed that treatment using 6% urea on the cocoa pod and save for 21 d decreased ADF by 8.2%. Urea provided ammonia after being hydrolyzed to produce stepwise ammonium carbamate and ammonium carbonate [41]. Such treatment reduces physical strength of primarily fibrous feed, disrupts the silicified cuticular barrier and cleavages of some lignin-carbohydrate bonds [41-42]. As a consequence of this, the degradability of urea-treated cocoa pod increased as shown by higher IVDMD values compared to the control. Apart from modifying structural carbohydrate of the cocoa pod, urea addition could contribute to feeding the microbes an additional nitrogen source for microbial protein synthesis purposes and subsequent utilization of the protein by the host animal; it will give better digestibility than other treatment [4]. Gas production and degradation rate of particle affected the IVDMD. A significant correlation between gas production and degradability contributed by many factors such as the composition of nutrient in which associated with microbial ability to adhere and degrade the fraction of feed materials [43].

4. Conclusion

In vitro degradability indicated nutrient utilization in the rumen which attributed by gas production kinetics parameters and VFA. In this study ammoniation treatment using 5% of urea had a positive and effect on the digestibility and ruminal fermentation of cocoa pod.

Acknowledgment
The author would like to thank for DIPA BPTBA LIPI 2015 as a source of financial support for this activity and also for Mr. Nana Hidayat and Andri Suwanto for their help in this activities and data collection.

References
[1] Rahmanto B 2004 Analisis usaha peternakan sapi potong rakyat *ICAERD Working Paper No. 59 Pusat Penelitian dan Pengembangan Sosial Ekonomi Pertanian, Departemen Pertanian*, (Jakarta: Pusat Penelitian dan Pengembangan Sosial Ekonomi Pertanian, Departemen Pertanian)
[2] Sarnklong C, Coneja J W, Pellikaan W and Hendriks W H 2010 Utilization of rice straw and different treatments to improve its feed value for ruminants: a review *Asian Australas. J. Anim. Sci.* **23** 680-92
[3] Kim Y I, Lee Y H, Kim K H, Oh Y K, Moon Y H and Kwak W S 2012 Effects of supplementing microbially-fermented spent mushroom substrates on growth performance and carcass characteristics of Hanwoo steers (a field study) *Asian Australas. J. Anim. Sci.* **25** 1575-81
[4] Laconi E B and Jayanegara A 2015 Improving nutritional quality of cocoa pod (*Theobroma cacao*) through chemical and biological treatments for ruminant feeding: *In vitro and In vivo Evaluation Asian Australas. J. Anim. Sci.* **28** 343-50
[5] Ditjenbun, 2014 http://regionalinvestment.bkpm.go.id/newsipid/commodityarea.php?ic=3&ia=34
[6] Puastuti W and Susana I W R 2014 Potensi dan pemanfaatan kulit buah kakao sebagai pakan alternatif ternak ruminansia *Wartazoa* **24** 151-9
[7] Puastuti W and Yulistiani D 2011 Utilization of urea and fish meal in cocoa pod silage based rations to increase the growth of Etawah crossbred goats. In: Ali A, Kamil KA, Alimon AR, Orskov, Zentek J, Tanuwiria UH *Proc 2nd Int Semin AINI Feed Saf Heal Food* (Jatinangor) July 6-7, 2011 (Bandung Indonesia: Padjadjaran University) p 463-9
[8] Tarka S M, Zoumas B L and. Trout G A 2014 Examination of effect cocoa shell with theobromin in lamb *Nutrition Report International* 1998 In: Puastuti, W dan Susana IWR. Potensi dan pemanfaatan kulit buah kakao sebagai pakan alternatif ternak ruminansia. *Wartazoa.* **24** 151-9
[9] Sutikno A I 1997 Cocoa pod for feeding of ruminant livestock *Wartazoa.* **6** 38-43
[10] Oluokun J A 2005 Intake, digestion and nitrogen balance of diets blended with urea treated and untreated cowpea husk by growing rabbit *African J Biotechnol.* **4** 1203-8
[11] Aji D P, Sri U and Suparwi 2011 Fermentasi kulit buah kakao (*Theobroma cacao* L.) menggunakan *Aspergillus niger* pengaruhnya terhadap kadar VFA dan N-NH3 secara *in-vitro J Ilmu Peternakan* **1** 774-80
[12] Puastuti W, Yulistiani D, Supriyati. 2009. Ransum berbasis kulit buah kakao diperkaya mineral: tinjauan pada kecernaan dan fermentasi rumen *in vitro"*. Dalam: Sani Y, Natalia L, Brahmantiyo B, Puastuti W, Sartika T, Nurhayati, Anggraeni A, Matondang R, Martindah E, Estuningsih SE, penyunting. Teknologi peternakan dan veteriner mendukung industrialisasi sistem pertanian untuk meningkatkan ketahanan pangan dan kesejahteraan peternak. *Pros Semnas Teknologi Peternakan dan Veteriner.* Bogor 13-14 Agustus 2009. (Bogor Indonesia: Puslitbangnak) p 442-7
[13] McDonald P, Edwards R A and Greenhalgh J F D 2002 *Animal Nutrition. 6th Edition* (London : UK Longman)
[14] Nguyen V N, Wanapat M, Khejornsart P and Kongmun P 2012 Nutrient digestibility and ruminal

fermentation characteristic in swamp buffaloes fed on chemically treated rice straw and urea *Trop. Anim. Health Prod.* **44** 629-36

[15] Sundstøl F, Coxworth E and Mowat D N 1978 Improving the nutritive value of wheat straw and other low quality roughages by treatment with ammonia *World Anim. Rev.* **26** 13-21

[16] Abate D and Melaku S 2009 Effect of supplementing urea-treated barley straw with lucerne or vetch hays on feed intake, digestibility and growth of Arsi Bale sheep *Trop. Anim. Health Prod.* **41** 579-86

[17] Schiere J B, Ibrahim M N M 1989 Feeding of urea-ammonia treated rice straw*: A compilation of miscellaneous reports produced by the Straw Utilization Project* (Sri Lanka: Pudoc, Wageningen)

[18] Chenost M and Kayouli C 1997 Roughage utilization in warm climates *FAO animal production and health* (Rome)

[19] Alexander G. Singh B, Sahoo A and Bhat T K 2008 *In vitro* screening of plant extracts to enhance the efficiency of utilization of energy and nitrogen in ruminant diets. *Anim. Feed Sci. Technol.* **145** 229-44

[20] AOAC Association of Official Analysis Chemist 2012 *Official methods of analysis. AOAC International 19th ed.* (Washington DC USA: Association of Official Analysis Chemist)

[21] Menke K H and Steingass H 1988 Estimation of the energetic feed value obtained from chemical analysis and *in vitro* gas production using rumen fluid. *Anim Res Dev.* **28** 7-55

[22] Ørskov E.R and McDonald I. 1979 The estimation of protein degradability in the rumen from incubation measurements weighted according to rate of passage. *J Agric Sci.* **92** 499-503

[23] Chen X B 1997 Neway-excel microsoft office: a utility for processing data of feed degradability and *in vitro* gas production (Aberdeen UK: Rowett Research Institute)

[24] Sofyan A, Sakti A A, Herdian H, Khairulli G, Suryani A E, Karti P D M H and Jayanegara A 2017 *In vitro* gas production kinetics and digestibility of king grass (*Pennisetum* hybrid) added by organic mineral and natural crude tannin *J Appl Anim Res* **45** 122-5

[25] Sofyan A, A A Sakti, M F Karimy, H Julendra, L Istiqomah, H Herdian, E Damayanti and A E Suryani 2015 Effectivity of probiotic, micromineral enriched yeast and their combination with *Azadirachta indica* leaves containing tannin on fermentability and digestibility of *Pennisetum* hybrid *JITV.* **20** 95-104

[26] Sun P, Wang J Q, Deng L F 2013 Effects of *Bacillus subtilis* natto on milk production, rumen fermentation and ruminal microbiome of dairy cows *Animal.* **7** 216-22

[27] Cohort 2008 *CoSTAT Version 6.400* Cohort Software 798 (Moneterey)

[28] Despal and H Abel 2017 Nutritional Properties of Cocoa Pods as Ruminant Feedstuff http://repository.ipb.ac.id/handle/123456789/53590

[29] Smith R H 1989 Nitrogen metabolism in the ruminant stomach In: Bock H D, Eggum B O, Low A G, Simon O, Zebrowska T, Low A G (Eds.): *Protein Metabolism in Farm Animals: Evaluation, Digestion, Absorption and Metabolism* (Berlin: VEB Deutscher Landwirtschaftsverlag) pp 165 – 203

[30] Arhab R, Laadjimi K, Driss D, Djabri B and Bousseboua H 2010 Evaluation of feed mixture interactions by using in vitro gas production method *Livest Res Rural Develop.* **22** http://www.lrrd.org/lrrd22 /11/arha22217.htm

[31] Van Soest P J 1994 *Nutritional Ecology of The Ruminant 2nd ed.* (Ithaca and London: Comstock Publishing Associates a Division of Cornell University Press)

[32] Istiqomah L, H Herdian, A Febrisantosa and D Putra 2011 Waru leaf (*Hibiscus tiliaceus*) as saponin source on *in vitro* ruminal fermentation characteristic *J.Indonesian Trop.Anim.Agric.* **36** 43-49

[33] Madigan M T, Martinko J M and Brock J P 2003 *Biology of Microorganisms* (Southern Illinois University Carbondale. 10th ed: Pearson Education, Inc)

[34] Wolin M J, Miller T L, Stewart C S 1997 Microbe-microbe interaction In: Hobson PN, Stewart CS, editors. *The Rumen Microbial Ecosystem. 2nd ed.* (London: Chapman & Hall UK) pp 467-

91
[35] Van Houtert M F J 1993 The production and metabolism of volatile fatty acids by ruminants feed roughages: a review. *Anim. Feed Sci. Technol.* **43** 189-225
[36] Herdian H, Istiqomah L, Febrisiantosa A dan D Setiabudi 2011 Pengaruh penambahan daun *Morinda citrifolia* sebagai sumber saponin terhadap karakteristik fermentasi, defaunasi protozoa, produksi gas dan metana cairan rumen secara *in vitro. JITV.* **16** 99-104
[37] Clark J H, Klusmeyer T H and Cameron M R 1992 Microbial protein synthesis and flows of nitrogen fractions to the duodenum of dairy cows *J. Dairy Sci.* **75** 2304-23
[38] Bach A, Calsamiglia S and Stern M D 2005 Nitrogen metabolism in the rumen *J. Dairy Sci.* **88** 9-21
[39] Kamra D N 2005 Rumen microbial ecosystem *Curr. Sci.* **89** 124-35
[40] Zain M 2009 Substitution of native grass with ammoniated cocoa pod in sheep diet *Media Peternakan.* **32** 47-52
[41] Schiere J B and Nell A J 1993 Feeding of urea treated straw in the tropics I a review of its technical principles and economics *Anim. Feed Sci. Technol.* **43** 135-47
[42] Van Soest P J 2006 Rice straw, the role of silica and treatments to improve quality *Anim. Feed Sci. Technol.* **130** 137-71
[43] Sandoval-Castro C A, Capetillo-Leal C, Cetina-Gongora R, Ramirez-Aviles L 2002 A mixture simplex design to study associative effects with an *in vitro* gas production technique. *Anim Feed Sci Technol.* **101** 191-200

IOP Conf. Series: Earth and Environmental Science 101 (2017) 012032 doi:10.1088/1755-1315/101/1/

Effect of feed supplement containing earthworm meal (*Lumbricus rubellus*) on production performance of quail (*Coturnix coturnix japonica*)

L Istiqomah, A A Sakti, A E Suryani, M F Karimy, A S Anggraeni, H Herdian

Research Unit for Natural Products Technology, Indonesian Institute of Sciences, (BPTBA-LIPI). Jl. Jogja-Wonosari Km. 31.5, Gading, Playen, Gunungkidul, 55861 D. I. Yogyakarta, Indonesia

Email: lusty.istiqomah@gmail.com

Abstract. The objective of this study was to evaluate the effect of feed supplement (FS) contained earthworm meal (EWM) on production performance of laying quails. Twenty weeks-old of 360 *Coturnix coturnix japonica* quails were used in a Completely Randomized Design (CRD) with three dietary treatments A= CD (control without FS), B= CD + 0.250 % of FS, and C= CD + 0.375 % of FS during 6 weeks of experimental period. Each treatment in 4 equal replicates in which 30 quails were randomly allocated into 12 units of cages. Variable measured were feed intake, feed conversion ratio, feed efficiency, mortality rate, hen day production, egg weight, and egg uniformity. Data were statistically analyzed by One Way ANOVA and the differences among mean treatments are analysed using Duncan's Multiple Range Test (DMRT). The results showed that administration of 0.375% FS based on earthworm meal, fermented rice bran, and skim milk impaired the feed conversion ratio and increased the feed efficiency. The experimental treatments did not effect on feed intake, mortality, hen day production, egg weight, and egg uniformity of quail. It is concluded that administration of feed supplement improved the growth performance of quail.

1. Introduction

In recent years, breeding of quail has taken an important place in alternative poultry production. Many factors such as hatchability and fertility rate, male/female ratio, genetics, age, egg quality, nutritional status and live weight of breeders affect quail's performance [1]. Consumers demand on quail meat and eggs (*Coturnix coturnix japonica*) increase annually hence the quail production establish significantly in the last decade. One way of increasing protein supply is to diversify. The Japanese quails have the potential to serve as an excellent and cheap source of animal protein for consumers [2]. The optimum performance of livestock depends largely on the quality and quantity of their dietary nutrients. To obtain high growth and reproductive rates of quail, the given feed should fit the nutritional needs. In the quail starter phase requires a higher protein content of 22-24% in the feed formula, whereas in the grower phase protein needs to be 20-22% [3].

Feed additives commonly added into feed formulation in order to improve the quail performance. Feed additives are non-nutritive substances used in poultry feed including antibiotics, enzymes, antioxidants, pellet-binders, antifungal, colored pigments and flavoring agents [4]. For many years feed additives have been widely used to increase animals' performance and recently it is used in poultry industry to improve growth, feed efficiency and layers performance [5]. Many studies have reported that earthworm meal can be used as one alternative source of animal protein in feed [6]. Earthworm meal contained high protein content (63.06%) as nutrient source for animal [7]. Essential amino acid of earthworm meal was dominated by isoleucine (1.98% of dry matter basis) [7]. Earthworm *L. rubellus* contained 'lumbricin I' which had antibacterial activity, included in peptide group which contained 62 amino acids [8]. Amino acid utilized as proteins are primary constituents of structural and protective tissues, including skin, feathers, bone, ligaments, as well as muscles and organs. Amino acids are used as poultry feed supplement for optimum performance and also used in livestock health care [9].

Sofyan et al. [10] reported that supplementation of feed supplement containing earthworm meal in broiler feed increased the final body weight, decreased feed intake, and improved feed conversion. Prayogi [11] stated that the addition of earthworm meals improved weight gain and lower feed conversion of quail. Rezaeipour et al. [12] reported that the addition of 10% earthworm meal *Eisenia foetida* increased weight gain, severe chest muscles, and protein intake of broiler chickens.

This study aimed to determine the effect of feed supplement contains earthworm meal on production performance of layer quail (*Coturnix coturnix japonica*).

2. Materials and methods

2.1. Preparation of feed supplement

Preparation of EWM refers to [13]. Earthworms are separated from the media and removed the dirt by washing. Then worms soaked in cold water 4 °C for 24 h. Formic acid 80% is added as much as 3% by weight of the worms. The worm is milled using a blender to become a paste and the paste of the worm is dried in an oven at 50 °C for 12 h and sieved to obtain a homogeneous particle size of ±40 mesh.

Fermented rice bran (FRB) is prepared according to [14]. *Rhizopus* sp. is isolated from tempeh and cultured in Potato Dextrose Agar (PDA) medium for 3-5 days of incubation. This isolate is propagated in PDB medium and incubated 5-7 days to obtain liquid inoculum. Rice bran is mixed with 5% of chitin-chitosan production waste (w/v) and 35% of warm water (v/v). The inoculums of *Rhizophus* sp. is added to the mixture (3.5% v/w) after the previous mixture temperature was 30-40 °C. The mixture is incubated for 7 days at room temperature (27 °C and 75% of RH). After incubation, fermented rice bran dried under the sun heat (2 days), prior to dry in oven (temperature 70 °C) for 2 hours until the moisture content was less than 10%. The fermented rice bran is ground and sieved to obtain a homogeneous particle size of ±60 mesh.

Skim milk is used in this study and categorized as food grade. After prepared the EMW with size of ±40 mesh, FRB with particle size of ±60 mesh, and skim milk then materials are mixed homogeneously with ratio 25:45:30 (w/w/w) and called as feed supplement. Nutrient content of EWM, FRB, and skim milk is performed on Table 1.

2.2. Experimental procedure

A total number of 360 quails *Coturnix coturnix japonica* aged 20 weeks were tested in a completely randomized design (CRD) with 3 treatments A= CD (control without FS), B= CD + 0.250 % of FS, C= CD + 0.375 % of FS. Each treatment in a 4 equal replicates in which 30 quails were randomly allocated into 12 units of multilevel cages.

Quails were kept under similar condition of management at temperature 27-28 °C with adequate water and feeding spaces provided throughout the experimental period lasting for 6 weeks. Light is provided for 24 h to avoid pilling and death. Through quail is known to be resistant to disease, anti-stress and antibiotics were administered through water at various times to check against possible disease outbreak. Good hygiene and biosecurity were ensured throughout the experimental period.

Feed supplement is given daily and mixed into the diet (0, 2.5, and 3.75 g of FS/kg of total diet). This study used commercial diet (CD) as diet contained 20% crude protein (CP). Nutrient content of diet based on laboratory analyses are presented in Table 2.

2.3. Data collection

Laying percentage (HDP) is calculated as number of eggs produced per hen divided by the number of days of the experimental period. Egg mass production per hen per day is calculated as laying percentage multiplied by average egg weight of the hen. Average egg weight is determined per hen. Feed intake is determined per replicate. Feed conversion ratio (FCR) is calculated as grams of feed consumed per gram of egg mass [5].

2.4. Data analysis

The commercial diet (CD) were analyzed for proximate composition according to the Association of Official Analytical Chemists [16]. Statistical analyses were performed with Costat [17]. The average data per replicate (30 hens) were treated as the experimental unit. Performance data were subjected to analysis of variance (ANOVA) and the differences among mean treatments were analyzed for significance using Duncan's Multiple Range Test [18].

3. Result and discussion

3.1. Quail growth performance

Earthworms are known as a protein rich source of cattle feed. The dry matter of earthworms has been found to be around 20 to 25% of the fresh weight. This contains around 60% of protein, 7-10% of fat, 8-10% of ash, 0.55% of Ca and 1% of phosphorus [19]. Nutrient composition of earthworm meal (EWM), fermented rice bran (FRB), skim milk, and feed supplement (FS) are presented in Table 1. Nutrient content in commercial diet (CD) with FS supplementation level are performed in Table 2.

Table 1. Nutrient composition of materials (DM basis)

	Materials			Feed supplement
Nutrient content	EWM	FRB	Skim milk	(FS)
Dry matter (%)	11.43 ± 0.04	92.46 ± 0.07	89.41 ± 0.14	8.01 ± 0.09
Ash (%)	5.15 ± 0.06	8.40 ± 0.01	8.21 ± 0.10	6.21 ± 0.11
Crude protein (%)	55.87 ± 0.06	11.69 ± 0.36	22.33 ± 1.40	24.13 ± 0.06
Fat (%)	16.39 ± 0.18	12.51 ± 0.28	0.19 ± 0.05	8.99 ± 0.22
Crude fiber (%)	0.17 ± 0.01	13.30 ± 0.89	NA[a]	NA[a]

[a] NA (not avalailable)

Table 2. Nutrient content (DM basis) of the diet used in the experiment

	Treatment		
Nutrient content	A	B	C
Moisture (%)	12.31 ± 0.23^{a}	11.82 ± 0.30^{ab}	11.49 ± 0.60^{b}
Crude protein (%)	20.54 ± 0.42	20.60 ± 0.72	19.76 ± 0.13
Crude fat (%)	3.72 ± 0.20^{ab}	3.94 ± 0.56^{a}	3.05 ± 0.19^{b}
Ash (%)	11.13 ± 0.28^{b}	12.58 ± 0.68^{a}	11.89 ± 0.17^{ab}

Note: Average in the same row with different superscript shown significantly differ (P<0.05)
A= CD (control without FS)
B= CD + 0.250 % of FS
C= CD + 0.375 % of FS

Table 2 showed that protein content in diet (20%) was in accordance with the requirements of feed quality for quail in layer phase around 20% [20]. Quails convert dietary protein content in diet into body protein (lean meat) and egg production, insufficient of protein concentration of feeds may diminish the economic and productive performance of quail. The utilization of dietary crude protein by poultry are affected by many factors such as sex, age, disease, nutritional status, protein balance, presence of anti-nutritional factors, feed waste and poultry genotype [21,22].

The effect of FS containing of EWM in dietary on quail performance are presented in Table 3. Administration of FS did not significantly affect (P>0.05) feed intake (FI). It is estimated that administration of EWM did not affect the feed palatability, hence it did not affect the feed consumption. Feed consumption is affected by environmental temperature, also the quality and quantity of ration [23]. Contrary, Sofyan et al. [10] reported that administration of feed supplement containing EWM decreased feed intake of broilers. Bahadori et al. [24] stated that 1-3% of earthworm (*E. fetida*) meal in diet decreased feed intake of broiler chicken compared to control diet without EWM. Japanese quail are less efficient to convert feed to body tissue than broilers, but quail can utilize low energy diets more efficiently than fowl [25]. Shoukry et al. [26] stated that Japanese quail consumed higher metabolizable energy per Kg body weight than chicken. This higher energy cost for egg production in Japanese quail is due to high losses of metabolizable energy from different way such as the energy costs of thermoregulation where, the mature Japanese quail expended about 6-10% of its metabolic rate to adjust a new level of metabolism under temperature fluctuation condition [27]. Another way is the calorific values of fat and carbohydrates in quail lays are highly energy egg compared to chicken [26]. Morever, metabolic body size of Japanese quails are larger than the chicken, thence the metabolizable energy is high losing. This condition indicates that Japanese quail are different in the energy efficiency compared to chicken.

Table 3. Production performance of quail

	Treatment		
Observed parameter	A	B	C
Feed intake (g/hen/d)	23.26 ± 0.66	22.86 ± 0.80	22.85 ± 0.91
Feed conversion ratio	2.97 ± 0.13^{a}	2.82 ± 0.08^{ab}	2.77 ± 0.14^{b}
Feed efficiency (%)	33.77 ± 1.47^{a}	35.52 ± 0.99^{ab}	36.23 ± 1.82^{b}
Mortality (%)	3.33 ± 2.72	1.67 ± 3.33	3.33 ± 3.85

Note: Average in the same row with different superscript shown significantly differ (P<0.05)
A= CD (control without FS)
B= CD + 0.250 % of FS
C= CD + 0.375 % of FS

The lowest FCR observed in this study, which is defined as the ratio between feed intake and egg production, occurred with 0.375% FS administration in the diet. Feed conversion cn be used as surrogate marker of the egg production coefficient, whereby a smaller value indicates a more efficient use of feed to produce an egg. Supplementation of EWM in diets could increase body weight and impaired feed conversion [11]. Istiqomah et al. [28] reported that administration of bio-supplement containing earthworm extract decreased feed conversion (FCR) of broiler. Administration of earthworm *E. foetida* meal in broiler feed decreased the feed conversion [12,24] and net protein ratio (NPR) [12]. In this study the FCR value was considerably similar with previous study 2.91 [2].

Administration of 0.375% FS significantly (P>0.05) increased the feed efficiency (P>0.05). Rezaeipour et al. [12] reported that protein efficiency percentage increased with increasing level of earthworm meal from 5% to 10% numerically but did not significantly different. Marks [29] suggested that the variation in feed efficiency is primarily the result of differences in metabolic efficiency and is approximately 40% genetically supplement. Murakami et al. [30] that dietary protein had no significant influence on feed efficiency when laying quails were fed diets with different protein levels.

Administration of feed supplement did not significantly (P>0.05) affect mortality of quail. Mortality rate were 1.67 and 3.33% for 6 weeks of experimental period. Health status of the hens was good and mortality during the experiment was acceptable (30 birds, 1.67 and 3.33%). Mortality is caused by pullorum disease (infection of *Salmonella pullorum*) with clinical signs such as white color of excreta, lower feed intake, shortness of breath, feathers puckered, and wings move weak. The overall mean livability of female layer Japanese quails were 97.12±0.22% (mortality rate 2.88%) [31].

3.2. *Performance of egg production*

Performance production of quails are shown in Table 4. There were no difference (P>0.05) on hen day production (HDP) among treatments. In this study, the protein content in diet administered by FS (Table 1) did not significantly different (P>0.05) with control diet, therefore the administration of FS did not affect the formation of egg and egg production during the experimental period. This result was probably due to adequacy of nutrient content between treatments that lead to a healthy quail, so it did not affect the process of egg and egg production is running normally. Karimy et al. [32] also reported that addition of earthworm meal (*L. rubellus*) did not affect egg production of quail.

Tuleun et al. [1] reported that hen day egg production (HDEP) on 20% CP diet was 81.67% for thirteen weeks study period. In this study, HDP ranging from 78.17 to 81.50% was considerably good. According to Mufti [33] eggs production are affected by lighting and the protein content in diet. Grindstaff et al. [34] also reported that the process of egg formation is influenced by the nutrients in the diets consumed by poultry.

Table 4. Effect of feed supplement containing EWM on hen day production, egg weight, egg mass, and egg uniformity during 6 weeks of experimental period

	Treatment		
Observed parameter	A	B	C
HDP (%)	78.17 ± 6.63	79.37 ± 5.94	81.50 ± 4.99
Egg weight (g/egg)	10.06 ± 0.19	10.24 ± 0.20	10.17 ± 0.31
Egg mass (g/hen/d)	7.86 ± 0.54	8.13 ± 0.27	8.29 ± 0.51
Egg uniformity (%)	65.45 ± 13.52	70.00 ± 4.33	70.45 ± 7.62

Note: Average in the same row with different superscript shown significantly differ (P<0.05)
A= CD (control without FS)
B= CD + 0.250 % of FS
C= CD + 0.375 % of FS

Based on Table 4, administration of FS did not significantly (P>0.05) affect egg weight of quail. The results obtained were still at the normal range of 6-16 g [35]. Bertechini [36] stated that normal range of quail's egg weight were 10-12 g. Tuleun et al. [2] reported that quail egg weight on 20% CP diet was 9.75 g and stated that egg weight depended greatly on daily protein intake. Quail egg weight, shape and colour can vary greatly among different females in a population, but are quite specific and consistent for one female [37]. The relationship between Japanese quail egg colour, shell quality and internal features is confirmed in some experiments inter alia carried out by Taha [38].

Administration of FS did not significantly (P>0.05) affect egg uniformity of quail. In this study, quail eggs showed lower uniformity, with high egg weight variation compared to Nepomuceno et al. [39]. According to Murakami et al. [30], the variation of egg quail weight is between 10 to 11 g, considering the deviation from 7 to 14 g.

Hen day production (HDP) among treatments was observed 6 weeks (figure 1).

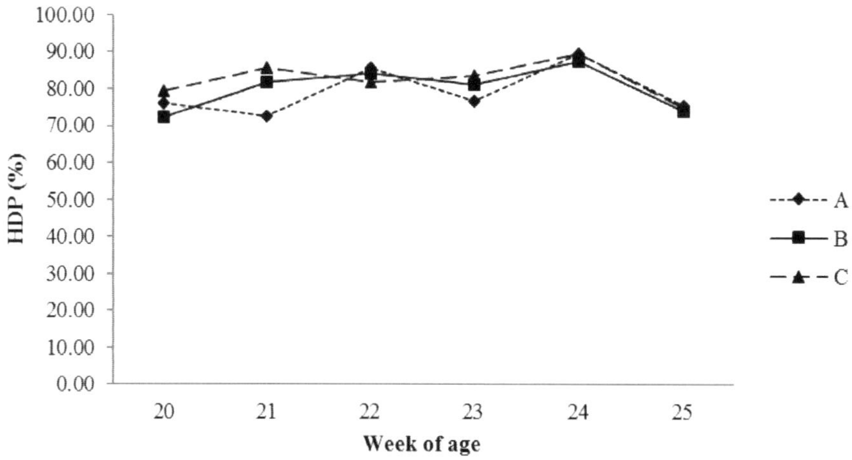

Figure 1. Weekly hen day production in A treatment (CD without FS), B treatment (CD + 0.250 % of FS), and C treatment (CD + 0.375 % of FS).

The result showed that peak production during 6 weeks of experimental period is observed in week 24 of age then decreased in week 25 of age. Quail hens attain peak production between 12-15 weeks and it is expected that as the birds increased in age, the number of small sized eggs would decrease consistently, while the medium and large sized eggs would increase consistently as well [40]. Japanese quails in the age group of 11 to 14 weeks resulted HDP of 76.93 ± 0.46% [31]. This optimum egg production could be achieved when body metabolism run well. This optimum metabolism are affected by environment factor (temperature, relative humidity, and lightning) and nutrition. At laying period, 22-24 °C is the required temperatures and the relative humidity should not exceed 60% during these production phase [36]. The differences observed for the peak of lay could be as a result of variations in the environments in which the experiments were conducted.

4. Conclusion

The result showed that administration of 0.375% of feed supplement (FS) based on earthworm meal, fermented rice bran, and skim milk impaired the feed conversion and increased the feed efficiency. However it did not affect hen day production, egg weight, and egg uniformity of quail. It is concluded that administration of feed supplement containing earthworm meal have potential to improve the growth performance of quail.

Acknowledgement

Authors would like to thanks to Manager of CV Holstein Indonesia, Boyolali, Central Java Mr. Itmamul Khuluq, researcher, and technician of BPTBA LIPI for supporting this experiment.

References

[1] Seker I 2003 The effect of some factors on the fertility rate and incubation results of hatching eggs in quails *YYÜ Vet Fak Derg* **14** 42-6
[2] Tuleun C D, Adenkola, A Y and Yenle F G 2013 Performance and erythrocyte osmotic membrane stability of laying Japanese quails (*Coturnix coturnix japanica*) fed varying dietary protein levels in a hot-humid tropics *Agric. Biol. J. N. Am.* **4** 6-13
[3] Panjaitan I, Sofiana A and Priambudiman Y 2012 Supplementation of cricket meal as a source of protein influence on performance of quail (*Coturnix coturnix japonica*) *J. Ilmiah Ilmu-Ilmu Peternakan* **XV** 8-14

[4] Bali S K D, Anupam K, Dipanwita P, Biswas S and Bhattacharyya D 2011 A comparative study on the antioxidant and antimicrobial properties of garlic and coriander on chicken sausage *International Journal of Meat Science* **1** 108-16
[5] Khan H S, Sardar R and Anjum M A 2007 Effects of dietary garlic on performance and serum and egg yolk cholesterol concentration in laying hens *Asian Journal of Poultry Science* **1** 22-7
[6] Sogbesan A O and Madu C T 2008 Evaluation of earthworm (*Hyperiodillus euryaulos*, clausen, 1914; oligocheata: eudrilidae) meal as protein feedtuffs in diet for *Heterobranchus longifilis* valenciennes, 1840 (teleostei, clariidae) fingerlings under laboratory condition. *Research Journal of Environmental Sciences* **2** 23-31
[7] Istiqomah L, Sofyan A, Damayanti E and Julendra H 2009 Amino acid profile of earthworm and earthworm meal (*Lumbricus rubellus*) for animal feedstuff *J. Indonesia Trop. Anim. Agric.* **34** 253-7
[8] Salzet M A, Tasiemski E and Cooper 2006 Innate immunity in lophotrochozoans: The annelids *Current Pharmaceutical Design* **12** 1-8
[9] National Organic Standards Board [NOSB] 2001 Methionine. Technical Advisory Panel Review for the USDA National Organic Program (United States: Organic Materials Review Institute) p 1- 20
[10] Sofyan A, Julendra H, Damayanti E, Sutrisno B and Wibowo M H 2010 Performance and Histopathology of broiler bhickens infected with *Salmonella pullorum* after administered by feed additives containing earthworm meal (*Lumbricus rubellus*) *Media Peternakan* **33** 31-5
[11] Prayogi H S 2011 The effect of earthworm meal supplementation in the diet on quail's growth performance in attempt to replace the usage of fish meal *Int. J. Poultry Sci.* **10** 804-6
[12] Rezaeipour V, Nejad O A and Miri H Y 2014 Growth performance, blood metabolites and jejunum jorphology of broiler chickens fed diets containing earthworm (*Eisenia foetida*) meal as a source of protein *Int. J. Adv. Biol. Biom. Res.* **2** 2483-94
[13] Istiqomah L, Herdian H, Damayanti E, Hayati S N and Julendra H 2012 Inhibitory of encapsulated earthworm extract (*Lumbricus rubellus*) on pathogenic bacteria in vitro *Media Peternakan* **35** 1-8
[14] Istiqomah L, Febrisiantosa A, Sofyan A, and Damayanti E 2010 Implementation of fermented rice bran as a flavor enhancer additive and its effect on feed utilization and beef cattle performance *Proceed. Int. Seminar on Tropical Animal Production Community Empowerment and Tropical Animal Industry* vol 5 (Yogyakarta) p 109-14
[15] Dauda G, Momoh O M, Dim N I and Ogah D M 2014 Growth, production and reproductive performance of Japanese quails (*Coturnix coturnix japonica)* in humid environment *Egypt. Poult. Sci.* **34** 381-95
[16] AOAC 1990 *Official Methods of Analysis* 15th ed. (Washington: Assoc. Off. Anal. Chem)
[17] Cohort 2008 *CoSTAT Version 6.400* Cohort Software 798 (Monterey)
[18] Gomez K A and Gomez A A 2007 *Statistical Procedures for Agricultural Research* 2nd ed (Jakarta: UI Press)
[19] Sharma S, Pradhan K, Satya S and Vasudevan P 2005 Potentiality of earthworms for waste management and in other uses – a review *J. American Sci.* **1** 4-16
[20] National Research Council (NRC) 1994 *Nutrient Requirements of Poultry* 9th ed. (Washington DC: The National Academies Press)
[21] Samadi F and Liebert F 2006 Estimation of nitrogen maintenance requirements and potential for nitrogen deposition in fast growing chickens depending on age and sex *Poultry Sci.* **85** 1421-9
[22] Jordão Filho J, Silvia J H V and Silvia C T 2011 Energy requirement for maintenance and gain for two genotypes of quails housed in different breeding rearing systems *R. Bras.*

Zootec. **40** 2415-22

[23] Leeson S and Summers J D 2005 *Comercial Poultry Nutrition* 3rd ed. (Nottingham: University Press)

[24] Bahadori Z, Esmaylzadeh L and Torshizi M A K 2015 The effect of earthworm (*Eisenia fetida*) and vermihumus meal in diet on broilers chicken efficiency and carcass components *Biological Forum – An International Journal* **7** 998-1005

[25] Shim K F and Vohra P 1984 A Review of the nutrition of Japanese quail *World's Poult. Sci.* 261-74

[26] Shoukry H M S, Mohammed F A and Sabri H 1993 Energy cost of egg production in Japanese quail (*Coturnix cotunix japonica*) *Al-Azhar J. Agric Res.* **18** 1-14

[27] Prinzinger R 1982 The energy costs of temperature regulation in hens: The influence of quick sinusoidal temperature fluctuations on the gaseous metabolism of the Japanese quail (*Coturnix coturnix japonica*) *Comp. Biochem. Physiol.* **71A** 469-72

[28] Istiqomah L, Hayati S N, Damayanti E, Julendra H, Sakti A A and Untari T 2013 Performance and meat quality of broilers infected with *Escherichia coli* and administered with bio Additive, probiotic, and antibiotic *Media Peternakan* **36** 14-20

[29] Marks H L 1991 Feed efficiency changes accompanying selection for body weight in chickens and quail *World'S Poultry Sci. J.* **47** 197-212

[30] Murakami A E, Furian A E, Tateishi, A, Kira K and Ribeiro R P 1994 Niveles de merionia para codornas Japanese (*Coturnix coturnix japonica*) *Em Postura.* 31 Reunla Annual da Sociedade Barsilera de Zooteenia Moringa PR (Brazil) 64

[31] Sangilimadan K, Rajini R A, Prabakaran R, Balakrishnan V and Murugan M 2012 Effect of dietary protein on layer Japanese quails (*Coturnix coturnix japonica*) in tropics *Tamilnadu J. Vet. Anim. Sci.* **8** 271-8

[32] Karimy M F, Istiqomah L, Sakti A A, Suryani A E and Khuluq I 2015 Egg production performance and blood profile quail (*Coturnix Coturnix japonica*) administered with feed containing earthworm meal (*Lumbricus rubellus*). *Proceed. National Seminar on Sustainability in Animal Husbandry* (Indonesia)

[33] Mufti M 1997 *Photoregulation Impact and Level of Dietary Protein during the Growth Period of the Performance of Quail Laying* (Bogor: Institut Pertanian Bogor, Thesis)

[34] Grindstaff J L, Demas G E and Ketterson E D 2005 Diet quality affects egg size and number but does not reduce maternal antibody transmission in Japanese quail *Coturnix japonica J. Anim. Ecol.* **74** 1051-58

[35] Sezer M 2007 Heritability of exterior eggs quality traits in Japanese quail *J. Appl. Biol. Sci.* **1** 37-40

[36] Bertechini A G 2012 The quail production. *Proceed. World's Poultry Congress* XXIV (Brazil) p 1

[37] Tolik D, Polawska E, Charutta A, Nowaczewski S and Cooper R 2014 Characteristics of egg parts, chemical composition and cutritive value of Japanese quail eggs – a review *Folia Biologica (Kraków)* **62** 287-92

[38] Taha A E 2011 Analyzing of quail eggs hatchability, quality, embryonic mortality and malposition in relation to their shell colors *Egypt. Poult. Sci.* **3** 807-15

[39] Nepomuceno R C, Watanabe P H, Freitas E R, Cruz C E B, Peixoto M S M and de Sousa M L 2014 Quality of quail eggs at different times of storage *Cienc. anim. bras.* **15** 409-13

[40] Olusuji A S 2011 *Effect of Age and Egg Size of Japanese Quail Hens on Hatchability and Post Hatch Performance of Quail Chicks Fed Different Dietary Protein Levels* (Nigeria: Ahmadu Bello University, Thesis)

Aflatoxin effect on erythrocyte profile and histopathology of broilers given different additives

M F Karimy[1], B Sutrisno[2], A Agus[3], A E Suryani[1], L Istiqomah[1] and E Damayanti[1]

[1]Research Unit for Natural Product Technology, Indonesian Institute of Sciences (LIPI), Jl. Jogja-Wonosari, km. 31,5, Gading, Kec. Playen, Kab. Gunungkidul, 55861 D.I. Yogyakarta, Indonesia

[2]Faculty of Veterinary Medicine, Jl. Fauna No. 2, Bulaksumur, 55281 Yogyakarta, Indonesia

[3]Faculty of Animal Science, Jl Fauna No. 3, Bulaksumur, 55281 Yogyakarta, Indonesia

E-mail: mfaizkarimy@gmail.com

Abstract. The aim of this study was to evaluate erythrocyte profile and microscopic changes effect of AF induces by low level (57.18 ppb) and chronic exposure (34 days) with administration of additive (*Lactobacillus plantarum* G7 and methionine). Aflatoxin-contaminated corn was prepared by inoculate *Aspergillus flavus* FNCC 6002 on corn. Total number of 576 broiler Lohman strain (MB202) unsexed DOC were allocated completely randomized into four treatments and 12 replicates, with 12 broiler chicks each. The treatments as follows: T1= aflatoxin-contaminated diet, T2= aflatoxin-contaminated diet + 1% of LAB (w/w), T3= aflatoxin-contaminated diet + 0.8% of methionine (w/w), and T4= aflatoxin-contaminated diet + 1% of LAB + 0.8% of methionine (w/w). The effect of treatments was evaluated using ANOVA and the difference among mean treatments were analyzed using DMRT. The result showed that administration of additives had no significant effect (P>0.05) on erythrocyte profile, liver, and bursa of Fabricius. The dose of additive in each treatment (T2, T3, T4) were insufficient to reduce adverse effect of chronic aflatoxicosis. It was concluded that the LAB dose for binding AF (57.18%) should be evaluated and the dose for methionine should be reduced for chronic treatment of aflatoxicosis.

1. Introduction

Mycotoxins are secondary toxic metabolites produces by fungi that occur as a contaminant in feed and feed ingredient universally [1]. More than 300 kinds of mycotoxins can cause toxicity sign in mammalian an poultry, aflatoxin (AF) and deoxynivalenol (DON) are two of the most common and harmful for the feed and animal industries [2,3]. Aflatoxins (AF) produced by *Aspergillus flavus* and *A. parasiticus* are a major concern in the poultry production, because it has association with illness and disease in poultry and domestic animals, immunosuppress effect, thermostable and hepatic lesions as the endpoint most associated with aflatoxicosis [4-6].

Erythron is defined by erythrocyte circulating in the body and erythropoietic tissue (bone marrow). On the normal conditions, erythron considers the balance of production, erythrocyte destruction by reticuloendothelial system (RES) in the liver, spleen and bone marrow. The method of erytrhon evaluation by using haematocrit values (Hct), haemoglobin (Hb), MCV, MCH, and MCHC examination [7].

The presence of small to moderate amount of mycotoxins in daily feed rations increases the susceptibility of animals to viral, bacterial and parasitic diseases, because mycotoxins can suppress immune system, even innate and adaptive immunity both of them [5]. AF can cause crucial macroscopic and microscopic changes of the liver, such as hepatomegaly, paleness, hydropic degeneration, fatty change, bile duct hyperplasia and periportal fibrosis [6].

Preventing management and decontamination technologies to minimize the effect of AF that contaminated feeds and feedstuffs is the important thing to develop. Procedure of decontamination focuses on degrading, destroying, inactivating, or removing AF by physical, chemical or biological methods has been developed [8,9]. The aim of this study was to evaluate erythrocyte profile and microscopic changes effect of AF induced induces by low level (57.18 ppb) and chronic exposure (4 days) with administration of additive (*Lactobacillus plantarum* G7 and methionine).

2. Material and methods

2.1. Material

Materials utilized in the present experiment contained raw material for diet formulation (corn, soybean meal, meat bone meal, palm oil, premix, NaCl, $CaCO_3$, and Ca_2PO_4), *Aspergillus flavus* FNCC 6002, dried culture of *Lactobacillus plantarum* G7, and methionine. The 576-days old chicks Lohman strain (MB202) were used in the present experiment.

2.2. Preparation of aflatoxin binder additive

Formulation of feed additive consisted of dried culture of *L. plantarum* G7 [10], methionine [11], and its combination. The dried culture of LAB was prepared using spray dryer [12]. *Lactobacillus plantarum* G7 was cultivated in de Man Rogosa Sharpe (MRS) Broth media at 37 °C for 18 hours. The culture was centrifuged at 4500 rpm for 10 minutes then the biomass/pellet was mixed with skim solution 20% w/v. The solution was homogenized before spray drying process. Spray dryer operating conditions as follows: inlet air temperature 120 °C outlet air temperature 68 °C, aspirator 90%, speed 25-30, and nozzle cleaner 2 (Buchi Mini Spray Dryer B-290). Dried culture obtained from spray dryer then added with skim as fillers and adjusted to $1x10^7$ CFU/g of cell density.

2.3. Experimental diet

An experimental diet composed of corn, aflatoxin-contaminated corn, soybean meal, meat bone meal, then formulated to meet nutrient requirements of broilers [13]. Experimental diet could be seen in Table 1. Aflatoxin-contaminated corn was prepared in several steps [14]. Corn was sterilized at 130 °C for 2 hours. *Aspergillus flavus* FNCC 6002 was cultivated in potato dextrose agar (PDA) media at 30 °C for five-seven days. Corn was inoculated by *A. flavus* and incubated for five days. After incubation, corn was dried under the sun heat then homogenized. Aflatoxin production from each treatment diet was determined using AgraQuant® Total Aflatoxin Assay (Romer Lab) by ELISA (Enzyme-Linked Immunosorbent Assay) reader (Thermo Scientific Multiscan Go). Aflatoxin dose in diet was 57.18 ppb=57.18 µg/kg.

Table 1. Composition and nutrient content of ingredients in the experimental diet

Ingredients	Composition (%)
Corn	59.0
Soybean meal	23.0
Meat bone meal	11.5
Palm oil	3.5
Premix	0.5
NaCl	0.5
$CaCO_3$	1.5
Ca_2PO_4	0.5
Total	100

2.4. In vivo experiment

A total number of 576 broilers Lohman strain (MB202) unsexed one-day-old broilers were allocated randomly into four treatments each. All chicks kept under a similar condition of management throughout the experimental period lasting for 35 days of age. The initial brooding temperature was 33 °C in the first week and reduced gradually two °C per week to 24 °C. Diets and water were provided *ad libitum* all over the experimental period. The experiment was designed in a completely randomized design with four treatments, 12 replicates with 12 broiler chicks each. The treatments as follows: T1= aflatoxin-contaminated diet, T2= aflatoxin-contaminated diet + 1% of LAB (w/w), T3= aflatoxin-contaminated diet + 0.8% of methionine (w/w), and T4= aflatoxin-contaminated diet + 1% of LAB + 0.8% of methionine (w/w).

Feed intake was recorded daily by subtracting the amount of offered feed with the residual for each treatment. Final body weight was assessed basis from initial day to the final day of the experiment. Mortality was recorded daily, and percentage of mortality was calculated. Feed conversion ratio (FCR) was calculated as total feed intake divided by final body weight of live chicken [15].

Vaccination was given at the age of 4 days old (ND-IB vaccine, Newcastle Disease, Infectious-Bronchitis-eye drop), eight days old (IBD vaccine, Infectious Bursal Disease), and 18 days old (ND Lasota vaccine, Newcastle Disease-oral). At the end of experimental period, five broilers per treatment were randomly selected and weighed. Blood samples were collected for erythrocyte profiles analysis, and then the necropsy was performed to observe pathological lesions, especially in livers and bursa of Fabricius.

2.5. Hematological examination

Five broilers (18 and 34 days old) from each treatment were selected at random and bled by wing vein puncture. The blood was collected into tube with 10% of EDTA as an anticoagulant [16]. The red blood cells (RBC) counts were determined by hemocytometer method. Haematocrit values were measured by the microhaematocrit method. Haemoglobin amounts were determined by the cyanmethaemoglobin method [8]. The mean corpuscular volume (MCV), mean corpuscular haemoglobin (MCH) and mean corpuscular haemoglobin concentration (MCHC) were calculated using formulation [7].

The formulation could be seen below:

$$\text{Mean corpuscular volume (MCV)} : \frac{PCV\ x\ 10}{RBC}\ (\mu 3/FL) \tag{1}$$

$$\text{Mean corpuscular hemoglobin (MCH)} : \frac{Hb\ x\ 10}{RBC}\ (\mu\mu/pg) \tag{2}$$

$$\text{Mean corpuscular hemoglobin concentration (MCHC)} : \frac{Hb\ x\ 100}{PCV}\ (\%) \tag{3}$$

2.6. Pathological examination

At the end of the experiment (34 days), five broilers from each treatment were selected at random and killed for pathological examination. All organs were examined for pathologic anatomic lesions. The samples of tissue, especially liver and bursa of Fabricius were collected and fixed in 10% normal buffered formalin.

Microscopically, hepatocellular degeneration in livers was graded as follows; Slight (degree 1): Mild hepatocellular swelling due to hydropic degeneration and fatty changes only in centrilobular areas. Moderate (degree 2): Clear hepatocellular swelling in both centrilobular and mid zonal areas. Severe (degree 3): Diffuse and severe hepatocellular swelling, cytoplasmic paleness and rupture, necrotic coagulatif and cholangioma [6] with modification.

2.7. Statistical analysis

The effect of treatments on broilers were evaluated using the analysis of variance of completely randomized design and the difference among mean treatments were analyzed using Duncan's Multiple Range Test [17].

3. Result and Discussion

The results of the present study are presented in Table 2 and 3. The data showed that the effect of treatments (T2, T3, and T4) on erythrocyte profile were not significantly different (P>0.05), especially at the 34 days of blood sampling (Table 3). From the erythrocyte profile evaluation, T1 treatment (AF) occured anemia microsyte-hyperchromic, which is marked by the a decline of MCV value at the 18 days of sampling (Table 1) but not significantly different (P>0.05) with another treatments (T2, T3, T4). At the 34 days of sampling, MCV and MCHC values were not shown anemia. MCV is an indicator for the size of erythrocyte and MCHC is indicator Hb concentration in the RBC.

Erytron is circulated RBC and the erythropoietic organ are called bone marrow. Erytron defense equilibrium between production, destruction of RBC by reticulo endothelial systems (RES) occurs in the liver, spleen and bone marrow. The condition influences the eritron is polycythemia, anemia, hydremia or hemodilution, and dehydration. Anemia is decrease of RBC, Hb or both of them in the blood circulation [7]. Base on the etiology of anemia at the present study, the anemia occurred at the 18 days of sampling all of treatments, but it did not occur at the 34 days of blood sampling.

The result of treatments (T1, T, T3, T4) on histopathological lesions of livers and bursa of Fabricius at 34 day old shown in the Table 4. From descriptive of liver histopathological changes the best result was found in T2 treatment (57.18 ppb of AF + 1% of LAB). Histopathological lesions in the livers observed in T2 are bile duct hyperplasia, edema, necrotic and dilated sinusoid (Figure 1-4). Bursa of Fabricius in the T2 treatment had focal follicular cyst (Figure 4), which is the same as in T4 for bursal organ.

Liver and immune system organs are considered to be target organs for AF and these organs are primarily affected in aflatoxicosis [6]. The particular aim of this study was induce crhonic aflatoxicosis in broiler (34 days) with lower AF level (57.18 ppb), which may naturally occur under field conditions. The study of Ortatatli et al.[6] showed that low doses of aflatoxicosis were between 50-100 ppb.

Table 2. Effect of treatments (T1, T2, T3, T4) on some hematological parameters for the first of blood sampling (18 days of age)

	1st blood profile sampling (18 days)					
	2.16 (N)	7.7(N)	26.3(N)	121(N)	35(N)	29(N)
Treatment	RBC (10^6/mm^3)	Hb (g/dL)	PCV (%)	MCV (fL)	MCH (pg)	MCHC (%)
T1	2.51 ± 0.59^{a}	7.70 ± 0.84^{a}	25.60 ± 4.22^{a}	99.63 ± 27.37^{a}	31.01 ± 8.91^{a}	30.35 ± 2.98^{a}
T2	2.09 ± 0.45^{ab}	6.70 ± 0.45^{a}	25.00 ± 2.12^{ab}	125.95 ± 37.09^{a}	33.70 ± 9.81^{a}	26.93 ± 2.64^{a}
T3	2.44 ± 0.29^{a}	6.50 ± 0.50^{a}	24.00 ± 2.35^{ab}	100.23 ± 20.11^{a}	26.95 ± 3.96^{a}	27.42 ± 4.60^{a}
T4	1.77 ± 0.48^{b}	6.60 ± 1.43^{a}	20.60 ± 3.58^{b}	135.43 ± 52.09^{a}	40.43 ± 18.18^{a}	32.20 ± 4.92^{a}

^{a,b} Values within columns with no common superscripts are significantly different (P < 0,05), according to Duncan's multiple range tests. N: normal value (21 days of age [18]); RBC: red blood cell; Hb: Hemoglobin; PCV: packed cell volume; MCV: mean corpuscular volume; MCH: mean corpuscular hemoglobin; MCHC: mean corpuscular hemoglobin concentration.

Table 3. Effect of treatments (T1, T2, T3, T4) on some hematological parameters for the second of blood sampling (34 days of age)

	2nd blood profile sampling (34 days)					
	2.16 (N)	7.7(N)	26.3(N)	121(N)	35(N)	29(N)
	RBC	Hb	PCV	MCV	MCH	MCHC
Treatment	(10^6/mm^3)	(g/dL)	(%)	(fL)	(pg)	(%)
T1	2.17 ± 0.24^a	9.16 ± 1.05^a	25.60 ± 4.22^a	130.01 ± 2.84^a	42.29 ± 1.39^a	32.53 ± 0.48^a
T2	2.11 ± 0.31^a	9.04 ± 1.61^a	25.00 ± 2.12^{ab}	131.29 ± 2.39^a	42.69 ± 1.56^a	32.51 ± 0.72^a
T3	2.28 ± 0.09^a	9.74 ± 0.61^a	24.00 ± 2.35^{ab}	132.38 ± 2.81^a	42.66 ± 1.28^a	32.22 ± 0.32^a
T4	2.23 ± 0.06^a	9.34 ± 0.29^a	20.60 ± 3.58^b	131.10 ± 2.50^a	41.94 ± 1.52^a	31.99 ± 0.81^a

^{a,b} Values within columns with no common superscripts are significantly different (P < 0,05), according to Duncan's multiple range tests. N: normal value (35 days of age [18]); RBC: red blood cell; Hb: Hemoglobin; PCV: packed cell volume; MCV: mean corpuscular volume; MCH: mean corpuscular hemoglobin; MCHC: mean corpuscular hemoglobin concentration.

The previous study reported that administration of 1% of LAB + 1.2% of methionine in aflatoxin-contaminated diet (6.1 ppb) resulted the lowest damage organ (64.4 %) of liver, kidney, spleen, thymus, bursa of Fabricius, pancreas, duodenum and ileal, but did not significantly affect the body weight gain and feed conversion of broiler [20]. The organs damage caused by AF-contaminated diet are hematopoietic organs, Immune system, and reticuloendothelial systems (RES) [21-23].

Target organs that considered of AF exposure are liver, kidney and the immune system organs and these are primarily affected in aflatoxicosis cases [6]. The particular aim of this study was to induce chronic aflatoxicosis in broilers (34 days) with low levels of AF (57.18 ppb) to the erythrocyte profile, liver, and bursa of Fabricius. The range of AF include low level of aflatoxicosis are 50 and 100 ppb and those may naturally occur under field conditions [6]. Based on SNI standard, the maximum level of AF in the diet of broilers is 50.0 μg/Kg (50.0 ppb) for all phases and broilers concentrate [24-26].

Table 4. Effect of treatments (T1, T2, T3, T4) on liver and bursa of Fabricius histopathology for 34 days of age

	Treatment			
Change/Lesion	T1	T2	T3	T4
Liver				
Bile duct hyperplasia	0/5	3/5	0/5	3/5
Periportal fibrosis	1/5	0/5	0/5	0/5
Hepatic necrosis, edema, dilated sinusoid	1/5	0/5	1/5	0/5
Bursa of Fabricius				
Focal folicular cyst and lymphocytic hyperplasia	0/5	2/5	0/5	1/5
No pathologic lesion	5/5	3/5	5/5	4/5

The values represent the number of chicks showing changes/number of chicks examined in each treatment group

In this study, the toxic effect of AF on erythrocyte profile, liver and bursa of Fabricius have been clearly observed by feeding 57.18 ppb of AF fed for 34 days. Whether 50 ppb of AF level causes aflatoxicosis in broiler as no significant differences was observed [6]. In this study all treatments has no significant differences (P>0.05) for binding or detoxification AF contamination diets event observed from erythrocyte profile and histopathological (liver and bursa of Fabricius). Level of LAB (1%) or methionine (0.8%) or combine between LAB and methionine not adequate for binding or detoxification the level of AF contaminated diets (57.18 ppb). LAB that has functioned as binder of AF and methionine that have function as detoxification agent need to be increased twice until three time level for looking for the best doses for binding and detoxification function.

Feed supplementation with methionine could prevent liver necrosis in the right dose because methionine supplementation could increase the glutathione (GSH) production. AFBI transformation in liver yielding a very radical AFBI epoxides. AFB1 transformation could be prevented by glutathione conjugation (y-glutamic-cysteinyl-glycine, GSH), catalyzed by glutathione-S-transferase (GST) become a mercapturic-AFBI molecule (8,9-dihydro-(S-Cysteinyl-(N-acetyl))-9-hydroxy aflatoxin Bl), that molecule is polar and could be taken out from the body via urine [11]. Over dosage of methionine supplementation could become toxic, so that the right dosage needs to be well managed for obtaining a good result [11].

The right doses of methionine for detoxification 1000 μg/Kg (1ppm) between 0.75%-1.25%, the 1.5% of methionine exactly cause hepatic necrosis [11]. GSH production has homeostatic mechanism in liver, methionine supplemented in feed could not always change the GSH, because production of GSH could not inhibited by cometionine, and its derivates like buthionine and neutral or non polar amino acid (glycine, alanine, phenylalanine, and proline) acidic amino acid (glutamic acid), basic amino acid (lysine), or synthesis (2-amino-2-norboman carboxylic acid and omethilaminoisobutiric acid) [27] and AFB1 bundled by GSH not always in linear curve, on equilibrium point GSH-AFB1 would get weaken [28], that is why 0.8% of methionine could be toxic for liver with the 57.18 ppb of AF, even though that dosage suitable for 1000 ppb (1 ppm) of AF. 1% of LAB could reduce 44.5% of lesions at 21 days and 31.1% of lesions at 35 days for 2.71 ppb of AF [20], could not adequat enough for binding 57.18 ppb of AF. This also applied to a combination between 1% of LAB and 0.8% of methionine could liver toxicity and got necrosis and interfere erytron as a homeostasis of hematological function.

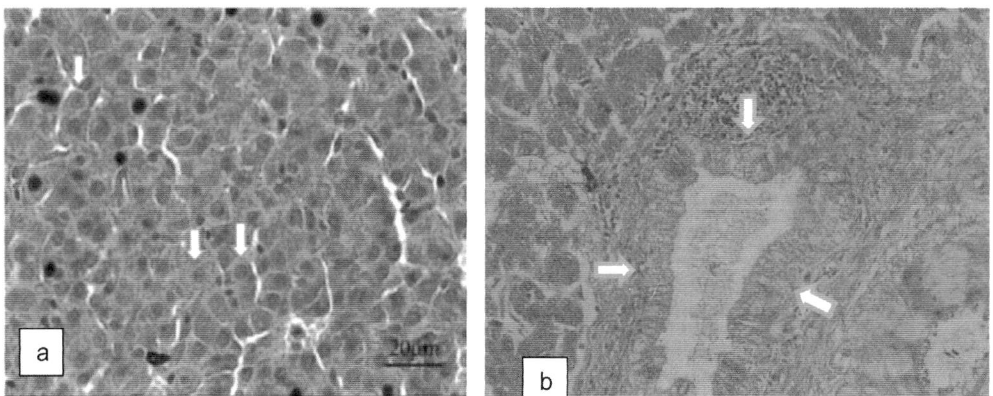

Figure 1. (a) Normal hepatocyte in chicken. The hepatic lobule has complete structure and the liver cell has regular shape (H & E, 400x) [19], (b) Mild to moderate bile duct hyperplasia occur on T2 (57.18 ppb of AF + 1% of LAB) and T4 (57.18 ppb of AF + 1% of LAB + 0.8% of methionine) (H & E, 1000x).

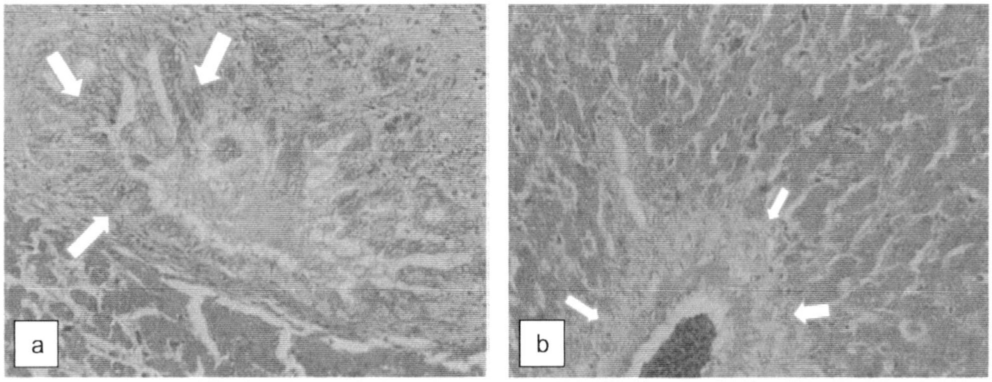

Figure 2. (a) Severe bile duct hyperplasia only occurs on T4 (57.18 ppb of AF + 1% of LAB + 0.8% of methionine) (H & E, 500x), (b) Periportal fibrosis only occurs on T1 (57.18 ppb of AF) (H & E, 1000x).

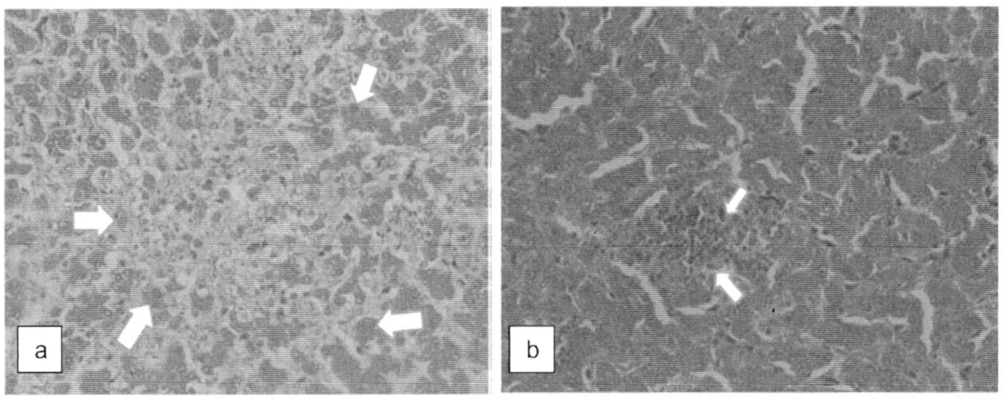

Figure 3. (a) Edema, dilated sinusoid and necrotic hepatocyte occurs on T1 (57.18 ppb of AF) and T3 (57.18 ppb of AF + 0.8% of methionine) (H & E, 1000x), (b) Focal lymphocytic infiltration in the liver occurs on all treatments (T1, T2, T3, and T4) (H & E, 1000x).

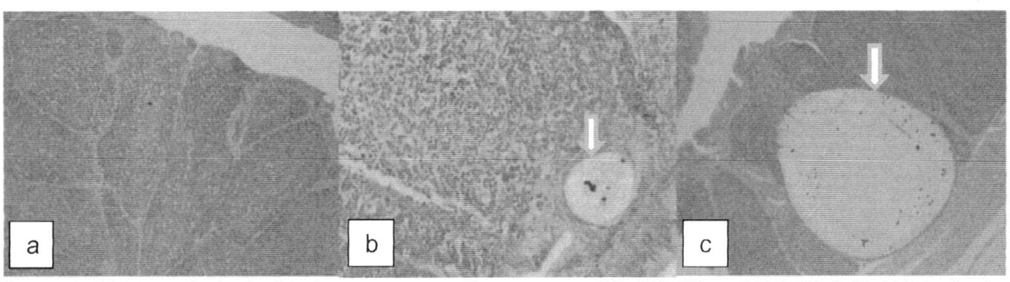

Figure 4. (a) No pathological lesion occurs on all treatments (T1, T2, T3 and T4) (H & E, 100x), (b-c) Bursal cyst and lymphocytic hyperplasia occurs on T2 (57.18 ppb of AF + 1% of LAB) two cases of five samples, and T4 (57.18 ppb of AF + 1% of LAB + 0.8% of methionine) one case of five samples (H & E, 500x (b) and H & E, 1000x (c)).

4. Conclusion

A number of all treatment doses (T2, T3, T4) is not adequate or not exact to reduce adverse effect of chronic aflatoxicosis. Needs to be evaluated of LAB dosage for binding AF (57.18%) and to reduce the dosage of treatments especially for methionine dosage for chronic treatments aflatoxicosis.

Acknowledgement
The authors gratefully acknowledge to all who have assisted and supported this research project (Faculty of Agricultural Technology UGM, Faculty of Animal Science UGM, Faculty of Veterinary UGM and technician in BPTBA LIPI). Particularly the Ministry of Research Technology and Higher Education (Kemenristekdikti) for funding this Insinas 2015 programed. Also special thanks to Ahmad Sofyan, Ph.D. for statistical support.

Reference

[1] Surai P F and Mezes M 2005 *Praxis Vet.* **53** 71-88
[2] Fink-Gremmels J 1999 *Vet. Quatery* **21** 115-20
[3] Streit E, Schatzmayr G, Tassis P, Tzika E, Marin D, Taranu I, Tabuc C, Nicolau A, Aprodu I, Puel O and Oswald I P 2012 *Toxins* **4** 788-809
[4] Ali E J 2014 *Int. J. Sci. Nat.* **5** 113-20
[5] Bondy G S and Pestka J J 2000 *J. Toxic. Env. Health* **3** 109-43
[6] Ortatatli M, Oğuz H, Hatipoğlu F and Karaman M 2005 *Res. Vet. Sci.* **78** 61-8
[7] Salasia S I O and Hariono B 2010 *Veterinary Clinical Pathology: Clinical Pathology Cases* (Yogyakarta: Samudra Biru) pp 9-50
[8] Oğuz H, Keçeci T, Birdane Y O, Önder F and Kurtoğlu V 2000 *Res. Vet. Sci.* **69** 89-93
[9] Parlat S S, Ozcan M and Oguz H 2001 *Res. Vet. Sci.* **71** 207-11
[10] Damayanti E, Suryani A E, Sofyan A, Karimy M F and Julendra H 2013 *J. Agric. Tech.* **35** 164-9
[11] Yunianta A, Agus A, Nuryono and Zuprizal 2010 *Proceed. ISTAP-5* (Yogyakarta: Universitas Gadjah Mada) p 215-20
[12] Istiqomah L, Hayati S N, Damayanti E, Julendra H, Sakti A A and Untari T 2013 *J. Anim Sci Tech.* **36** 14-20
[13] National Research Council 1994 *Nutrients Requirements of Poultry* (Washington DC: National Academy Press) pp 19–34
[14] Abbas H K, Zablotowicz R M, Bruns H A and Abel C A 2006 *Biocontrol Sci. Technol.* **16** 437–49
[15] Timmerman H M, Veldman A, van den Elsen E, Rombouts R M and Beynen A C 2006 *Poult. Sci.* **5** 1383-8
[16] Grimes S E 2002 *A Basic Laboratory Manual for The Small-Scale Production and Testing of I-2 Newcastle Disease Vaccine: Collection of Blood from Chickens* (Thailand: Food and Agriculture Organization of the United Nations (FAO)) pp 31-7
[17] Gomez K A and Gomez A A 2007 *Statistical Procedure for Agricultural Research* (Jakarta: UI Press Jakarta) pp 8–13
[18] Glomsky C A and Pica A 2011 *The Avian Erythrocyte It's Phylogenetic Odyssey* (USA: Science Pub. Edenbridge Ltd) p 421
[19] Wang C, Zhang T, Cui X, Li S, Zhao X and Zhong X 2013 *Evidence-Based Compl. Altern. Med.* 1-9
[20] Istiqomah L, Damayanti E, Julendra H, Suryani A E, Sakti A A and Anggraeni A S 2017 *Proceed. 3rd International Biology Conference* (USA: AIP Pub.) **1854,** 020017 1-7
[21] Dafalla R, Yagi A I and Adam S E I 1987 *Vet. Human Toxic* **29** 222–5
[22] Qureshi M A, Brake J, Hamilton P B, Hagler W M and Nesheim S 1998 *Poultry Sci.* **77** 812–19
[23] Ortatatli M and Oguz H 2001 *Res. Vet. Sci.* **71** 59–66
[24] SNI standard 2006 Standard of broiler starter diets **SNI 01-3930-2006**
[25] SNI standard 2006 Standard of broiler finisher diets **SNI 01-3931-2006**
[26] SNI standard 2009 Standard of broiler concentrate diets **SNI 3148.5:2009**
[27] Yee T A, Ookhtens M and Kaplowitz N l984 *J. Biol. Chem.* **259** 9355-8
[28] Fernandez C, Ruiz C G, Colell A, Yi J R and Kaplowitz N 1996 *Am J. Physiol.* **270** G969-75

Decolorization and Degradation of Batik Dye Effluent using *Ganoderma lucidum*

Diah Pratiwi, A.W. Indrianingsih, Cici Darsih, Hernawan

Research Unit for Natural Product Technology, Indonesian Institute of Science, Yogyakarta, Indonesia 55861

E-mail: dee.diahpratiwi@gmail.com

Abstract. Batik is product of traditional Indonesia culture that developed into a large textile industry. Synthetic dyes which widely used in textile industries including batik. Colour can be removed from wastewater effluent by chemical, physical, and biology methods. Bioremediation is one of the methods that used for processing colored effluent. Isolated White-rot fungi *Ganoderma lucidum* was used for bioremediation process for batik effluent. G. *lucidum* was developed by *G. lucidum* cultivation on centers of mushroom farmer Media Agro Merapi Kaliurang, Yogyakarta. The batik effluent was collected from a private small and medium Batik enterprises located at Petir, Rongkop, Gunungkidul Regency. The aim of the study were to optimize decolorization of Naphtol Black (NB) using *G. lucidum*. The effect of process parameters like incubation time and dye concentration on dye decolorization and COD degradation was studied. *G. lucidum* were growth at pH 5-6 and temperature 25°C at various Naphtol Black dye with concentration 20 ppm, 50 ppm, and 100 ppm for 30 day incubation time. The result from this study increased decolorization in line with the increasing of COD degradation. Increasing percentage of decolorization and COD degradation gradually increased with incubation time and dye concentration. The maximum decolorization and COD reduction were found to be 60,53% and 81,03%. *G. lucidum* had potential to decolorized and degraded COD for NB dye effluent on higher concentration.

1. Introduction

Batik is product of traditional Indonesia culture that developed into a large textile industry. In the past, dyes and supporting materials used in the process of batik use natural ingredients, but over its the development into a industry then many batik using materials synthesis. Of course this affects the quality and quantity of effluent generated from the process batik production.

Nowadays, water pollution is a major problem faced by the world. It is one of the main cause of death and diseases all over the world [1]. Removal of various pollutants such as heavy metals and dyes from industrial effluents has become a critical issue due to strict environmental regulations [2]. In particular, the dyes wastewater such as azo dyes are the potential polluting synthetic dyes with an extensive range of structures as well as colors as they are used widely in batik industry. Wastewater containing dyes from batik industries affects the aesthetic value of surface water and reduces light penetration, endangering aquatic lives and hindering photosynthesis [3]. Dyes can also cause skin irritation, allergic dermatitis and some of them have been reported to be carcinogenic and mutagenic [4].

Azo dyes are synthetic aromatic compounds [5]. The aromatic group around the azo bond helps to stabilize the azo group. Aromatic azo compounds are usually stable and produce strong colors [6]. Naphthol Black (NB) [Disodium;(6Z)-4-amino-3-[(4-nitrophenyl)diazenyl]-5-oxo-6-(phenyl-hydrazinylidene)naphthalene-2,7-disulfonate] is the most commonly used benzidine-based azo dye. It is usually found in the effluents discharged from batik industries. It has a complex structure and high solubility in water. Due to all these nature, NB was selected as a model anionic dye in this study

Some methods can be applied to remove dyes from the waste water, such as membrane filtration, electrocoagulation, electrochemical process, reverse osmosis, irradiation, oxidation, precipitation, and adsorption [7]. Each method has its advantages and disadvantages. The application of traditional physicochemical methods has been restricted due to high cost and low efficiency. In recent years, biological decolorization methods have been considered as a useful, specific, less energy, and environmentally benign, since it results in bioconversion of pollutants to stable and non-toxic end products [8,9]. White-rot fungi (WRF) have received much attention due to their ability to degrade synthetic dyes, giving rise to non-toxic compounds. Some WRF has been reported to degrade rather azo dyes, such as *Phanerochaete chrysosporium*, *Pycnoporus cinnabarinus*, *Pleurotus ostreatus*, and *Basidiomycetous Fungi* [8]. But there not much progress has been made in decolorization of textile effluents such as batik industries. Batik effluents constitute high chemical and biological oxygen demands, suspended solids, and toxic compounds. Thus, there is need to search other microorganisms for treating batik effluents, which are capable of growing at alkaline pH and having sturdy ligninolytic activities.

G. lucidum is a representative of white-rot fungi. *G. lucidum* is highly liable to grow on and degrade agro-industrial lignocellulosic biomass. *G. lucidum* can grow on alkali lignin and expresses high laccase activity [10]. This work was aimed at evaluating *G. lucidum* for their ability to decolorize NB dyes in liquid cultures.

2. Material and Methods

2.1. Batik Dye Effluent

The batik effluent was collected from a private small and medium Batik enterprises located at Petir, Rongkop, Gunungkidul Regency. Synthetic dyes used in Batik enterprises is NB that included azo dyes.

2.2. Organism and Inoculum

G. *lucidum* was developed by *G. lucidum* cultivation on centers of mushroom farmer Media Agro Merapi Kaliurang, Yogyakarta. Taxonomy this mushroom has been verified at Pharmaceutical Biology Divison, Faculty of Pharmacy, Universitas Gadjah Mada. The strain is maintained on PDA (Potato Dextrose Agar) Sigma Aldrich at 30°C. The mycelium was transferred to PDA plates and allowed grow for 7th days. Liquid culture experiments were conducted in 500 ml Erlenmeyer flasks containing 230 ml of liquid medium PDB (Potato Dextrose Broth) and Batik dye effluent at different concentration. Liquid medium PDB containing 20 g/L dextrose and 4 g/L potato extract. Each flask was inoculated with 3 number of agar plug cut from PDA plates and incubated for 30 days.

2.3. Enzyme Assays

The extraction using the centrifugation enzyme method [11]. One gram sample is mixed with 50 mm phosphate buffer PH 6,5 with a comparison of 10/1 (v/w), then shake on speed 150 rpm for 1 hour. After that, samples centrifuged at 7000 x g for 20 minutes. Supernatant liquid separated and used to analyze the activity of the enzyme.

Laccase activity was measured at 30 °C using 1 mM ABTS as substrate. The enzyme analysis mixture contained 0,5 ml of 0,5 M sodium acetate buffer (pH 5), 0,1 ml of 0,1 mM ABTS (Roche Diagnostics), and 0,4 ml enzyme filtrate. The increase in absorbance of the solution was monitored at 420 nm ($E_{420} = 36\ mM^{-1}\ cm^{-1}$) using Halo UV-VIS Spectrophotometer RB-10.

2.4. Decolorization

Decolorization was monitored by a spectrophotometer and expressed as the percentage of removed dye. Percentage of decolorization was calculated as follows:

$$Decolorization\ (\%) = \frac{(Abs\ 0 - Abs\ t)}{Abs\ 0}\ x\ 100$$

Where the absorbance at time 0 is Abs_0 and Abs_t is the absorbance at time t, and λ_{max} is measured at the maximum visible wavelength of dye.

2.5. COD Degradation

Chemical Oxygen Demand is an amount of oxygen consumed by the organic compound and inorganic matter which was oxidized to the water. COD reflect the pollution degree of water and comprehensive index of relative content of organics [12]. Degradation of batik dye effluent in term of COD removal was also investigated. This COD analysis was done at Universitas Gadjah Mada using the volumetric method.

3. Results and discussions

3.1. Mycelial Growth

The Growth of the mycelial already filled all surface of the media on the 7 day, reaching 10,18 cm in diameter. The thickness of mycelia also increased in line with the incubation time. Mycelium was white, but after 20 day the color changed to yellowish or yellow and till 30 day mycelial color began to change to brownish. There factors that affecting growth mycelial biomass are medium, pH and strain. Medium play an important role, followed by pH and type of strain [13]. The growth of *G. lucidum* mycelial to fully covered medium depends on diameter surface, amount and nutrients that contained at medium. This parameter growth has been reported by Kapoor and Sharma, that the mycelial growth depends on such factor as culture media, pH, temperature, and nutrient component. These factors greatly influenced the expansion of fungi in each field and laboratory condition [14].

(a) (b)

(c)

Figure 1. The Growth of Mycelial at (a) day 7; (b) day 20; and (c) day 30

3.2. Effect of Dye Concentration and Incubation Time on Decolorization

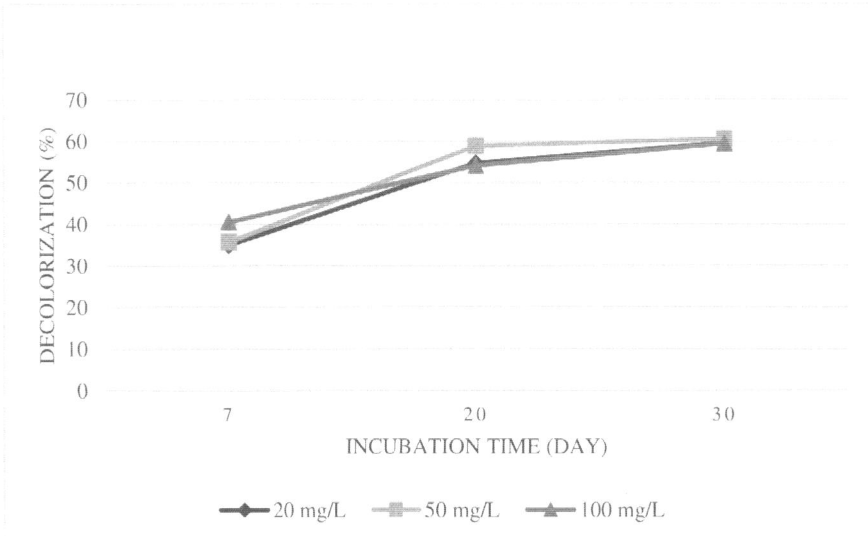

Figure 2. The Effect of Dye Concentration and Incubation Time on Decolorization of Batik Effluent by *G. lucidum*

The percentage of decolorization increased with time incubation. These results are accordance with research conducted by Sharma et.al (2015) [15], where an increasing percentage of decolorization gradually increased with incubation time up to a point till it became constant. In this study, the incubation time where decolorization became constant is not reached during the 30 day incubation time or reached after 30 day. The optimum decolorization of batik effluent in this study reaching 58,79 % at 20 day on 50 ppm concentration. The maximum color removal of textile dye using *G. Lucidum* found to be 81,4% at 5 day operation [16]. Textile effluent Arzoo textile (ART) and Crescent textile (CRT) was maximally decolorized 49,5 ± 0,71% and 36,2 ± 0,56% 0n 10th day of incubation. The ability of white rot fungi to degrade textile industrial effluent is correlated directly to its ability to degrade different dyes present in effluent [17]. Decolorization of Reactive Blue 19 by Ganoderma Sp. achieved 95% after 5 day at pH 7 and 25°C [18]. Decolorization of synthetic dyes by *G. lucidum* taken time 7 to 15 days [19]. There are two factors that influenced decolorization process by fungi, one is related to fungal growth conditions and other is characteristics of dye solution or effluent [20].

The amount usage of dyes, affect the concentration of effluent textiles and dyeing industries. The effects of dye concentration on decolorization of batik effluent by *G.lucidum* was studied at different concentrations from 20 mg/L to 100 mg/L till 30 day. The result of this study revealed that the decolorization was increased with the increasing dye concentration. This is contrast with the results of previous study stated that dye concentration affects the efficiency of decolorization, high dye concentration decreased the decolorization rates [20]. Decolorization was decrease with increasing dye concentration [21]. The removal of dyes increases with decrease in initial dye concentration. At high concentration, the chemical and other pollutants present in the dye effluent inhibit the growth of microorganism [16]. But, this result of study are in line with Revankar and Lele which states that time required to complete decolorization increased over the increased of concentration. That could be possible to sequentially adapt the biomass to higher concentrations and thus improve the performance [22]. Decolorization ability of white rot fungi can be substantially increased by carefully optimizing the operational conditions such as nutrient content of media culture, age of fungus and environmental condition [18]. *G. lucidum* had potential to decolorized and degraded COD for NB dye effluent on higher concentration.

3.3. Effect of Dye Concentration and Incubation Time on COD Degradation

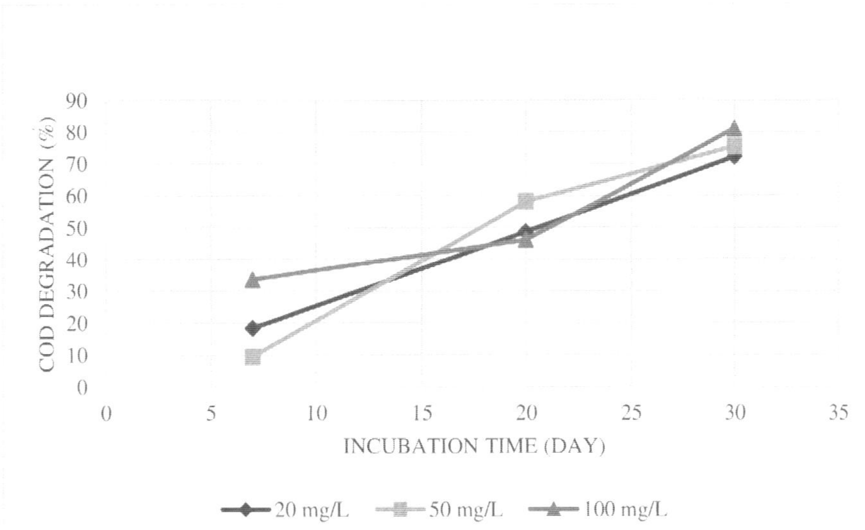

Figure 3. The Effect of Dye Concentration and Incubation Time on COD Degradation of Batik Effluent by *G. lucidum*

Complete decolorization of effluent dyes does not mean that the effluent is completely degraded [12], so degradation of batik dye effluent regarding of COD removal were also investigated. COD removal increased with incubation time. Long incubation times giving *G. lucidum* more time to degrade the COD contained in batik effluent. The maximum percentage of COD degradation achieving 81,03% at 30 day incubation time. Higher concentration of batik dye effluent increased COD removal. This result different with a general study which stated that the higher concentration of dye in the effluent decrease the ability of COD removal. It possible that this caused the presence of elements contained in batik effluent which is a nutrient for *G. lucidum* to growth. An increased of dye concentration increase the elements that be a nutrient for *G. lucidum*. The medium that fungi are mostly grown with dyes or dye effluent to develop a biosorbent containing living fungi mainly composed of carbon source, nitrogen source, and other nutrients [20]. From this study, the result shows that increased decolorization in line with the increasing of COD degradation.

3.4. Enzyme Activity

A method has been patented for decolorization dye effluent using WRF to absorb, degrade, and remove the color of dye effluent[23]. Biodegradation mechanism of textile dyes by fungi because they can produce the lignin modifying enzymes, laccase, manganese peroxidase (MnP) and Lignin peroxidase (LiP) [24]. *G. lucidum* has potential to produce laccase as main ligninolytic enzyme at optimazed parameters were incubation period of 16 days, pH 5, and temperature of 30°C [25].

The concentration of dye affect the enzyme activity, increased of dye concentration decreased the enzyme activity. An increasing of enzyme activity gradually increased with incubation time up to optimum condition. The result of the study showed that when enzyme activity higher then the percentage of decolorization and COD removal lower, neither the opposite when the percentage of decolorization and COD removal higher then the enzyme activity lower. There is posibble mechanisms of fungal decolorization and degradation of dyes, i.e. adsorption, biodegradation, mineralization, enzymatic degradation, and utilization as a carbon source [26]. Decolorization was carried out by a variety of WRF and in most cases laccase was responsible for color removal [17]. In this study, COD removal and decolorization may not affected by laccase activity but other extracellular enzyme such as LiP and MnP. Extracellular enzyme such as LiP and MnP are involved in biodegradation of dyes as well as in color removal. The extent of dye removal depends on the degree of dye complexity, nitrogen avability, and enzymatic in the medium [26]. Other mechanisms of fungal decolorization such as adsorption may due happen on this study. Adsorption as mechanisms of decolorization a wide range dye wastewater by *Ganoderma sp.*[27]. Revankar and lele stated that to ensure that the dye decolorization was not as a result of adsorption of dye on fungal mycelium, the mycelial cultures were filtered, re-suspended in methanol, extracted and centrifuged. If there was no color detected within the extract supernatant, this indicated that decolorization of effluent was achieved because dye degradation by ligninolytic enzyme systems of *Ganoderma sp.*[22].

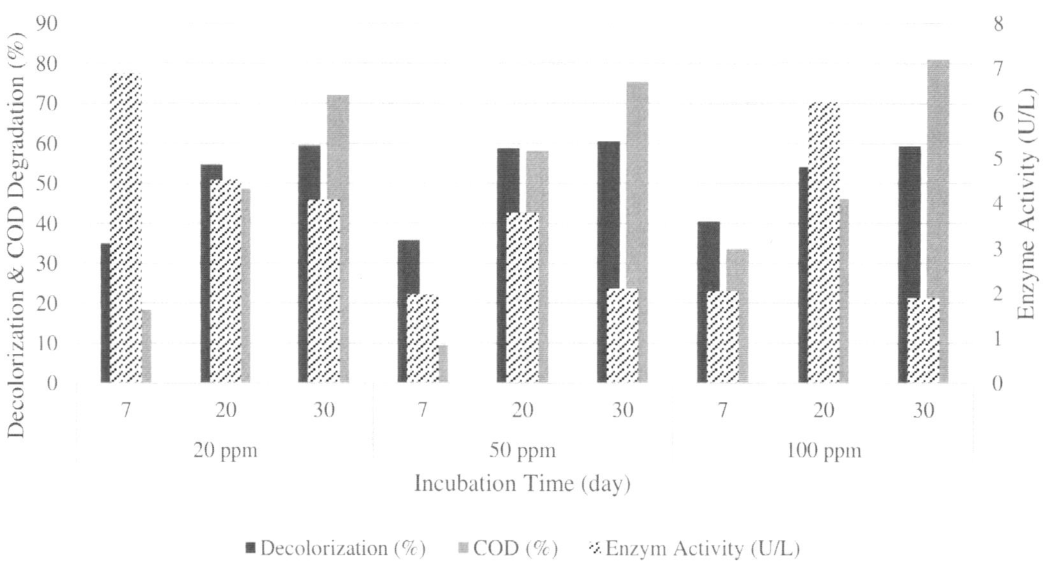

Figure 4. Decolorization, COD degradation, and Enzyme Activity of Batik Dyes Effluent by *G. lucidum*

4. Conclusion

G. lucidum as one of WRF that tremendous potential for biodegradation of a variety textile effluent. In these study *G. lucidum* were growth at pH 5-6 and temperature 25°C at various NB dyes concentration: 20 ppm, 50 ppm, and 100 ppm for 30 day incubation time. Increased decolorization in line with the increasing of COD degradation. Increasing percentage of decolorization and COD degradation gradually increased with incubation time and dye concentration. The maximum decolorization and COD reduction were found to be 60,53% and 81,03%. *G. lucidum* had potential to decolorized and degrade COD for NB dye effluent on higher concentration. Laccase activity may not affected the decolorization and COD removal of batik NB dyes effluent by *G. lucidum*. Need to do more study to find out the optimum concentration of NB as well as to know really mechanisms of deocolorization and COD degradation.

References

[1] Shadeera Rouf and M. Nagapadma 2015 *Int. J. Sci. Eng. Researc* **6** 538–44
[2] Li C, Cui J, Wang F, Peng W and He Y 2016 *Desalination Water Treat.* **57** 14060–6
[3] Farzana M H and Meenakshi S 2015 *Int. J. Biol. Macromol.* **72** 900–10
[4] Gong R, Li M, Yang C, Sun Y and Chen J 2005 *J. Hazard. Mater.* **121** 247–50
[5] Raval N P, Shah P U, Ladha D G, Wadhwani P M and Shah N K 2016 *Desalination Water Treat.* **57** 9247–62
[6] Anon Modeling of Fixed Bed Column Studies for Adsorption of Azo Dye on Chitosan Impregnated with a Cationic Surfactant - Modeling-of-Fixed-Bed-Column-Studies-for-Adsorption-of-Azo-Dye.pdf
[7] Jain R and Sikarwar S 2008 *J. Hazard. Mater.* **152** 942–8
[8] Kuhad R C, Sood N, Tripathi K K, Singh A and Ward O P 2004 *Advances in Applied Microbiology* vol 56 (Elsevier) pp 185–213
[9] Diwaniyan S, Kharb D, Raghukumar C and Kuhad R C 2010 *Water. Air. Soil Pollut.* **210** 409–19
[10] Sitarz A K, Mikkelsen J D, Højrup P and Meyer A S 2013 *Enzyme Microb. Technol.* **53** 378–85
[11] Gassara F, Brar S K, Tyagi R D, Verma M and Surampalli R Y 2010 *Biochem. Eng. J.* **49** 388–94
[12] Yang Q, Liu Z and Yang J 2009 *J. Water Resour. Prot.* **01** 286–9
[13] Qianling Zhou, Wei Yang, Jun Fang Lin and Li Qiong Guo 2015 *Eur. J. Med. Plants* **6** 17–25
[14] Pooja Kapoor and B.M. Sharma 2014 *Int. J. Sci. Environ. Technol.* **3** 1515–24
[15] Sharma A, Shrivastava B and Kuhad R C 2015 *3 Biotech* **5** 621–31
[16] Selvakumar S, Manivasagan R and Chinnappan K 2013 *3 Biotech* **3** 71–9
[17] Asgher M, Noreen S and Bhatti H N 2010 *Water Environ. Res. Res. Publ. Water Environ. Fed.* **82** 357–61
[18] Fazli M M, A.R. Mesdaghinia, K. Naddafi, S. Nasseri, M. Yunesian, M. Mazaheri Assadi, S. Rezaie and M. Yunesian 2010 *Iran J Env. Health Sci Eng* **7** 35–42
[19] Lingan Rajendran , Ramjegathesh Rajendran , Ramu Senthil , Venkatesan Shanthiyaa , Thiruvengadam Raguchander and Ramasamy Samiyappan 2014 *Wudpecker J. Agric. Res.* **3** 130–5
[20] Fu Y and Viraraghavan T 2001 *Bioresour. Technol.* **79** 251–62
[21] Murugesan K, Nam I-H, Kim Y-M and Chang Y-S 2007 *Enzyme Microb. Technol.* **40** 1662–72
[22] Revankar M and Lele S 2007 *Bioresour. Technol.* **98** 775–80
[23] Shen, H.-P., D.-G. Mou, K.-K. Lim, P. Feng and C.-H. Chen Microbial decolorization of wastewater
[24] Ertugay N and Acar F N 2017 *Arab. J. Chem.* **10** S1158–63
[25] Zill-e-Huma Aftab and Shakil Ahmad 2015 *Pak J Phytopathol* **27** 95103
[26] Singh H 2006 *Mycoremediation* (Hoboken, NJ, USA: John Wiley & Sons, Inc.)
[27] Mou D-G, Lim K K and Shen H P 1991 *Biotechnol. Adv.* **9** 613–22

Cellulase and Xylanase Production from Three Isolates of Indigenous Endophytic Fungi

Yopi, W Tasia, and R Melliawati

Research Center for Biotechnology, Indonesian Institute of Sciences (LIPI), Cibinong, Bogor 16911, West Java, Indonesia.

E-mail : *yopi@bioteknologi.lipi.go.id, yopi001@lipi.go.id*

Abstract. Cellulases and hemicellulases have good potential to be used in energy production, in pulp, paper, textile industries, as well as in animal feed industries. Moreover, its utilization in food industries also cannot be ignored, among others, cellulase and xylanase roles in bakery, wine, and fruit and vegetables juice production. One of the potential enzyme source is endophytic fungi. Object of this study is to explore the potency of endophytic fungi isolated from medicinal plants as source of cellulolytic and xylanolytic enzymes. HL.47F.216 is endophytic fungi isolated from traditional medicinal plants ironwood tree was determined as xylanase producer. HL.51F.235 from pin-flower tree is cellulase producer, while CBN.6F.29 which produces both xylanase and cellulase is originated from Madagascar periwinkle. HL.47F.216 showed 2.5 cm in clear zone diameter and its xylanase activity was 0.262 U/mL with optimum condition pH 7 at 50°C. HL.51F.235 showed 2.4 cm clear zone diameter and 0.239 U/mL of cellulase activity at pH 5 and 70°C. CBN.6F.29 showed 2.8 cm and 0.394 U/mL (pH 5, 40°C) for its cellulase activity, while 2.3 cm and 0.439 U/mL (pH 8, 70°C) for its xylanase activity. Xylanase from HL.47F.216 and CBN.6F.29 showed low molecular masses of 20 kDa and 37-50 kDa, respectively. Molecular masses for cellulases from HL.51F.235 and CBN.6F.29 were 25 and 50 kDa for HL.51F.235 and 100 kDa for CBN.6F.29. Based on macroscopic and microscopic identification, fungal isolate CBN.6F.29 is a member of Class *Coelomycetes*, while HL.47F.216 was *Acremonium* sp. and HL.51F.235 was *Aspergillus nigri*.

Keywords : Bioprocess, Cellulase, Endophytic fungi, Xylanase

1. Introduction

Recent studies focus on cellulases and xylanases are highly increase in number. It is due to their high demand in worldwide markets for food and beverage industries. Cellulase decompose cellulose, a recalcitrant polymer formed from D-glucose residues linked by β-1,4-glycosidic bonds. On the other hand, xylan, a complex polysaccharide consisted of a backbone of xylose residues linked by β-1,4-glycosidic bonds is degraded by xylanase.

Cellulases are employed for their capability to increase the yield of potato pulp maceration, guava juice clarification, and tomato juice filtration (Ali 2003). Kuhad *et al.* (2011) emphasized their usefulness by mentioning their applications in food industry. These enzymes can release antioxidants from fruit and vegetable pomace, improve starch and protein extraction yields, modify the viscosity of fruit purees, control bitterness of citrus fruit, and take roles in texture and quality of bakery products improvement.

Beg *et al.* (2001) stated that xylanases are also remarkable in baking industries, increasing the specific bread volume. In biscuit making, it is recommended for making cream crackers lighter (Harris

& Ramalingam 2010). Both cellulase and xylanase from more than one microorganism are better than when obtained from a single microorganism when applied to improve the extraction of olive oil (Bhat 2000).

Microorganisms are frequently used as source of enzymes because they can be cultured in large quantity but in a short time period. Moreover, microbial proteins are often more stable than enzymes from other sources, such as animal or plant origin (Archarya & Chaudhary 2012). Fungi are well-known as agents of decomposition of organic matter in general and of cellulosic substrate in particular (Jahangeer et al. 2005). Therefore, commercial enzymes that are on the market are often fungal enzymes, including cellulases and xylanase (Polizeli *et al.* 2005). Fungal cellulases can be produced in large amounts and often are less complex than bacterial cellulases. This makes it easier to clone the fungal genes and to produce the corresponding enzymes through recombination method for advanced up-scaling (Archarya & Chaudhary 2012).

Endophytic fungi from root, rod, or leaves of plants have potential for cellulase and xylanase production in industries since they have adapted in cellulosic and hemicellulosic materials (Robl *et al.* 2013, Sunitha *et al.* 2013). They spend the entire or part of their life cycle colonizing intercellular and/or intracellular spaces inside the healthy tissues of the host plants, typically causing no apparent symptoms of disease (Ravindran et al. 2010). Foliar fungi from medicinal plants even may play a role in the function of leaves as traditional medicines (Maria *et al.* 2005). This information emphasizes the great potency of medicinal plants endophytes as enzymes producers for food industry (Patil *et al.* 2015). The aim of the present study is to determine cellulase and xylanase secretion from indigenous endophytic fungi isolated from medicinal plants in Indonesia.

2. Materials and Methods

2.1 Microorganisms

Unidentified endophytic fungi belonging to the Research Center of Biotechnology, Indonesia Institute of Science, were used. CBN.6F.29 was isolated from leaf of Madagascar periwinkle (*Catharanthus roseus*) in Cibinong. HL.47F.216 was isolated from the middle rod of ironwood tree (*Memecylon excelsum* Blume), while HL.51F.235 originated from leaf of pin-flower tree (*Glochidion borneese*). Both were sampled from Mount Halimun National Park.

2.2 Enzyme Production

Pre-culture of the isolates was carried out in Potato Dextrose Broth for 72 h with shaking at 110-120 rpm. Ten percent (10%) of each pre-culture along with potato extract (10%) and sugarcane bagasse (2.5%) was employed as a culture of the isolates. The latter was incubated for 7 d with shaking at 110-120 rpm at room temperature (25-28°C). To obtain crude enzyme extract, twice centrifugation at 13,000 rpm and 4°C for 5 min each was conducted. Subsequently, filtration with filter paper was carried out if required.

2.3 Screening of Cellulase and Xylanase Activity

Cellulase and xylanase activities were first qualitatively observed by the Congo red assay. Carboxymethyl cellulose (CMC) and xylan from beechwood were used as substrate for cellulase and xylanase activity determination, respectively. Crude enzyme was spotted on double layer agar pH 6 supplemented with 0.5% substrate. Incubation for 3 d was conducted at 37°C prior to the Congo red assay. To get better visualization of clear zones, acetic acid (5%) was then flooded to the plates. Each sample was subjected to protein concentration determination at 280 nm and protein visualisation using SDS-PAGE and zymogram analysis (Laemmli 1970). Polyacrylamide gels 10% were employed on vertical electrophoresis for 1.5-2 h at 200 W 40 mA 100 V. Staining in SDS-PAGE analysis was done by 0.02% Coomassie Brilliant Blue G-250 and proteins marker (*Bio-Rad, Precision Plus Protein Standards Dual Color*) were used as an indicator of protein molecular mass. Blue bands showed the protein profile. Gels for zymogram analysis were soaked in 2.5% Triton X-100 for 30 min at 25 rpm, followed by rinsing with milliQ water for 5 min, which was done twice. The gel was then soaked in 50 mM buffer phosphate pH 6 for an hour at 45°C 50 rpm. The gel was rinsed with milliQ water and stained with 0.1% Congo red for 30 min. 1 M NaCl was employed in gel destaining, while 5% acetic acid was used for clearer visualization. Protein showed as unstained clear bands against a dark-purple background.

2.4 Optimization of cellulase and xylanase reaction

Activities of xylanase and cellulase were quantified by incubating 250 μL xylan or CMC (0.5%), respectively, with 250 μL enzyme sample, followed by the DNS assay for determination of the reducing sugars (Miller 1959). The pH value was optimized in the range of pH 4-10 using citrate (pH 4 and 5), phosphate (pH 6 and 7), and glycine (pH 8-10) buffer as substrate solvent. Optimization of temperature then was conducted in the range of 30°C-100°C.

2.5 Fungal identification

Morphological identification of fungal isolates was conducted macroscopically and microscopically by *Indonesia Culture Collection* (Barnett & Hunter 1998; Domsch *et al.* 1980; Samson *et al.* 1995). Fungal isolates were grown on Potato Dextrose Agar and identified using light microscope *Olympus BX53* with bright field exposure and magnification scale up to 1000X using immersion oil. Camera *Olympus DP 26* which connected to computer was used for microscopic documentation.

3. Result and Discussion

3.1 Screening of cellulase and xylanase activities

The Congo red assay showed that HL.47F.216 and HL. 51F.235 have different abilities and enzyme profile. The latter was shown as a CMC degrader, while HL.47F.216 is able to utilize xylan. HL.47F.216 formed a 2.5 cm clear zone in diameter and HL.51F.235 formed a clear zone with 2.4 cm diameter, indicating their xylanolytic and cellulolytic ability, respectively. CBN.6F.29 was shown to have both xylanolytic and cellulolytic features. Its xylanase formed a bigger clear zone than HL.47F.216, 2.8 cm. However, its clear zone diameter on CMC plate did not differ much to the one of HL.51F.235, 2.3 cm (Figure 1). The initial screening showed that the endophytes have a massive ability to degrade cellulose and hemicellulose since they produced clear zones of >2 cm. These abilities may be gained due to the adaptation of the fungi to their habitat, which is in a set of lignocellulosic materials in the body of plants (Bhagobaty & Joshi 2012).

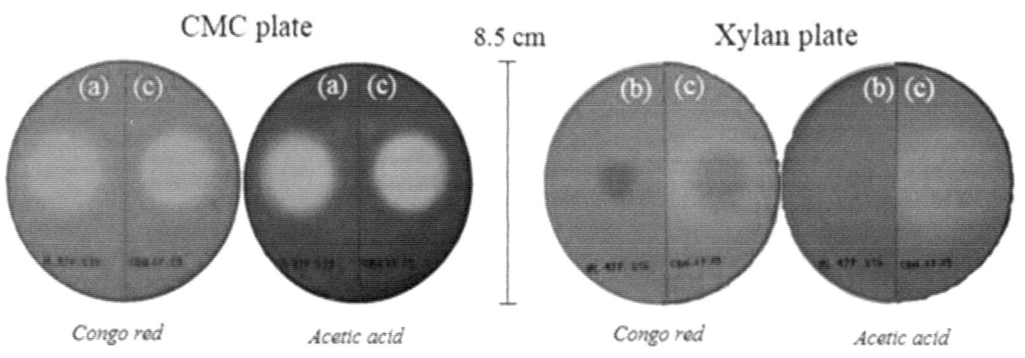

Figure 1. (a) HL.51F.235 (b) HL.47F.216 (c) CBN.6F.29. HL.51F.235 (2.4 cm) and CBN.6F.29 (2.3 cm) could degrade CMC, while HL.47F.216 (2.5 cm) and CBN.6F.29 (2.8 cm) have xylanolytic abilities

Results of zymogram analysis as depicted in Figure 2, show that 25 kDa and 50 kDa proteins are responsible for CMC decomposition by HL.51F.235. On the other hand, a higher molecular mass was shown for the cellulase from CBN.6F.29, approximately 100 kDa. The enzyme responsible for the

IOP Conf. Series: Earth and Environmental Science **101** (2017) 012035 doi:10.1088/1755-1315/101/1/012035

decomposition of xylan appeared on the analysis at 20 kDa in the crude enzyme sample of HL.47F.216, while it shows a mass between 37 and 50 kDa for CBN.6F.29. SDS-PAGE analysis showed that protein pattern from these samples have distinct differences between substrates, indicating true endocellulases or xylanases.

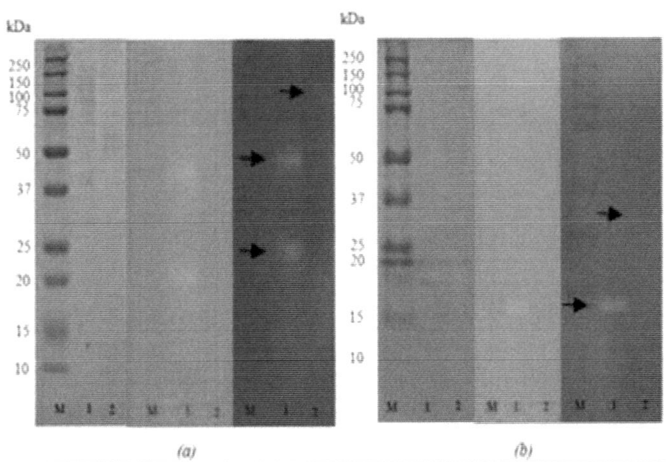

Figure 2. SDS-PAGE and zymogram analysis on 10% polyacrylamide gel. Blue bands in SDS-PAGE gel showed profile of protein from the sample. Unstained bands indicated particular protein role in CMC or xylan degradation. *(a)* cellulase, M: Marker; 1: HL.51F.235; 2: CBN.6F.29 *(b)* xylanase. M: Marker; 1: HL.47F.216; 2: CBN.6F.29.

The molecular masses of the xylanases or cellulases identified in this study are in good agreement with prior researches. The molecular mass of HL.51F.235 cellulase is similar to those from *Trichoderma reesei* and *Phanerochaete* reported in Nayebyazdi *et al.* (2012) ranging from approximately 25 to 50 kDa. Ncube (2013) also reported a molecular mass of 20-43 kDa for *Aspergillus niger* cellulases, while Ritter *et al.* (2013) and Li *et al.* (2011) stated in a review that the molecular weight of fungal cellulase can span a wide range (30-250 kDa). The present study corresponds to the latter as CBN.6F.29 cellulase was detected at approximately 100 kDa.

The molecular mass of xylanases from the endophyte isolates was at 20-50 kDa, in agreement to previous studies (Polizeli *et al.* 2005). *Aspergillus aculeatus* F50 xylanases were 18 kDa, 26 kDa, and 52 kDa (Fujimoto *et al.* 1995), *Hypocrea lixii* xylanase was 29 kDa (Sakthiselvan *et al.* 2014), and *Neocallimastix frontalis* xylanases were 45 kDa and 70 kDa (de Segura & Fevre, 1993) in mass. A xylanase with low molecular weight (19 kDa) was also found in *Aspergillus fumigatus* (Silva *et al.* 1999).

3.2 Optimization of pH value and temperature

Enzyme activity assays using the DNS method were conducted for 15 min reaction time. These data (Figure 3) showed that the optimal pH for cellulase activity was 5 for crude enzyme samples from both isolates, and declined as the pH value rose. The patterns of optimal temperature of cellulase activity from HL.51F.235 and CBN.6F.29 were quite different. HL.51F.235 was rather high, 70°C, whereas cellulase from CBN.6F.29 reacts optimally at moderate temperature, 40°C. On the other hand, xylanase from HL.47F.216 and CBN.6F.29 have higher optimal pH. Moreover, they still have considerable activities at pH 10. Xylanase from CBN.6F.29 was alkaline as its optimal pH was 8 at 70°C. Xylanase from HL.47F.216 was optimal at pH 7 at 50°C. In general, HL.47F.216 had a lower xylanase activity compared to CBN.6F.29.

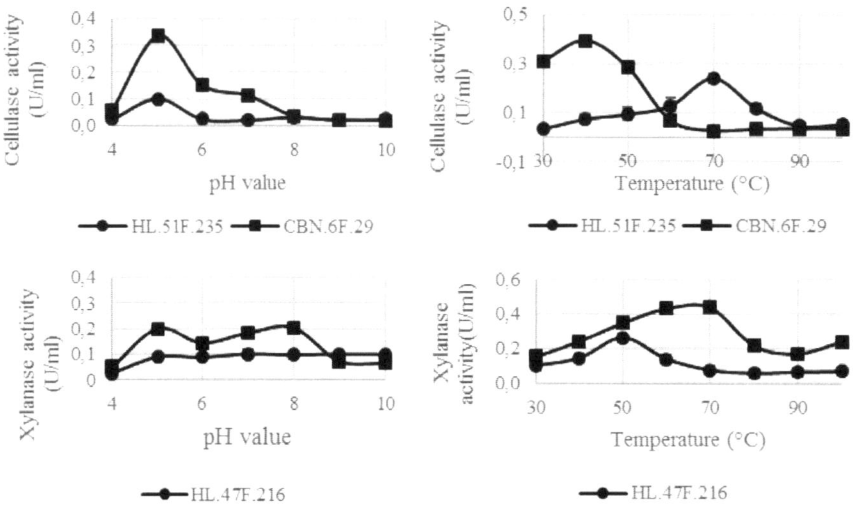

Figure 3. Optimization of cellulase and xylanase activity with respect to pH value and temperature from HL.47F.216, HL.51F.235, and CBN.6F.29 for 15 m using CMC and beechwood xylan as substrates.

A summary of enzymes activities of the endophytes presented in Table 1. Cellulase activity of HL.51F.235 and CBN.6F.29 cultivated on sugarcane bagasse were 0.239±0.0061 U/mL and 0.394±0.0224 U/mL, respectively. Samples from HL.47F.216 have an activity of 0.262±0.0172 U/mL, while CBN.6F.29 shows 0.439±0.0003 U/mL for xylanase.

Table 1. Enzymes activities and properties determined for crude culture supernatants from the three endophytic fungi cultivated on media containing sugarcane bagasse

No	Isolate	Optimal Condition		Enzyme Activity (U/mL)	
		pH	Temperature	Cellulase (CMCase)	Xylanase
1	HL.47F.216	7	50 °C	-	0.262±0.0172
2	HL.51F.235	5	70 °C	0.239±0.0061	-
3	CBN.6F.29	5	40 °C	0.394±0.0224	-
		8	70 °C	-	0.439±0.0003

3.3 Fungal isolates identification

Macroscopically, fungal isolate CBN.6F.29 looked like cotton on PDA media. It was white yet no color on reversionary colony. Its hyphae was examined under the microscope and found that it was transparent (hyaline), septate (monocytic) with no clamp connection structure, and 3-5 μm in width. This fungal isolate did not formed reproduction structure neither sexual nor asexual during sterile incubation (Figure 4). Based on these information, CBN.6F.29 was identified as member of Class Coelomycetes. Taxon was showed on Table 2.

Coelomycetes is a general term for asexual fungi and it has three genera, *Phyllosticta, Phomopsis,* and *Phloeospora*. Coelomycetes occupy many ecological niches and may be pathogens of terrestrial or aquatic plants, endophytes, or saprobes. Some endophytic species may be latent pathogens, existing as symptomless endophytes, but expressing themselves as pathogens once the host defense system weaken. This Class of fungi are common in tropical and temperate regions. It is used as biological control agents, bioremediator, and for pharmaceutical uses (Wijayawardene *et al.* 2012).

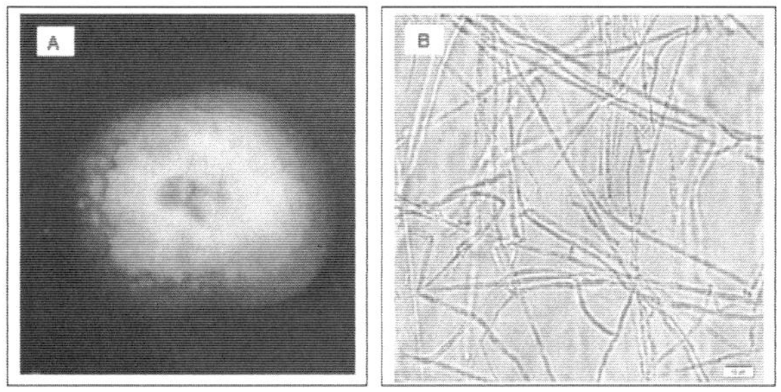

Figure 4. (A) Macroscopic appearance of fungal isolate CBN.6F.29, (B) Anamorphic structure of conidia head (magnification 1000X)

Center part of fungal isolate HL.47F.216 was pale pink cotton-like, rather wet, and slimy. Its reversionary colony was found to be creamy white. Hyphae was found transparent, septate (monocytic), 3-5 μm in diameter with no clamp connection structure. It had single/ramified conidiophore. Fialid formed at substrate and pustles mycelium. Its huddle conidia formed slimy head conidia. Conidia was short oval and lengthwise sized 3-5 μm x 1-2 μm. Clamydiospore was not found during incubation on PDA. *Acremonium* sp. was considered to be alike with HL.47F.216 based on the information investigated.

Table 2. Taxon of fungal isolates identified macroscopically and microscopically

Takson	CBN.6F.29	HL.47F.216	HL.51F.235
Kingdom	Fungi	Fungi	Fungi
Division	Eumycota	Eumycota	Eumycota
Subdivision	Deuteromycotina	Ascomycotina	Ascomycotina
Class	Coelomycetes	Ascomycetes	Ascomycetes
Order	-	Hypocreales	Eurotiales
Family	-	Hypocreaceae	Trichocomaceae
Genus	-	*Acremonium*	*Aspergillus*
Subgenus	-	*Acremonium* sp.	*Aspergillus* section *Nigri*

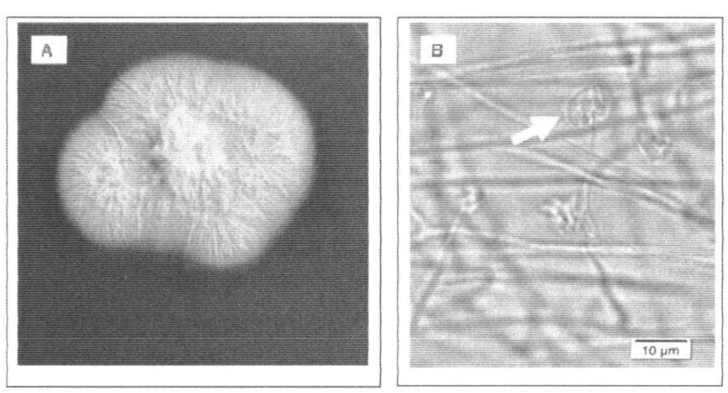

Figure 5. (A) Macroscopic appearance of fungal isolate HL.47F.216, (B) anamorphic structure of conidiophore, fialid, and group of conidia (magnification 1000X)

Figure 6 showed that fungal isolate HL.51F.235 was like black powder and its reversionary colony was no color. It had transparent hyphae, septate (monocytic), 4-6 μm in width, and clamp connection structure was not found. Its conidiophore was in color (hyaline), had smooth wall and its rounded ends formed vesicle. Vesicle was round to semi-round and 33-55 μm in diameter. Fialid was directly formed at vesicle or at metula (biseriate). Sclerotia structure was not found during incubation on PDA. Fungal isolate HL.51F.235 was determined as *Aspergillus* section *nigri* with taxon showed on Table 2.

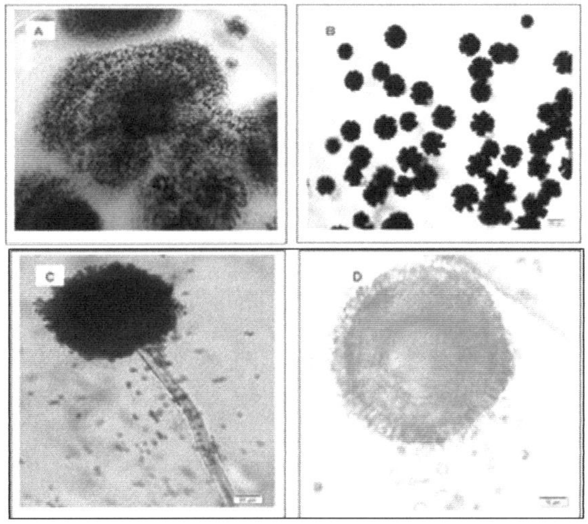

Figure 6 (A) macroscopic appearance of fungal isolate HL.51F.235, B) Anamorphic structure of conidia (magnification 100X) (C) Stipe conidiophore and conidium head (magnification 400X) (D) vesicle and conidia (magnification 1000X)

4. Conclusions

Three Isolated endophytic fungi, HL.47F.216, HL.51F.235, and CBN.6F.29 could produced cellulase and xylanase. Further detail investigations are needed to improve the activities and to make their production economically feasible. Besides, several biochemical and toxicological tests are required to ensure that the endophytes from the medicinal plants are food-grade fungi. Hence, morphological, biochemical, and genetic identification along with finding the efficient conditions in enzymes production are going to be the future aims of the study.

5. Acknowledgements

This research was supported by DIPA Biorefinery Project, Unggulan LIPI Pangan & Obat, Research Center for Biotechnology, Indonesian Institute of Sciences (LIPI).

References

[1] Acharya S, Chaudhary A. 2012. Bioprospecting thermophiles for cellulase production: A review. *Braz. J. Microbiol.* 844-856.
[3] Ali SG. 2003. Biotechnological application of alkaline cellulase in food technology. *Ass. University Bull Environ Res* **6**: 117-124.
[4] Barnett HL, Hunter BB. 1998. Illustrated genera of imperfect fungi 4th ed. New York (USA): Prentice Hall, Inc.
[5] Beg QK, Kapoor M, Mahajan L, Hoondal G. S. 2001. Microbial xylanases and their industrial applications: a review. *Appl Microbiol Biot* **56**: 326-338.
[6] Bhagobaty RK, Joshi SR. 2012. Enzymatic activity of endophytic fungi on five medicinal plant species of the pristine sacred forests of Meghalaya, India. *Biotechnol Biopro Eng* **17**: 33-40.
[7] Bhat MK. 2000. Cellulases and related enzymes in biotechnology. BiotechAdv 18: 355-383.
[8] De Segura BG, Fevre M. 1993. Purification and characterization of two 1,4-β-xylan endohydrolases from the rumen fungi *Neocallimastix frontalis*. *Appl Environ Microbiol* **59**: 3654-3660.
[9] Domsch KH, Gams W, Anderson TH. 1980. Compendium of soil fungi. Vol 1. London (UK):

Academic Press.

[10] Fujimoto H, Oor T, Wang S, Takizawa T, Hidaka H, Murao S, Arai M. 1995. Purification and properties of three xylanases from *Aspergillus aculeatus*. *Biosci Biotech Biochem* **59**: 538-540.

[12] Harris AD, Ramalingam C. 2010. Xylanases and its application in food industry: A review. *J Exp Sci* **1**: 1-11.

[13] Jahangeer S, Khan N, Jahangeer S, Sohail M, Shahzad S, Ahmad A, Khan SA. 2005. Screening and characterization of fungal cellulases isolated from the native environmental source. *Pak J Bot* **37**: 739-748.

[14] Kuhad RC, Gupta R, Singh A. 2011. Microbial cellulases and their industrial applications. *Enz Res*, 1-10.

[15] Laemmli, U. K. 1970. Cleavage of structural proteins during the assembly of the head of bacteriophage T4. *Nature* **227**(5259): 680-685.

[16] Li D, Li A, Papageorgiou A. 2011. Cellulases from thermophillic fungi: Recent insights and biotechnological potential. *Enz Res*, 1-9.

[17] Maria GL, Sridhar KR, Raviraja NS. 2005. Antimicrobial and enzyme activity of mangrove endophytic fungi of southwest coast of India. *J Agr Tech* **1**: 67-80.

[18] Miller GL. 1959. Use of dinitrosalicylic acid reagent for determination of reducing sugar. *Anal Chem* **31**: 426-428.

[19] Nayebyazdi N, Salary M, Ghanbary MAT, Ghorbany M, Bahmanyar M. 2012. Investigation of cellulase activity in some soil borne fungi isolated from agricultural soils. *Ann Biol Res* **3:** 5705-
5713.

[20] Ncube T. 2013. Development of a fungal cellulolytic enzyme combination for use in bioethanol production using *Hyparrhenia* spp. as a source of fermentable sugars. *PhD thesis.* Limpopo: University of Limpopo.

[21] Patil MG, Pagare J, Patil SN, Sidhu AK. 2015. Extracellular enzymatic activities of endophytic fungi isolated from various medicinal plants. *Int J Curr Microbiol Appl Sci* **4**: 1035-1042.

[22] Polizeli MLTM, Rizzati ACS, Monti R, Terenzi HF, Jorge JA, Amorim DS. 2005. Xylanases from fungi: properties and industrial applications. *Appl Microb Biot* **67**: 577-591.

[23] Ravindran C, Naveenan T, Varatharajan G. 2010. Optimisation of alkaline cellulase production from marine derived fungi, *Chaetomium* sp., using agricultural and industrial wastes as substrates. *Bot Mar* **53**: 275-282.

[24] Ritter CET, Camassola M, Zampieri D, Silveira MM, Dillon AJP. 2013. Cellulase and xylanase production by *Penicillium echinulatum* in submerged media containing cellulose amended with sorbitol. *Enz Res*, 1-9.

[25] Robl D, Delabona PS, Mergel CM, Rojas JD, Costa PS, Pimentel IC, Vicente VA, Pradella JGC, Padilla G. 2013. The capability of endophytic fungi for production of hemicellulases and related enzymes. *BMC Biot* **13**: 94.

[26] Sakthiselvan P, Naveena B, Partha N. 2014. Molecular characterization of a xylanase-producing fungi isolated from fouled soil. *Braz J Microbiol* **45**: 1293-1302.

[27] Samson RA, Hoekstra ES, Frisvad JC, Filtenborg O. 1995. Introduction to food borne fungi. 4th ed. Netherlands: Ponsen & Looyen.

[28] Silva CHC, Puls J, de Sousa MV, Filho EXF. 1999. Purification and characterization of a low molecular weight xylanase from solid-state cultures of *Aspergillus fumigatus*. *Rev Microbiol* **30**: 114-119.

[29] Sunitha VH, Devi DN, Srinivas C. 2013.*W J Agr Sci* **9**: 1-9.

[30] Tong CC, Cole AL, Shepherd MG. 1980. *Biochem J* **191**: 83-94.

[31] Wijayawardene NN, McKenzie EHC, Chukeatirote E, Wang Y, Hyde KD. 2012. Coelomycetes. *Cryp Mycol* **33**: 215-244.

IOP Conf. Series: Earth and Environmental Science **101** (2017) 012036 doi:10.1088/1755-1315/101/1/012036

Screening of agricultural wastes as a medium production of catalase for enzymatic fuel cell by *Neurospora crassa* InaCC F226

Pugoh Santoso[1] and Yopi[1]

[1]Indonesia Institute of Sciences, Biocatalyst and Fermentation Laboratory, Indonesia

E-mail adress: pugoh_santoso@ymail.com
yopi@bioteknologi.lipi.go.id

Abstract. Explorations of local microorganisms from Indonesia that can produce of catalase are still limited. *Neurospora crassa* is a fungus which resulting of two kinds of catalase, namely catalase-1 and catalase-3. We studied the production of catalase by *Neurospora crassa* (no. F226) from Indonesia Culture Collection (InaCC) in Solid State Fermentation (SSF). Among four screened agro wastes (corn cob, rice straw, oil palm empty fruit bunches, and bagasse), rice straw and oil palm empty fruit bunches (OPEFB) were remarked as the most promising substrate suited for the excellent growth and adequate production of catalase. Based on the result, the method of solid state fermentation was suitable to production of catalase. It is caused that the medium served to maintain microbial growth and metabolism. The filamentous filament is more suitable for living on solid media because it has a high tolerance to low water activity, and it has a high potential to excrete hydrolytic enzymes that caused of its morphology. The filamentous filament morphology allows the fungus to form colonies and penetrate the solid substrates in order to obtain nutrients. The results showed that the highest catalase activity was obtained on rice straw and oil palm empty fruit bunches medium with catalase activity of 39.1 U/mL and 37,7 U/mL in 50% moisture content medium, respectively. Optimization of humidity and pH medium in the rice straw were investigated which is the highest activity obtained in 30% moisture content and pH medium of 6. The catalase activity was reached in the value of 53.761 U/mL and 56.903 U/mL by incubated 48 hours and 96 hours, respectively.

1. Introduction

Recently, energy from petroleum become a serious problem for the whole countries in the world, including Indonesia. Decreasing world's petroleum stock make its price is very high. So, it is urgent to search other alternative energy sources [1]. The bio-refinery is a prospective key to solve the problem. The utilization of biomass in the bio-refinery concept to production of energy has been a focus on by researcher in the world. In the bio-refinery concept, the biomass has been converted to simple sugar, such as a glucose, which is used as a building block to result of bioethanol or biochemical by microbes or chemical treatment. Because there are several problems in the conversion of glucose to bioethanol by microbes, such as inhibitor furfural and HMF, or pollution of environmental by chemical treatment, the researchers have been looking for the best method to the conversion of the glucose to energy [2; 3].

The most effective way to solve the problem is the development of cell bio-electricity by the concept of the enzymatic fuel cell. An enzymatic fuel cell is a device that can directly change the chemical into electrical energy through an electrochemical [4]. Chemical reactions in the enzymatic fuel cell may occur via direct electron transfer (DET), which electron transfer occurs directly between the enzyme and the electrode or through an intermediary in the form of mediated electron transfer (MET). The reaction between glucose with glucose oxidase will have resulted of a gluconolactone and hydrogen peroxide. The hydrogen peroxide that resulted from this reaction will be changed by the oxidoreductase groups, such as catalase, being water and oxygen. Two reactions in this process can be used as a principle of cell bio-electricity enzymatic based on biomass in the resulting of energy.

Catalase (hydrogen peroxide oxidoreductase, EC1.11.1.6) is one of the antioxidant enzymes, which deals with removal of the oxidative damage in cells. It has a double function; it catalyses the decomposition of hydrogen peroxide into oxygen and water (catalase activity) and also oxidizes electron donors such as ethanol, methanol, or phenols (peroxidative activity) [5;6]. It is classified into four groups. The first is monofunctional-heme (typical) catalases. This group is heme catalases, containing iron-protoporphyrin IX as a prosthetic group in their active sites. Monofunctionalheme (typical) catalases composed of two classes based on the size of the subunits: small-subunit catalases (<60 kDa) and large-subunit catalases (>75 kDa) and they have molecular masses of 200–340 kDa [7]. Various of microorganisms, such as bacteria, fungi and yeast, can be resulting of this enzyme. Diaz *et al* [8] reported that the fungi of *Neusrospora crassa* resulted in two kinds of catalase, namely catalase-1 and catalase-3 and it found in conidia at a very high concentration, tolerant in the high solvent and stable in the large of range pH.

Up to now, the research of catalase, particularly in Indonesia, using indigenous microbe of Indonesia are still limited. By seeing of region of Indonesia as a tropical country with rich of biodiversity can be used to produce of catalase. Indonesia is a mega biodiversity country and the world second biggest biodiversity after Brazil. These bio-diversities include microbial diversity in soil and water. Today, Indonesia has isolated of various microbes, and it saved in the Indonesia Culture Collection (InaCC). InaCC has more than 10,000 isolates of species of bacteria, actinomycetes, fungi, yeast and microalgae [9]. Our previous research using *Neurospora crassa* InaCC F226 and production it in the solid medium of Vogel resulted in a high activity and stable as long as 2 hours in the temperature of 30-40^{0}C [10]. Our current research is focused on the screening of an agricultural biomass that can be used as a medium to the production of catalase effectively in the solid state fermentation (SSF).

2. Material and methods

2.1. Materials
All of the chemicals were used from Merck and Sigma chemical unless otherwise stated.

2.2. Microorganisms and inoculum preparation
Neurospora crassa (no. F226) were obtained from the Indonesia Culture Collection. This fungus was isolated from the leaf litter at the Salak mountain, Bogor-Indonesia. It was grown from stocks of conidia on potato dextrose agar (PDA) slant at room temperature for 3 days and stored at 4°C. The spores were suspended by the addition of 5 mL of distilled water to the slant medium.

2.3. Physical treatment of agro-industrial residues
The corn cob, bagasse, rice straw and oil palm empty fruit branches used in this research were kindly provided by Biomaterial Laboratory of Indonesia Institute of Sciences, while the bagasse obtained from the sugar cane farmer in Kediri-West Java. These materials were ground with a machine to make the particle size being 100 mesh [11].

2.4. Screening of agricultural wastes to medium production of Catalase

Catalase production was produced in Solid State Fermentation (SSF) using corn cob, rice straw, oil palm empty fruit bunches (OPEFB) and bagasse with modification of Ayesha Sadaf & S. K Khare method [11]. The biomass was ground by machine until the size of biomass is 100 mesh, and each of agricultural wastes weighed as much as 2 grams. These medium were added of Vogel's medium, 0.2% glucose, 0.5% yeast extract and 1% of glycerol [12]. The humidity and pH medium were adjusted being 50% and 5.5, respectively, and it was sterilised at temperature 121°C by 15 minutes. It was inoculated with 0.2 mL of *Neurospora crassa* InaCC F226 in the sterile of distilled water and incubated at room temperature for 120 hours. The sampling was done at 72 hours, 96 hours and 120 hours.

2.5. Moisture content optimization in the selected of medium production of catalase

The moisture content of medium production was studied in the range of 20-60% with interval 10% [4]. The media of rice straw (2 gram) was added with Vogel's medium, 0.2% glucose, 0.5% yeast extract and 1% of glycerol [12]. The Vogel's medium pH was adjusted to 5.5 before sterilization at 121^{0}C for 15 minutes. The *Neurospora crassa* InaCC F226 has added aseptically into the medium as much as 0.2 mL and fermentation was allowed to proceed for 96. The sampling was done at 72 hours and 96 hours.

2.6. Optimization of pH medium in the selected of medium production

The optimum of pH medium for catalase production was determined using various pH medium ranging from 4.0 to 6.0 with interval 1. The Vogel's medium was adjusted by 10% HCl to make the acidic medium. The rice straw (2 gram) with 30% moisture content was added of Vogel's medium and sterilization at 121^{0}C for 15 minutes. 0,2 mL of inoculum was aseptically inoculated into the medium and fermentation was allowed to proceed for 120 hours. The sampling was done at 72 hours, 96 hours, and 120 hours.

2.7. Extraction and assay enzyme

Neurospora crassa InaCC F226 grown in the agro-residues medium containing Vogel's medium and incubated in the various times (72-120 hours). The medium containing *Neurospora crassa* InaCC F226 was added with 20 mL of Na-phosphate buffer (pH 7.0; 50 mM) containing 3% hydrogen peroxide. The medium was stirred at temperature 4^{0}C for over night (16 hours), and the supernatant containing crude enzymes of catalase was separated from biomass and cells using centrifugation (Kubota 6200) at 13000 x g in 4^{0}C for 15 minutes [10]. The commonly used procedure to investigate catalase activity conducted from modification of Santoso's method [10]. The hydrogen peroxide was used as substrate and decrease of hydrogen peroxide concentration was measured spectrophotometrically at 240 nm. The catalase activity was measured at a controlled temperature (35^{0}C) in 50 mM sodium phosphate buffer (pH 7.0). The free enzyme as much as 0.3 mL were mixed in 1.2 mL of 0.6 mg/mL hydrogen peroxide in buffer solution. The decrease of absorbance concentration hydrogen peroxide at 240 nm was monitored. Enzyme activity was determined using the initial rate of the reaction and the extinction coefficient for H_2O_2 of 39.4 M^{-1} cm^{-1} [13]. One enzyme unit was defined as the amount of enzyme that catalyzes the decomposition of 1 μmol hydrogen peroxide per minute at 35^{0}C

2.8. Measurement of protein concentration

To measure of protein concentration on solution was used Bradford's method [14]. A sample of 0.01 mL was added into 0.99 mL of Bradford solution in the test tube. It was incubated at room temperature for 10-15 min. The quantity of protein in solution was determined by spectrophotometer at wave length 595 nm using bovine serum albumin as standard protein

3. Result and discussion

3.1. Screening of agricultural biomass as a medium production of catalase

Agricultural biomass waste (Fig 1) can be used as a growth medium to produce of catalase by *Neurospora crassa* InaCC F226. There are two biomasses that it resulted in a good of catalase activity, namely Rice straw and oil palm empty fruit bunches (OPEFB). The highest catalase activity reaches in incubation of *Neurospora crassa* at 96 hours with activity value of 39.082 U/mL (Rice Straw). Incubation using rice straw at 72 hours and 120 hours yielded catalase activity of 33.016 U/mL and 34.306 U/mL (Figure 2), respectively. Production of catalase with oil palm empty fruit bunches gave the highest activity at incubation of 48 hours with an activity of 37.690 U/mL. The catalase activity began to decrease in incubation at 96 and 120 hours with activity 34.120 U / mL and 26.574 U/mL, respectively.

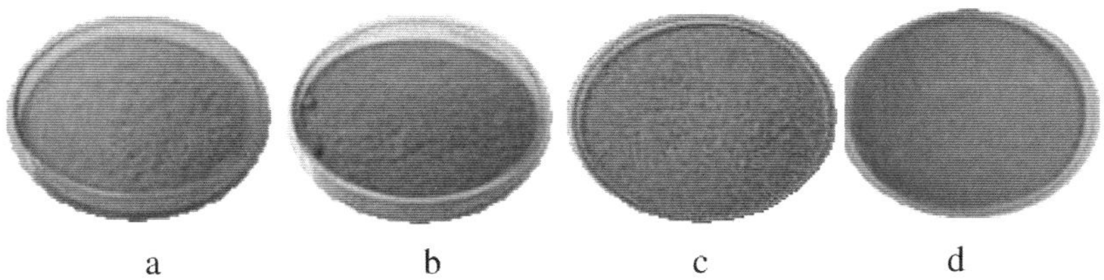

Figure 1. The various of agricultural biomass; a. corn cob b. TKKS; c. rice straw; d. bagasse

Utilization of corn cob and bagasse as a medium production of catalase still provides a pretty good catalase activity. It can be seen in Figure 1 that the highest catalase activity was produced by using bagasse in incubation of 96 hours (37.648 U/mL) and it was still given a good enough activity at 48 hours incubation with catalase activity 35.529 U/mL. The catalase activity at 120 h incubation provides an activity of 31.988 U/mL. Incubation of catalase using corn cobs yielded catalase activity at 96 h incubation (35.448 U/mL), while incubation at the 120 hours and 48 hours of catalase activity began to decrease with activity value of 32.183 U/mL and 30.947 U/mL, respectively.

Based on the result, the method of solid state fermentation (SSF) was suitable to production of catalase. Solid state fermentation (SSF) is defined as the growth of microorganisms on the dampened solid substrate, where the medium serves to maintain microbial growth and metabolism [19]. Generally, the filamentous filament is more suitable for living on solid media because it has a high tolerance to low water activity, and it has a high potential to excrete hydrolytic enzymes that caused of its morphology. The filamentous filament morphology allows the fungus to form colonies and penetrate the solid substrates in order to obtain nutrients.

Increased production of hydrolytic enzymes in solid-state fermentation is due to increased aerial hyphae growth. Aerial hyphae play an important role in the respiration process during fungi growth. Rahardjo *et al.* [15] reported that the aerial hyphae of *Aspergillus oryzae* grown on wheat flour substrate gave 75% oxygen intake. The amount of oxygen consumed is proportional to the amount of α-amylase produced by *Aspergillus orizae*. This means that the aerial hyphae can accelerate the oxygen uptake so that the formation of the aerial mycelium will be abundant. Abundant aerial formation of mycelium may increase the production of hydrolytic enzymes. In addition to high enzyme productivity, using solid media fermentation also provides high product stability and lower catabolic repression [16;17; 18; 19] reported that tanase production on solid media achieved the highest activity at 96 hours of incubation time while for media Liquid, the highest tanase activity is produced after 120 hours of incubation time. This means the use of solid media is more efficient for the production of enzyme tanase than liquid media. Aguilar *et al.* [20] also reported that the protease activity in the media was submerged 10 times higher than in solid media. With the low proteolytic activity of proteases in the media solid, then tanase production becomes high.

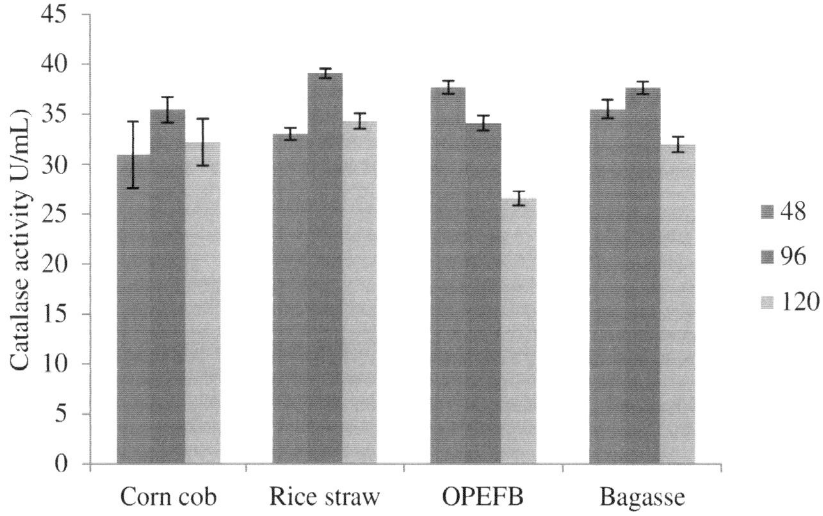

Figure 2. Production of catalase in the various of agricultural biomass

3.2. Optimazation of moisture content and pH medium production of catalase

Humidity of medium was studied in this research 20; 30; 40; and 60%. The catalase activity decreases in the moisture content and incubation time. The results in Figure 3 show the activity on a medium with 30% moisture content resulting high catalase activity of 53.761 U/mL (incubation 48 hours). Then, the moisture content of 40% produced an activity of 44.454 U/mL, catalase activity began to decrease in humidity 20; 50 and 60% with significant values of 26.768 U/mL; 26.848 U/mL; And 27.432 U/mL (incubation 48 h). The catalase activity with an incubation time of 96 hours resulted in activity of 28.701 U/mL (20%); 30.931 U/mL (30%); 31.146 U/mL (40%); 29.958 U/mL (50%); And 28.232 U/mL (60%).

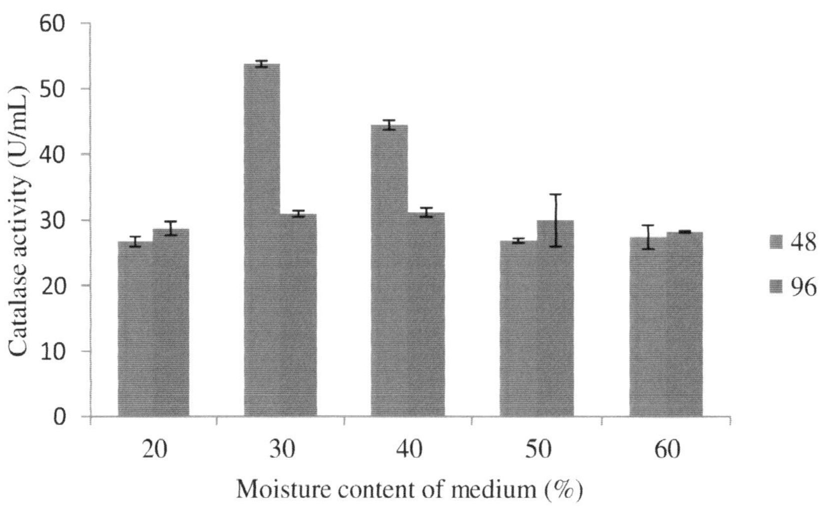

Figure 3. The effect of moisture content in medium production of catalase

The acidity (pH) on medium production can be increasing of catalase activity. It can be seen from Figure 4 that the highest catalase activity obtained on pH medium of 6 by the activity of 56.903 U/mL in 96 hours incubation while the incubation at 48 hours and 120 hours resulted in the activity 56.514 U/mL and 55.989 U/mL, respectively. Meanwhile, the catalase activity on pH medium of 5 increased in the incubation of 48 hours and 96 hours by the activity of 47.225 and 55.385 U/mL, respectively, but there was a small decrease in the incubation of 120 hours with activity 55.799 U/mL. The incubation of *Neurospora crassa* in the pH medium of 4 was resulting in the activity 45.630 U/mL for incubation of 48 hours, 49.107 U/mL for 96 hours, and 54.281 U/mL for 120 hours. pH values can affect structure and function of the protein so that enzymes present in the system will be inactive if microbes grow at too high or too low pH. In addition, if the pH growth is too low, it will disrupt the proton gradient between environments Outside and inside the cell so as to interfere with the cell's ability to produce energy [20].

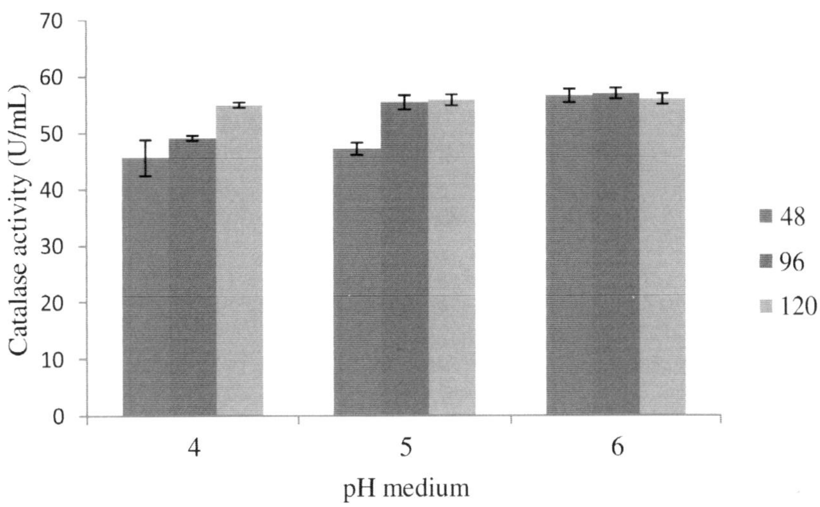

Figure 4. The effect of pH mediun on catalase activity

4. Conclusion

Agricultural waste (corn cob, rice straw, oil palm empty fruit bunches and bagasse) can be used as solid state fermentation to the production of catalase. The results showed that the highest catalase activity was obtained on rice straw medium and oil palm empty palm fruit bunches (OPEFB) with catalase activity of 39.1 U/mL (rice straw, 50% moisture content) and 37.7 U/mL (OPEFB , 50% moisture media). Variations of medium humidity and pH of the medium resulted in a 30% moisture content of rice straw medium by resulting catalase activity of 53.8 U/mL (48 hours incubation) and catalase activity at pH 6 resulting 56.903 U/mL on 96 hours incubation.

Acknowledgment

This research was supported by DIPA Tematik Puslit Biotechnology-LIPI, 2017. We would like to express our thank to Dra. Ruth Melliawati and Nuryati who has for their help in the research.

References

[1] Enny Hawani Loebis, Djarot S Hamiseno, Tri Panji 2008 Hydrolysis optimation process of palm oil empty fruit brunches to glucose by chemical and enzyme to production of ethanol. Bogor Agricultural University [thesis]

[2] D. Hernández, B. Riaño, M. Coca, M.C. García-González 2014 Saccharification of Carbohydrates In Microalgal Biomass by Physical, Chemical and Enzymatic Pretreatments as A Previous Step for Bioethanol Production. doi: 10.1016/j.cej.2014.10.049

[3] Leif J Jönsson, Björn Alriksson and Nils-Olof Nilvebrant 2013 Bioconversion of lignocellulose: inhibitors and detoxification. doi: 10.1186/1754-6834-6-16

[4] Bayu Cakra Buana, Laksmi Ambarsari, Akhirudin Maddu 2016 Design and characterisation enzymatic fuel cell based on glucose oxidase and horseradish peroxide. Bogor Agricultural University [Thesis]

[5] Arabaci G 2011 Partial purification and some properties of catalase from dill (Anethum graveolens L.). *J. Biol. Life Sci.* **2** 11–15

[6] Zhenxiao Yu, Hongchen Zheng, Xingya Zhao, Shufang Li, Jianyong Xu, Hui Song 2016 High level extracellular production of a recombinant alkaline catalase in *E. coli* BL21 under ethanol stress and its application in hydrogen peroxide removal after cotton fabrics bleaching. *Bioresource Technology.* 303—310

[7] Xinhua Fu, Wei Wang, Jianhua Hao, Xianglin Zhu, Mi Sun 2014 Purification and characterization of catalase from marine Bacterium *Acinetobacter sp.* YS0810. Hindawi. doi: 10.1155/2014/409626

[8] Diaz Adelaide, Rangel Pablo, Oca De M Y, Llidias Fernando, Hansberg W 2001 Molecular and kinetic study of Catalase-1, a durable large catalase of *Neurospora crassa. Free Radical Biology & Medicine.* **31** 11 1323-1333

[9] LIPI 2013 Bioresources untuk pembangunan ekonomi Indonesia. [Laporan Tahunan]: Cibinong (ID): Indonesia Institute of Sciences

[10] Santoso Pugoh, Ambarsari Laksmi, Suryani, Yopi 2016 Purification and characterization of catalase from indigenous fungi of *Neurospora crassa* InaCC F226. *International journal on Advanced Science, Engineering and Information Technology.* **6** 502-507. doi: 10.18517/ijaseit.6.4.885

[11] Ayesha Sadaf and S.K. Khare 2014 Production of Sporotrichum thermophile xylanase by solid state fermentation utilizing deoiled Jatropha curcas seed cake and its application in xylooligosachharide synthesis. *Bioresource Technology.* **153** 126–130

[12] Vogel Hj 1964 Distribution of lysine pathways among fungi: evolutionary implication. Am Nat. **98** 435-446

[13] Didem Sutay Kocabes, Ufuk Bakir, Simon E, V Philips, Michael J, Mc Pherson, Zumrut B, Ogel. 2008 Purification, characterization, and identification of a novel bifunctional catalase-phenol oxidase from *Scytalidium thermophilum. Appl Microbiol Biotechnol.* **79** 407-415. doi: 10.1007/s00253-008-1437-y

[14] Bradford M 1976 Rapid and sensitive method for the quantitation of microgram quantities of protein utilizing the principle of protein-dye binding *Journal Analytical Biochemistry.* **72** 248–254

[15] Rahardjo YSP, Weber FJ, Comte EP le, Tramper J, Rinzema A 2002 Contribution of aerial hyphae of *Aspergillus oryzae* to respiration in a modelsolid state fermentation system. *Biotechnol Bioeng* **78** 539-544

[16] Aguilar CN, Augur C, Favale-Torres E, Viniegra-Gonzalez G 2002 Culturesconditions dictate protease and tannase production in submerged and solidstate cultures of *Aspergillus niger* Aa-20. *Appl Biochem Biotechnol* **103** 1-3 407-414

[17] Holker U, Hofer M, Lenz J 2004 Biotechnological advantages of laboratory scalesolid state fermentation with fungi. *Appl Microbiol Biotechnol* **64** 2 175-186

[18] Rana NK, Bhat TK 2005 Effect of fermentation system on the production andproperties of tannase of *Aspergillus niger* van Tieghem MTCC 2425. *JGeneral Appl Microbiol* **51** 4 203-212

[19] Yunita Arian Sani Anwar, Hasim, Made A 2006 Production and characterization of tanin acid hydrolase enzyme from *Aspergillus niger*. Bogor Agricultural University [Thesis]

[20] Aguilar CN, Augur C, Favale-Torres E, Viniegra-Gonzalez G 2002 Culturesconditions dictate protease and tannase production in submerged and solidstate cultures of *Aspergillus niger* Aa-20. *Appl Biochem Biotechnol* **103**1-3 407-414
[21] Ray B 2000 Fundamental Food Microbiology. Ed ke-3 (Boca Raton: CRC Press)

IOP Conf. Series: Earth and Environmental Science **101** (2017) 012037 doi:10.1088/1755-1315/101/1/012037

Synthesis and characterization of dialkanolamides from castor oil *(Ricinus communis)* as nonionic surfactant

M Anwar[1*] and T D Wahyuningsih[2]

[1]Research Unit for Natural Product Technology, Indonesian Institute of Sciences (BPTBA-LIPI), Jl. Jogja-Wonosari Km. 31,5 Gading, Playen, Gunungkidul, Yogyakarta

[2]Department of Chemistry, Faculty of Mathematics and Natural Sciences, Universitas Gadjah Mada, Sekip Utara Bulaksumur, Yogyakarta 55281

E-mail: muslih.anwar@gmail.com

Abstract. Nonionic surfactant of dialkanolamide derivates was synthesized and characterized from castor oil *(Ricinus comunnis)*. Ricinoleic acid was isolated from castor oil by hydrolysis in alkaline (KOH) condition at 65 °C. Oxidation of ricinoleic acid by dilute potassium permanganate ($KMnO_4$) in alkaline condition at 75-90 °C gave dicarboxylic acid which was then reacted with ethanolamine at 140-160 °C for 6 hours. The product was recrystallized with isopropanol, and the structure elucidation was performed by FTIR, ^{1}HNMR spectrometer, and GC-MS with silylation method. Characterization of surfactants was carried out by surface tension measurement (capillary rise method), Critical Micelle Concentration (CMC) based on turbidity method and calculation of Hydrophilic-Lipophilic Balance (HLB) value with Griffin method and Bancroft rule. The result showed that ricinoleic acid in castor oil is 86.19 % and it is oxidation give an azelaic acid and octanedioic acid in 53.25 %. Amidation of a dicarboxylic acid and ethanolamine at 140-160 °C for 6 hours yielded of N^1,N^9-bis(2-hydroxyethyl)nona diamide in 49.35 %. Surfactant characterization indicates that dialkanolamide derivates can be used as a surfactant due to its ability to reduce the surface tension of ethanol with CMC at 1.2 g/L, HLB value is 5.58 and can be used as emulsifier water in oil (W/O).

1. Introduction

Surfactants are being used in a broad variety of applications, among in the household cleaning, agricultural crop protection, cosmetics, pharmacy, remediation, petroleum additive, enhanced oil recovery [1], food additive and processing [2]. Based on a 2016 Acmite Market Intelligence report [3], The global market of surfactants reached at approximately US$30.65 billion in 2015. Increasing by 4.4% annually, the market is expected to reach US$39.69 billion by 2021. So far, the most surfactant made from edible oil or petroleum. The production of surfactant from edible oils has increased seriously affect on food security. Several non-edible oils such as jatropha [4], castor, thumba oil and sal fat [5] were reported for surfactant preparation. One of the non-edible oil that is abundant in Indonesia is castor oil. The largest fatty acid composition of castor oil is ricinoleic acid, which is approximatelly 87-90% of the total fatty acids [6]. Ricinoleic acid is an 18-carbons unsaturated fatty acid which having an hydroxyl group on the 12th carbon atom. The hydroxy group and the double bond are easily oxidized to yield a broad range of products depend on the type of oxidant and reaction conditions. By condensing the fatty acids, methyl esters, or triglycerides with some alkanolamines produced alkanolamides [7]. These compounds are commonly used as biosurfactants or biolubricants, rush inhibitors, ink formulation, in the pharmaceutical industry and anti-inflammatory [8].

The chemical properties of alkanolamide derivatives vary greatly, depending on the length of the hydrocarbon chain and the substituent properties in the nitrogen atom [9]. Alkanolamides has a large function, especially as a stable emulsifier, due to the amide bond are not easy to decompose in an acidic condition [10] and easy to biodegrade [11]. Fatty alkanolamides can be synthesized from alkanolamines with fatty acids, fatty acid esters, and fatty acyl halide [11-12]. By breaking the double bond on the ricinoleic acid with cleavage oxidation produced a dicarboxylic acid. The dicarboxylic acid (azelaic acid) is a compound of nine carbon atoms with two carboxylate groups at both ends. Amidation of the azelaic acid produced dialkanolamide compound (N1, N9-bis (2-hydroxyethyl) nonadiamide) (Scheme 1). In addition, the products were characterized by Fourier transform infrared spectroscopy (FTIR), GCMS and H[1]NMR.

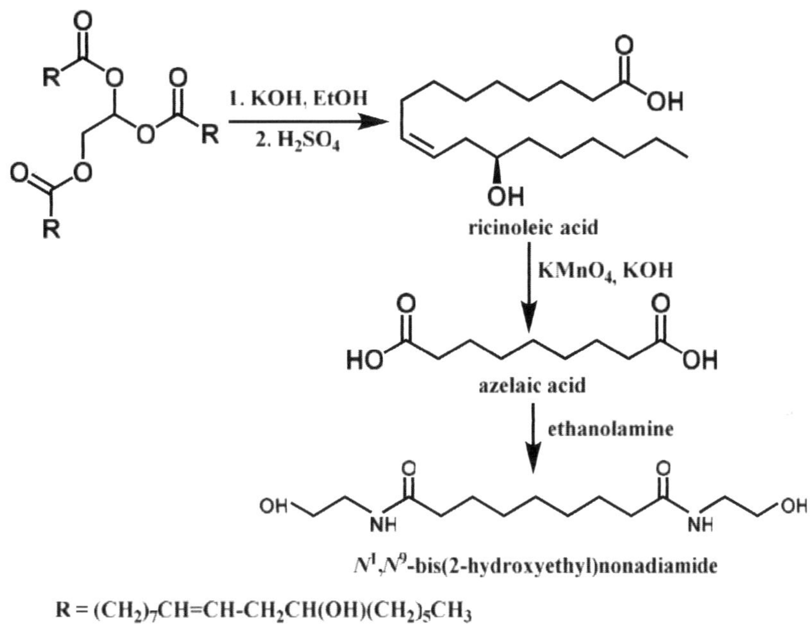

Scheme 1. Synthesis of the nonionic surfactant from castor oil

2. Experimental Section

2.1. General

Castor oil was purchased from Brataco and N, O-bis (trimethylsilyl) trifluoroacetamide (BSTFA) from Aldrich Chemical Co. Potassium hydroxide, potassium permanganate, ethanol, petroleum ether, diethyl ether, isopropanol, ethanolamine, acetonitrile, sulfuric acid, and sodium sulfate anhydrous were purchased from Merck with an analytical grade and were used without further purification. The synthesis products were identified by infrared spectrophotometer (FTIR, Shimadzu-8201PC), spectrometers Nuclear Magnetic Resonance Proton (^{1}H NMR, JEOL JNM-ECA 500, 500 MHz, $CDCl_3$, TMS) and Gas Chromatography (GC Hawlet Packard-5890), Gas Chromatography - Mass Spectrophotometer (GC-MS, Shimadzu QP-5000, and Shimadzu QP-2010). GC analysis of ricinoleic acid methyl esters was carried out using HP-5 column, 30 m x 0.025 mm id. The samples for GCMS analysis were prepared by using BSTFA as silylation agent and injected to Rtx-5MS column, 30 m x 0.025 mm id, 40 mL/min of flow rate and 60 kPa of inlet pressure. Silylation process: 200 μL sample (0.3 g in 0.6 mL acetonitrile) and 100 μL BSTFA was heated at 60-65 °C for 60 min.

2.2. Procedure

2.2.1. Hydrolysis of Castor Oil
Castor oil (14 g, 14 mmol) was introduced into KOH solution (11 g, 196 mmol) in 40 mL of ethanol and were refluxed for four h. The mixture was poured into an ice cube and acidified with sulfuric acid solution (20%v/v) under stirring until pH 2. The ricinoleic acid was produced by separated them from aqueous layer with diethyl ether. The extract was dried over anhydrous sodium sulfate, and the solvent was evaporated under vacuum to give yellow liquid of ricinoleic acid with 82.48% of yield. ν_{max} : 3348 (-OH, str.), 3008 (C_{sp^2}-H)) and 1712 (C=O acid, str.) cm^{-1}. The GC chromatogram as its ester form showed 4 peaks with the highest at t_R 8.3 minutes (86.19%)

2.2.2. Cleavage Oxidation
The mixture of fatty acid hydrolyzed from castor oil (12 g, 40 mmol) was dissolved in 90 mL potassium hydroxide solution (3.5 g). With a strong stirring, 316 g (200 mmol) potassium permanganate dissolved in 350 mL of distilled water and the salt of fatty acid mixture solution was added carefully at 75 °C for 30 minutes until permanganate color was disappeared. The mixture was amounted to 3 M sulfuric acid solution, and heated at 90 °C for 15 min to form a brown precipitate and filtered in a hot condition. The filtrate was cooled in an ice bath until a precipitate is formed and recrystallized with hot water. The solid white product was 4.07 g with m.p 104.2-108 °C. ν_{max} : 2939 and 2854 (C_{sp^3}-H, str.) , 1697 (C=O carboxylic acid, str.) and 1411 (-CH_2-, deformation) cm^{-1}. The GC chromatogram as its ester form showed 2 peaks dominant at t_R 12.77 (35.7%; m/z 110, 97, 69, 55 and 41(base peak)) and 14.08 minutes (45.91%; m/z 152, 124, 111, 83, 60, 55 (base peak) and 41).

2.2.3. Dialkanolamide synthesis
The oxidation product (2.85 g) and ethanolamine (2.88 g, 45 mmol) were heated for 6 hours at 160 °C. The formed product was dissolved in 50 mL of isopropanol and cooled in an ice bath to yield solid N^1, N^9-bis(2-hydroxyethyl)nonadiamide (2.05 g) with m.p. 119 °C. ν_{max}: 3266 (-O-H, str.), 2916 (-C-H, str.), 2855 (-C-H, str.), 11643 (C=O amide, str.), 1558 (N-H, bend) and 1049 (-C-N) cm^{-1}. The GC chromatogram (reverse phase) with silylation shows 1 peak (100% abundance) at t_R = 20.8 min. Mass spectrum (EI) : m/z 403, 375, 35, 286, 196, 172, 166, 101, 73 (base peak), 44, and 41. δH (500 MHz, CDCl3) 1.21 (6H, s, -CH_2-), 1.44 (4H, t, -CH_2-), 2.03 (4H, t, -CH_2-), 3.08 (4H, m, -CH_2-), 3.36 (4H, m, -CH_2-), 7.78 (2H, s, NH).

2.2.4. Surfactant Characterization Test
Critical Micelle Concentration (CMC) was evaluated by using Lamotte 2020 Portable Turbidity Meter with AMCO as a standard and surface tension by capillary rise method according to Cabral [13]. The HLB (Hydrophilic-Lipophilic Balance) value is calculated using the equations proposed by Davies-Rideal with the Bancroft rule [14] :

$$HLB = 7 + \Sigma H + \Sigma L \quad (1)$$

With, Σ H = Amount of x hydrophilic group values, Σ L = Amount of x hydrophobic group value

3. Result and Discussion
The synthesis of dialkanolamide derivates from castor oil consisted of several steps. First, castor oil was hydrolysis to obtain its fatty acids. Second, the mixture of fatty acids which most consisted by ricinoleic acid (86,19%) was strongly oxidized to break down the alkene bonds and to produce its corresponding acid groups. The final steps, the oxidation products were reacted with ethanolamine to give N^1, N^9-bis(2-

hydroxyethyl)nonadiamide. The aim of this research was to produce some nonionic surfactant from castor oil and characterized its physicochemical properties.

Saponification of castor oil was accomplished in ethanol as a solvent and acidified to produce a yellowish liquid of fatty acids with 82.48% of conversion. Elucidation of the product was done by FTIR and GC analysis. The infrared spectrum showed the appearance of a broad peak at 3348 cm^{-1} as a stretching of O-H group and shifting of carbonyl bond to 1712 cm^{-1}. GC chromatogram showed four dominant peaks with the highest abundance was a ricinoleic acid with 86.19% at 8.3 min of retention time.

The oxidation of fatty acid with concentrated potassium permanganate at 75-90 °C is performed to cleavage the double bond on the ricinoleic acid to its dicarboxylic form. This reaction is carried out under an alkaline condition and with high stirring so that the ricinoleic acid can be mixed well into water layer. The infrared spectra exhibited the appearing absorbance of the carbonyl group of carboxylic acids at 1697 cm^{-1}. The absence of alkene absorption at 3008 and 725 cm^{-1} indicated that ricinoleic acid was successfully broken down to produce azelaic acid (nonadioic acid). The GC chromatogram showed that the oxidation product was given two dominant peaks at retention time 12.77 and 14.08 min respectively were octanedioic (36%) and azelaic acid (46%) with a total yield of 53.25%.

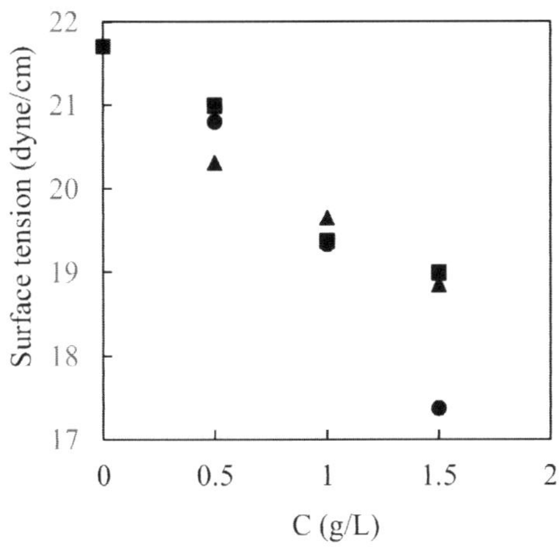

Figure 1. Surface tension ethanol with the addition of ricinoleic acid (■); N^1,N^9-bis(2-hydroxyethyl) nonadiamide (▲) and tween 80 (●)

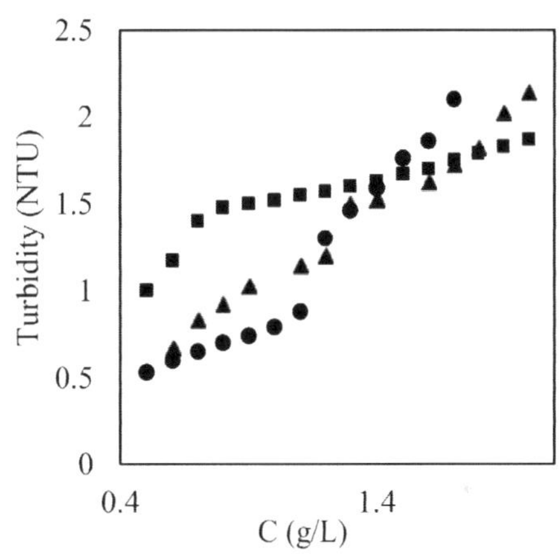

Figure 2. Turbidity of ricinoleic acid (■); N^1,N^9-bis(2-hydroxyethyl)nonadiamide (▲) and tween 80 (●)

Alkanolamides can be synthesized from carboxylic acid derivatives with hydroxylamine. The synthesis dialkanolamides carried out by heating ethanolamine and oxidation products with a molar ratio 3:1 at 140-160 °C for 6 hours. Recrystallization with isopropanol resulted in white solids with 119 °C of m.p. The appearance of peaks at 1049, 1558 and 3266 cm^{-1} were respectively characterized as C-N, N-H, and N-H/O-H bonds of alkanolamides. The shifting of carbonyl peak from 1697 to 1643 cm^{-1} indicated that amides group was presented with the product. Those data confirmed with MS of the silylated product gave a fragment at m/z 403 occurred when molecular ion (m/z 562) released two trimethylsilyls and a methyl radical. This fragment furthermore released two methyl radicals into m/z 375, which is further fragmented resulting in m/z 196, 44, and 41. The McLafferty rearrangement of molecular ion produced fragments with m/z 315, 286, 172, 166, 101, and 73. The purity of product confirmed by appearing only one dominant peak on GC chromatogram at 20.8 min of retention time. The GC chromatogram showed that recrystallization with isopropanol eliminated impurities from the product. ^{1}H-NMR spectrum confirmed the single product with the appearance protons with a chemical shift at 7.78 ppm as two hydrogen atoms of secondary amides. Multiplet signals at 3.08-3.36 ppm equivalent to eight protons from the methine group of ethanolamides. Moreover, fourteen methine protons from long chain fatty acid appeared at 1.21 -2.03 ppm. Based on the elucidation structure with FTIR, GCMS, and NMR, it can be concluded that reaction between oxidation products of castor oil fatty acid and ethanolamine than recrystallized with isopropanol produced N^1,N^9-bis(2-hydroxyethyl)nonadiamide with the percentage of yield was 49.35%.

Surfactant characterization was evaluated by using capillary rise method and turbidimetry to determine surface tension and CMC, respectively. The surface tension of ethanol was decreased in the greater addition concentration of N^1, N^9-bis(2-hydroxyethyl)nonadiamide. It has the same results with ricinoleic acid and tween 80 (Fig 1.). At a concentration of 0.5 g/L, N^1, N^9-bis(2-hydroxyethyl)nonadiamide was able to provide a greater decrease in surface tension than ricinoleic acid and tween 80, but when the concentration increased to 1 g/L and 1.5 g/L, tween 80 gave a better result than others.

The CMC value was determined by measuring the average turbidity of the sample at various concentration. The value of CMC was evaluated where the addition concentration of the sample dramatically change the turbidity gradient (Fig. 2). The critical micelles concentration of N^1, N^9-bis(2-hydroxyethyl)nonadiamide formed at a concentration of 1.2 g/L, whereas tween 80 will form CMC at a concentration of 1.1 g/L. In the other hand, ricinoleic acid can not be a good surfactant at a concentration of 0.5 - 2 g/L due to no critical micelles concentration in the solution.

HLB (Hydrophilic-Lipophilic Balance) is a value that describes the power of hydrophilicity or lipophilicity in some substances. In this empirical method, the surfactant is rated by using a number based on its solubility in water. The HLB value for a compound is fixed under all reaction conditions, so in this study, the determination of HLB was performed using the Griffin method with Bancroft rule. The calculation results obtained HLB value of N^1, N^9-bis(2-hydroxyethyl)nonadiamide, tween 80, and ricinoleic acid were 5.58, 3.45 and 2.93, respectively. Theoretically, the range of HLB values 3-4 is a non-dispersed type of surfactant in water, whereas the range of HLB 4-6 values can be utilized as emulsifiers for water emulsions in oil (emulsifier W/O) [15].

4. Conclusion

N^1, N^9-bis(2-hydroxyethyl)nonadiamide can be prepared from castor oil with several steps such as hydrolysis, cleavage oxidation with concentrated potassium permanganate and amidation with ethanolamine. Based on elucidation analyses as described above, it was concluded that all compounds were successfully synthesized. Surfactant characterization showed that N^1, N^9-bis(2-hydroxyethyl) nonadiamide could be used as an emulsifier for water emulsion in oil (W/O).

Acknowledgement

This publication was funded by DIPA of Indonesian Institute of Sciences of the Republic of Indonesia.

References

[1] Raffa P, Broekhuis A A and Picchioni P 2016 *J Petrol Sci Eng* **145** 723–733
[2] Romsted L S 2014 *Surfactant Science and Technology : Retrospects and Prospects* (Florida : CRC Press) chapter 11 pp 301
[3] Acmite Market Intelligence, Market Report: World Surfactant Market, Retrieved July 30, 2017 http://www.acmite.com/market-reports/chemicals/global-surfactant-market.html
[4] Kumar S, Kumar S and Mandal A 2017 *AIChE J* **63** 2731–2741
[5] Kamalakar K, Satyavani T, Mohini Y, Prasad R B N and Karuna M S L 2014 *J Surfact Deterg* **17** 637–645
[6] Binder R G, Applewhite T H, Kohler G O and Goldblatt L A 1962 *J Am Oil Chem Soc* **39** 513–517
[7] Kumar D and Ali A *Fuel* 2015 **159** 845–853
[8] Mudiyanselage A Y, Yao H, Viamajala S, Varanasi S and Kana Yamamoto 2015 *Ind. Eng. Chem. Res* **54 (16)** 4060–4065

[9] Bilyk A, Bistline R G, Piazza G J, Fearirheller S H and Hass M J 1992 *J. Am. Oil Chem. Soc.* **69** 488-491
[10] Muargard T, Remaud-Siemon M, Petre D, and Monsan P 1997 *Tetrahedron* **53** 5185-5194
[11] Nabel A N, El-Tabl A S, Aiad I A, Zakareya K and Moustafa A H J 2013 *Surfact Deterg* **16** 67-73
[12] Adewuyi A, Oderinde R A, Rao B V S K and Prasad R B N 2012 *J Surfact Deterg* **15** 89–96
[13] Cabral J P S and Smith D A R W J 1992 *J. Colloid Interface Sci.* **149** 1 1
[14] Kiil Sören, and Kontogeorgis Georgios M 2016 *Introduction to applied colloid and surface chemistry* (John Wiley & Sons) chapter 5 pp 116
[15] Ohshima H 2016 *Encyclopedia of Biocolloid and Biointerface Science 2V Set* (John Wiley & Sons chapter 45 pp 571

Development and in vitro evaluation of carboxymethyl chitosan based drug delivery system for the controlled release of propranolol hydrochloride

Hernawan[1]*, Septi Nur Hayati[1], Khoirun Nisa[1], Anastasia Wheni Indrianingsih[1], Cici Darsih[1], Muhammad Kismurtono[1]

[1] Research Unit for Natural Products Technology, Indonesian Institute of Sciences, Yogyakarta, Indonesia

E-mail: hern001@lipi.go.id

Abstract. Propranolol hydrochloride is a nonselective β-adrenergic drug and has been used as angina pectoris, antihypertensive, and that of many other cardiovascular disorders. It has a relatively short plasma half-life and duration of action are considered too short in certain circumstances. Thus, it's fascinating to elongate the action. The tablet formula was based on extended-release by a propranolol hydrochloride based carboxymethyl chitosan matrix. Here we used direct compression technique with internal wet granulation to prepare the tablets. The tablets were evaluated for physical properties (hardness, weight variation test, friability) and in vitro release studies. There was no interaction observed between propranolol hydrochloride and excipients. Dissolution profiles of each formulation were followed zero order model. In conclusion, these results strongly suggest that in appropriate proportions carboxymethyl chitosan with internal granulation is suitable for formulating propranolol hydrochloride controlled release.

1. Introduction

Controlling kinetics of drug release from pharmaceutical formulations is important to achieve the therapeutic effects [1]. The optimum therapeutic response of many medications can be observed once adequate blood levels are achieved and maintained with minimal fluctuations. Controlled release products became popular for the oral administration of such drugs because they give more consistent blood levels [2]. Controlled release with zero order kinetics maintains a plasma concentration of drugs at a constant level.

Propranolol hydrochloride or 1-isopropyl amino-3-(1-naphthyl oxy)-2-propanol hydrochloride is a nonselective β-adrenergic blocking agent, and has been used as angina pectoris, antihypertensive drug and that of many other cardiovascular disorders treatment [2][3][4]. It has relatively short plasma half-life so that patients are habitually asked to require the drug once every 6 to 8 h in divided daily doses. Frequent drug administration may reduce patient compliance and therapeutic efficacy [5]. Propranolol hydrochloride controlled release tablet with zero order kinetics is needed, to improve patient compliance and therapeutic efficacy.

Several methods were describing approaches for controlled release formulations of propranolol hydrochloride. Patra et al. [5] developed a bilayer tablet of propranolol hydrochloride using super disintegrant and immiscible water polymers such as ethylcellulose, Eudragit RLPO, and Eudragit RSPO which showed dissolution kinetics followed the Higuchi model via a non-Fickian diffusion. Sanghavi et al. [2] prepared matrix tablets using hydroxypropyl methylcellulose (HPMC) which exhibited first order release kinetics and Pandey et al. [3] and Chaturdevi et al. [6] develop a floating tablet using HPMC. Javadzadeh et al. [7] promote liquid-solid tablets which showed zero order release kinetics. Velasco-De-Paola et al. [8] described dissolution kinetics of controlled release tablets prepared using eudragit. Mohammadi-Samani et al. [9] described controlled-release dosage forms using ethyl cellulose coating, orodispersible tablets made by employing microcrystalline cellulose have been reported Deshpande and Ganesh [10], and Huang et al. [11].

Our previous research, prepared matrix tablets of propranolol hydrochloride by external wet granulation using carboxymethyl chitosan (CMCh) which exhibited Peppas-Korsmeyerkinetics [12]. Here on this study, we describe a preparation of controlled release matrix by internal wet granulation and study of propranolol hydrochloride controlled release from carboxymethyl chitosan-based matrix tablets aiming to find out release mechanism.

2. Methodology

2.1. Materials
Propranolol hydrochloride was provided by Kimia Farma (Jakarta, Indonesia). Carboxymethylchitosan (CMCh), CMC-Na, hydrochloric acid, lactose, magnesium stearate, xanthan gum and buffer phosphate 7.2. Distilled water was used throughout all experiments, and the other materials and chemicals procured for the studies were of analytical grade

2.2. *Instrumentation*
Tablets were prepared using (Reickermann Korsch, Germany) single punch tablet machine. Infrared spectra were recorded with FTIR Spectrometer Shimadzu Prestige-21. X-ray diffractogram was performed using X-Ray diffractometer (Shimadzu). The drug content was analyzed using Spectrophotometer UV-Vis Dynamica RB 10.

2.3. Preparation of Matrix Tablets
The controlled-release tablets were prepared by direct compression method Controlled release matrix-embedded tablets were made separately by internal wet granulation process using different proportions of CMCh (2%, 4%, and 6%) and denoted respectively as Fi1, Fi2, and Fi3. The formulations are given in Table 1. Propranolol hydrochloride, CMC-Na, xanthan gum, and lactose were mixed and homogenized in a cube mixer 90 rpm for 10 minutes. After completion, 50 ml CMCh solution was added to the mixture to form granules. The mixture granules were passed through 14 mesh, further dried at 40-60oC for 48 hours and mixed 1.33% of magnesium stearate. Tablets were compressed at 300 mg weight on a single punch tablet machine fitted with 5-mm round-shaped punches with compression pressure 10 kN and dwell time 2 s.

Table 1. Composition of Various Tablet Formulations

Formulation Code	Drug (mg)	CMCh (mg)	CMC-Na (mg)	Xanthan Gum (mg)	Lactose (mg)	Mg-Stearate (mg)	Total Weight (mg)
F_i1	120	6	24	100	46	4	300
F_i2	120	12	24	100	40	4	300
F_i3	120	18	24	100	34	4	300

2.4. Compatibility Studies
The interaction between the propranolol hydrochloride and polymer was investigated by using FTIR spectroscopy and X-ray powder diffraction.

2.4.1. Fourier-Transform Infrared Spectroscopy
Compatibility studies of drug, polymers, and the physical mixture were carried out using FTIR spectrophotometer in the scanning range of 300–4,000 cm-1 by KBr disc method.The samples were previously ground and mixed thoroughly with potassium bromide, at 1:5 (sample: KBr) ratio, respectively. An infrared transparent KBr matrix was prepared by compressing the powders at a pressure of 5 tons in a hydraulic press for 5 min.

2.4.2. X-ray powder diffraction
X-ray diffractogram of drug, excipient and matrix tablets were performed using X-Ray diffractometer (Shimadzu). The samples were exposed to x-ray radiation (Cu Kα) at a wavelength of 1.5406 A. The rate of the scanning was 0.6o/min at a range of 5–50o 2θ. Samples were measured on a flat background quartz plate in an aluminum holder

2.5. Evaluation of Tablets

2.5.1. Physical Properties of Tablets
Tablets were characterized for weight variation, hardness, and friability. The drug content of the manufactured tablets was determined. From each batch, ten tablets were taken, weighed, and finely ground. An adequate amount of this powder equivalent to 10 mg of drug was accurately weighed, dissolved and diluted with distilled water and analyzed using UV-Vis spectrophotometer at 290 nm. Hardness was determined by using a tablet hardness tester (Stokes Monsanto, Germany).Friability test was conducted using Erweka abrasive friability tester (Erweka TA 20, Germany).

2.5.2. In vitro Drug Release
The in vitro drug release was performed by dissolution test on the USP dissolution apparatus (Erweka DPT6R, Germany) which was operated at 50 rpm. The dissolution medium used was 900 mL of simulated gastric fluid (SGF) for the first 2 hours and followed by simulated intestinal fluid (SIF) for the remaining hours. The temperature was maintained at 37±0.5 °C. Five milliliters of the sample was taken using a syringe filter periodically and replaced with an equal volume of prewarmed (37±0.5 °C) fresh dissolution medium. The drug content in each sample was analyzed after dilution using UV-Vis spectrophotometer at 290 nm (Dynamica RB 10).

2.5.3. Drug Release Data Modeling
To identify the release mechanisms of propranolol hydrochloride, we tested concerning the release data. The release data were fitted to five kinetic models (zero order (Eq. 1), first order (Eq. 2), Higuchi models (Eq. 3), Peppas-Korsmeyer (Eq. 4) and Hixson-Crowell (Eq. 5)) [12][13][14][15]

$$R = k_1 t \qquad \text{(Eq.1)}$$

$$\text{Log UR} = (k_2 t)/2.303 \qquad \text{(Eq. 2)}$$

$$R = k_3\sqrt{t} \qquad \text{(Eq. 3)}$$

$$\text{LogR} = \text{Logk}_4 + n\text{Logt} \qquad \text{(Eq. 4)}$$

$$\sqrt[3]{UR} = k_5\, t \qquad \text{(Eq. 5)}$$

Where R and UR are respectively the released and unreleased percentages, t is time; k1, k2, k3, k4, and k5 are constants rate of zero-order, first-order, Higuchi models, Peppas-Korsmeyer, and Hixon-Crowell model, respectively

3. Results and Discussion

The interaction between propranolol hydrochloride and excipients often leads to identifiable changes in the infrared (IR) spectra of the drug excipient mixture in the formulation. The IR spectra of drug excipient mixture were compared with the standard frequency of propranolol hydrochloride. Carboxymethyl chitosan gives the peak in IR spectrum nearby 3441 cm^{-1} due to the hydroxyl group. Propranolol hydrochloride provides the peaks in IR spectrum nearby at 2962 cm^{-1} due to the presence of a secondary amine group, 3286 cm^{-1} due to the hydroxyl group (secondary), the aryl alkyl ether display a stretching band at 1103 cm^{-1} and the peak at 771 cm^{-1} due to α-substituted naphthalene [5].The IR spectra of the mixture showed broadening of the peak at 3325 cm^{-1} due to the extensive hydrogen bonding. There is no significant interaction between the propranolol hydrochloride and the excipients used in the formulations, indicating by the characteristic peaks for the functional groups in the propranolol hydrochloride struct ure are retained. The overlay IR spectrum is shown in Fig.1,

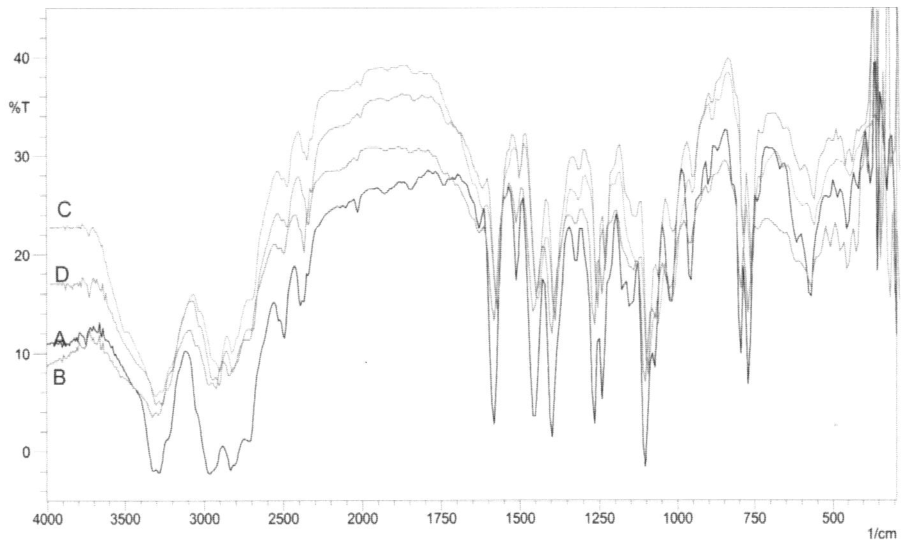

Figure 1. The overlay IR spectra of pure propranolol hydrochloride (A), Fi1 (C), Fi2 (B), and Fi3 (D)

The X-ray powder diffractograms of pure drug and formulations (Fig. 2) were measured for comparison. Pure propranolol hydrochloride showed a typical pattern of a crystalline substance which showed characteristic sharp refraction peaks at 12.54°, 16.76°, 17.23°, 19.92°, 23.68° and 25.12°. This result corresponds to X-ray diffraction patterns of polymorphs two crystalline forms mixture of propranolol hydrochloride as reported by Bartolomei et al. al [4] and Javadzadeh et. al [8]. The sharp peaks of the drug were also present in formulations, where it was mixed with excipients. This result indicated there was a physical interaction between propranolol hydrochloride and the excipient, but crystalline nature of propranolol hydrochloride was retained even after formulation [4][5][6].

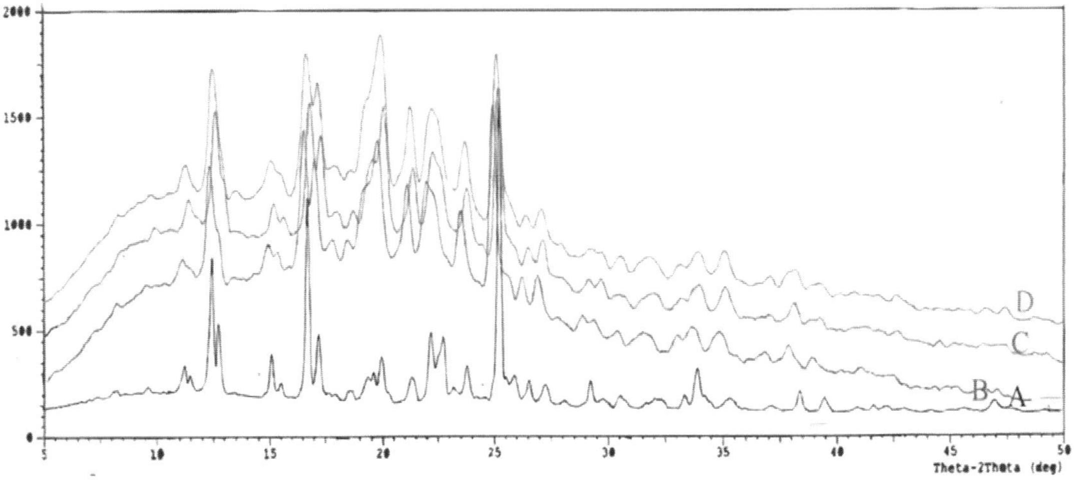

Figure 2. The overlay of X-ray powder diffractograms of pure propranolol hydrochloride (A), Fi1 (B), Fi2 (C) and Fi3 (D)

Polymorphic changes of the drug are important factors that may affect the dissolution rate and bioavailability [16]. Polymorphism also influences the pharmaceutical implication of the presence of meta stable crystalline forms in the bulk powder and the therapeutic efficacy of the drug. Also, crystal structure might affect tablet porosity, density, disintegration and aggregation mechanism, as well as the plastic and elastic properties of dosage forms [8]. Compatibility studies reveal that the drug is compatible with all excipients used in the formulation.

Physical Properties of Tablets

The physical properties (weight, hardness, drug content, water content and friability) of the tablets for all formulation were noted (Table 2). The average weight variations of all the samples were in the range 300± 2.5 mg compliant with the precise requirements of weight uniformity, are not more than 10%. The drug content was found to be close to 99.63±1.54% of the label claim for propranolol in all formulations which was within acceptable limits. The friability tablets were found to within losing weight or friability less than 0.5%-1% [4,14]. The low friability reveals that the matrix tablets are hard and compact.

Table 2. Physical Properties of Compressed Tablets

Formulation Code	F_i1 (2% CMCh)	F_i2 (4% CMCh)	F_i3 (6% CMCh)
Weight variation	302.00±0.66	302.66±0.93	301.50±0.50
Hardness (kg/cm^2)	6.60±2.50	6.30±2.30	5.70±2.70
Friability (%)	0.16±0.02	0.19±0.02	0.13±0.14
Water content (%)	2.00±0.50	2.00±0.35	2.00±0.68

Release Rate Studies

In-vitro dissolution studies were conducted in triplicate of each of the formulations such as F_i1, Fi2, and Fi3. The results for the fitting of the kinetic model for propranolol hydrochloride release from the tablet are shown in Table 3. The values for the release rate constants (k), the correlation coefficients (R^2), and are considered. The correlation coefficient (R^2) was used as an indication of the best fit, for each of the models.

Table 3. Modeling of Dissolution

Kinetic models	F_i1(2% CMCh)		F_i2(4% CMCh)		F_i3(6% CMCh)	
	R^2	k	R^2	k	R^2	k
Zero-order	0.972	2.52×10^{-3}	0.932	2.34×10^{-3}	0.932	3.27×10^{-3}
First-order	0.970	1.97×10^{-2}	0.929	1.98×10^{-2}	0.929	1.94×10^{-2}
Higuchi	0.952	1.84×10^{-2}	0.869	1.70×10^{-2}	0.891	2.22×10^{-2}
Peppas-Korsmeyer	0.961	2.52×10^{-3}	0.875	2.34×10^{-3}	0.880	3.27×10^{-3}
Hixon-Crowell	0.739	2.97×10^{-2}	0.930	3.14×10^{-2}	0.930	3.12×10^{-2}

The results for various models investigated for formulations containing carboxymethyl chitosan ranked in order of Zero order > First order > Peppas-Korsemeyer > Higuchi > Hixon-Crowell. In vitro release data of propranolol hydrochloride from carboxymethyl chitosan based matrix tablet best fitted to zero order model. Carboxymethyl chitosan in appropriate proportions with internal wet granulation is suitable for formulating propranolol hydrochloride controlled release tablets which exhibit zero order kinetics. The ability of the carboxymethyl chitosan to hydrate and form a gel layer around the drug is well-known and is essential to sustaining and controlling the release of a propranolol hydrochloride from the matrix. Zero order release kinetics is the ideal delivery of antihypertensive drugs. Zero order release kinetics refers to the process of constant drug release from a drug delivery device [15].

4. Conclusions

Carboxymethyl chitosan has successfully extended the release of propranolol hydrochloride. No interaction was observed between propranolol hydrochloride and excipients, and crystalline nature of propranolol hydrochloride was not disturbed. Dissolution profiles showed that of each formulation demonstrated that controlled releases of propranolol hydrochloride were following zero order models. It is concluded that in appropriate proportions carboxymethyl chitosan with internal granulation is suitable for formulating propranolol hydrochloride controlled release which exhibits zero order kinetics.

Acknowledgement

The authors would like to acknowledge the Ministry of State for Research and Technology Republic of Indonesia for Improvement the Ability of Researchers and Engineers Incentives Project I.160. 2012

References

[1] Tamanna T, Bulitta JB and Yu A 2015 *J. Mater. Sci. Mater. Med.* **26(2)** 117
[2] Sahoo J, Murthy PN, Biswal S, Sahoo SK and Mahapatra K 2008 *AAPS PharmSciTech.* **9(2)** 577–582
[3] Pandey S, Devmurari V, Paridhi S and Mahalaxmi R 2010 *Der Pharmacia Lettre* **2(1)** 75–86
[4] Bartolomei M, Bertocchi P, Cotta M, Santucci N and Valvo L 1999 *J. Pharm. Biomed. Anal.* **21** 299–309
[5] Patra CN, Kumar AB, Pandit HK, Singh SP and Devi MV 2007 *Acta Pharm.* **57(4)** 479–489
[6] Chaturvedi K, Umadevi S and Vaghani S 2010 *Sci. Pharm.* **78(4)** 927–939
[7] Javadzadeh Y, Musaalrezaei L, and Nokhodchi A 2008 *Int. J. Pharm.* **362(1–2)** 102–108
[8] Velasco-De-Paola MV, Santoro MI and Gai MN 1999 *Drug Dev. Ind. Pharm.* **25(4)** 535–541
[9] Mohammadi-Samani S, Adrangui M, Siahi-Shadbad MR and Nokhodchi A 2000 *Drug Dev. Ind. Pharm.* **26(1)** 91–94
[10] Deshpande KB and Ganesh NS 2011 *Int. J. Res. Pharm. Biomed. Sci.* **2(2)** 529–534.
[11] Huang YB, Tsai YH, Yang WC, Chang JS, Wu PC and Takayama K 2004 *Eur. J. Pharm. Biopharm.* **58(3)** 607–614
[12] Hernawan, Nurhayati S, Nisa K, Indrianingsih AW, Darsih C and Kismurtono M 2013 *Indones.*

J. Chem. **13(3)** 242–247
[13] Dash S, Murthy PN, Nath L and Chowdhury P 2010 *Acta Pol. Pharm.* **67(3)** 217–223
[14] Mahapatra AK, Sameeraja NH and Murthy PN 2015 *AAPS PharmSciTech.* **16(3)** 579–88
[15] Singhvi G and Sing M 2011 *Int. J. Pharm. Stu. Res.* **2(1)** 77-84
[16] Censi R and Di Martino P 2015 *Molecules* **20** 18759-18776

Improving Overall Equipment Effectiveness Using CPM and MOST: A Case Study of an Indonesian Pharmaceutical Company

Dousmaris Omega, Aditya Andika

Department of Industrial Engineering, Bina Nusantara University, Jakarta, Indonesia

E-mail: aditya.a@binus.ac.id

Abstract. This paper discusses the results of a research conducted on the production process of an Indonesian pharmaceutical company. The company is experiencing low performance in the Overall Equipment Effectiveness (OEE) metric. The OEE of the company machines are below world class standard. The machine that has the lowest OEE is the filler machine. Through observation and analysis, it is found that the cleaning process of the filler machine consumes significant amount of time. The long duration of the cleaning process happens because there is no structured division of jobs between cleaning operators, differences in operators' ability, and operators' inability in utilizing available cleaning equipment. The company needs to improve the cleaning process. Therefore, Critical Path Method (CPM) analysis is conducted to find out what activities are critical in order to shorten and simplify the cleaning process in the division of tasks. Afterwards, The Maynard Operation and Sequence Technique (MOST) method is used to reduce ineffective movement and specify the cleaning process standard time. From CPM and MOST, it is obtained the shortest time of the cleaning process is 1 hour 28 minutes and the standard time is 1 hour 38.826 minutes.

1. Introduction

Intense competition in the business world encourages every company to improve the quality in terms of products / services, performance of work and the company itself. Companies must be able to make continues improvements to develop their business. Particularly in the pharmaceutical field which is related to health. All kinds of drugs are produced should undergo any sterile process to produce good medicine.

Company X (the real name of the company is withheld due to confidentiality reasons) is a manufacturing company that produces pharmaceutical products, such as herbal remedies, supplements, and pharmaceutical drugs. In the production process, each machine affects the Overall Equipment Effectiveness (OEE) metric. Company X is facing a problem related to the machines. The problem is the availability rate of the machines.

World-class standard of OEE as stated by Frost and Sullivan is 85% composed of three factors that are Availability (90%), Performance (95%), and Quality (99%) [1]. The OEE metric achieved by Company X was 60% in 2015 for the area of production of liquid (fluid). Company X decided that the target for OEE metric in 2016 is 70%.

The purpose of this research is to increase OEE through improving availability, as one of OEE's factor, in order to achieve the OEE metric target for 2016. The current availability rate is below 70%

while the performance rate and quality rate is already above 90%. To increase availability, this study uses CPM and MOST. MOST has a vital role in the efforts to reduce or eliminate idle and/or down time, improve the working methods, standardize the time as well as enhance the overall capacity planning [2]. Researches on OEE Improvement by using MOST is not widespread. As far as the authors know, there are only few other researches that use MOST to improve OEE, especially in the pharmaceutical setting. Therefore, the results of this research can be used for future reference for other researches.

This study starts with the observation of production process on each machine to get the current OEE. Then, determine the machine to be focused and causation that may occur. The next step is to repair the flow of activities and division of tasks by using CPM. The final result of the improvement is in the form of standard time as the latest time limit by MOST and allowance.

2. Literature Review

2.1. Overall Equipment Efficiency

To determine the effectiveness of the equipment there is a calculation method that can be used is Overall Equipment Effectiveness [3]. Overall Equipment Effectiveness (OEE) is generally used as an indicator of the performance of the use of equipment.

OEE measurement is important to show which areas have a bottleneck in the production line. Moreover, it can be used as a measuring tool to evaluate and improve the proper way to increase the productivity of machinery/equipment. The value of the OEE itself is obtained from the multiplication of three main factors, namely availability rate, performance rate and quality rate, as follows [4]:

$$Availability\ rate = \frac{Required\ Availability - downtime}{Required\ Availability} x100\% \quad (1)$$

$$Performance\ rate = \frac{Output\ x\ Optimal\ cycle\ time}{Operating\ time} x100\% \quad (2)$$

$$Quality\ rate = \frac{Processed\ input - Quality\ defects}{Production\ Input} x100\% \quad (3)$$

$$OEE = Availability\ Rate\ x\ Performance\ Rate\ x\ Quality\ Rate \quad (4)$$

Availability ratio measures the overall time when the system is not operating due to the damage to the equipment, production preparation and adjustment. Performance ratio is measured as the ratio of the actual operating speed of the equipment with ideal speed based on design capacity. Quality ratio focused on how much quality loss in the form of defective products which occurred in connection with the equipment.

2.2. Improved Quality Technique

2.2.1. Pareto Diagram

Problem areas can be defined by the technique developed by the economist Vilfredo Pareto to describe the concentration of wealth. In the analysis of Pareto, the interesting part is to identify and measure on a common scale that was then ordered in descending order, as the cumulative distribution. This technique is commonly called the 80-20 rule [5].

2.2.2. Cause-Effect Diagram

This method defines the occurrence of events which are usually undesirable or problem, namely the effect, as "fish head" and then identify the factors, which is the cause, as "fish bone" [5]. The main cause is usually divided into five or six major categories, namely humans, machines, methods, materials, environments, and administration, each subdivided into several causes.

2.3. Critical Path Method

The critical path is a series of activities of a project that cannot be delayed the implementation time and show relationships that are interrelated to one another [6]. The more critical path exists in a project, the more activities that should be monitored intensively. This strategy is also frequently used in shortening project duration [7]. In general, network diagram can be drawn using the activity on node (AON) and activity on arrow (AOA) as can be seen in Figure 1.

2.3.1. Activity on Node (AON)

The activities are represented by points / nodes. The arrows simply explain the relationship of dependence between activities.

2.3.2. Activity on Arrow (AOA)

The activities are represented by arrows connecting the two circles that representing two events. The tail of the arrow is the beginning of the activity, while its end signaling as the end.

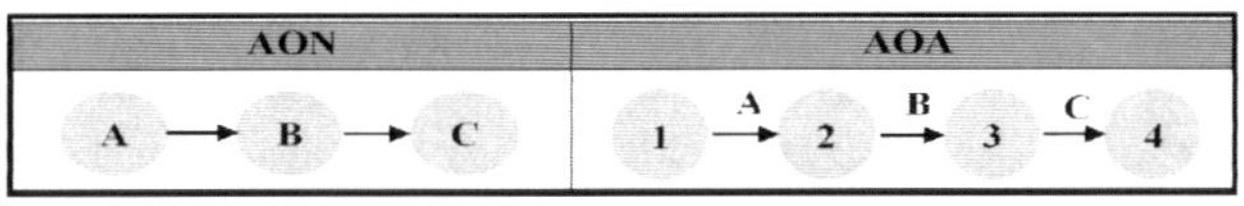

Figure 1. Comparison of AON and AOA

The process used to determine the time schedule for each activity is a two-pass consisting of a forward pass (ES and EF) and the backward pass (LS and LF), as can be seen in Figure 2. Forward pass and backward pass using a notation to indicate the schedules of activity on the network project clearly as follows [6]:

- Earliest Start (ES): the earliest time an activity can begin by assuming all its predecessors have been completed.

$$ES = \text{Max } \{\text{EF all immediate predecessors}\} \quad (5)$$

- Earliest finish (EF): The earliest finish time for an activity amount of the earliest start time (ES) and time of the activity itself.
-

$$EF = ES + \text{Time Activities} \quad (6)$$

- Latest Finish (LF): the last time an activity can be completed so as not to delay the completion of the entire project.

$$LF = \text{Min (LS from all activities that directly follow)} \quad (7)$$

- Latest Start (LS): Last time an activity can be initiated so as not to delay the completion of the entire project.

$$LS = LF - \text{Time Activities} \quad (8)$$

Early Start (ES)	Duration	Early Finish (EF)
	Activity	
Latest Start (LS)	Duration	Latest Finish (LF)

Figure 2. Critical Forward and Backward Pass Notation

- Slack is the time owned by an activity to be postponed without causing delays in the project implementation as a whole. Activities with slack = 0 is referred to as a critical activity and are on the critical path. Mathematically it can be written as follows:

$$\text{Slack} = \text{LS} - \text{ES or LF} - \text{EF} \quad (9)$$

2.4. Motion and Time Study

Motion and time study aim to eliminate unnecessary work and design most effective methods and procedures while providing methods of measuring work to determine a performance index for an individual or group of workers, department or entire plant. Motion and time study consists of four parts and two parts [8]:

- Motion study or work methods design: to find the preferred method of conducting work.
- Time study or work measurement: to obtain the standard time to perform a specific task.

2.5. Maynard Operation and Sequence Technique

Maynard Operation and Sequence Time (MOST) is a motion time system that is mostly used in industrial setting to set the standard time of workers' tasks [2]. This technique is based on the order of sub activities or movements. MOST quantifies work content by evaluating the movement of objects [9]. Each type of movement has different sequence. Therefore it is necessary to separate the model sequence of events in this method. Table 1 illustrates the sequence of movements in the MOST [5]:

Table 1. Sequence Model of The Basic Most System

Activity	General Move	Controlled Move	Tool Use
Sequence Model	ABGABPA	ABGMXIA	ABGABP*ABPA
Parameters	A-Action distance; B-Body Motion; G-Gain Control; P-Place	M-Move Controlled; X-Process Time; I-Alignment	*F-Fasten; L-Loosen; C-Cut; S-Surface treat; M-Measure; R-Record; T-Think

On the use of MOST method of measuring standard time, there is little difference that is not looking for a normal time needed first. To get the standard time, we can simply by adding the allowance to the normal time has been obtained [10].

$$Standard\ time = normal\ time\ x\ \left(\frac{100\%}{100\% - \%\ allowance}\right) \quad (10)$$

Time units in the MOST technique are based on Time Measurement Unit (TMU). The following is the conversion value between normal time and TMU [2]:

- 1 hour = 100,000 TMU
- 1 minute = 1,667 TMU
- 1 second = 27.8 TMU
- 1 TMU = 0.00001 hour = 0.0006 minute = 0.036 second

MOST assists in process standardization by evaluating the standard time per cycle and ignores unnecessary activities [11]. It is used to identify and analyze value added and non-value added activities [12]. By applying MOST, resources can be deployed efficiently [13].

2.6. Allowance
Allowance granted to three things: the personal needs, relieve fatigue and barriers that cannot be avoided [10].

3. Methodology

The research process consists of several steps. The first step is to do observations on the company and see the problem.

Observation is made on the production of liquid medicine. Based on observations, existing problem is Overall Equipment Effectiveness percentage is still low on a particular machine. Therefore the objective to be achieved is to increase the value of OEE with improvements on the cause of the problems.

Data that can be collected in this study are primary data and secondary data. Primary data were successfully observed directly are the cleaning work processes, the element of work each process, and the time of each work process. While secondary data obtained from the company are the output of production and log sheet each production.

In addition to see the percentage of OEE, the result of OEE is also to see the bottleneck area. From that result, it will be easier to evaluate and improve the existing machines.

Based on data collected, Pareto Diagram analysis is conducted to define which process takes the longest time. Then, Cause-Effect Diagram is developed to analyze the causes that make that certain process consume the longest time. Afterwards, Critical Path Method is utilized to determine the division of tasks. Maynard Operation and Sequence Technique are conducted to define the normal time of the process.

4. Results and Discussions

4.1. Overall Equipment Effectiveness
To see a machine can be said to be effective in a company, any company applying a calculation method that is Overall Equipment Effectiveness. This method uses the parameters of time and quantity which results in a percentage.

Table 2. OEE Comparisons between Filler and Mixer Machine

PERIOD	JUN 2015	JUL 2015	AUG 2015	SEPT 2015	OCT 2015	Average
%OEE of Filler	62,43%	42,13%	63,75%	55,69%	58,76%	56,55%
%OEE of Mixer	72,71%	80,27%	74,67%	61,50%	76,82%	73,19%

Based on the data recapitulated the filler machine and the mixer machine (Table 2), it can be seen that the effectiveness of the filler machine has decreased in each month, while the mixer machine has increased. So the filler machine is the one that needs to be improved.

4.2. Pareto Diagram and Cause-Effect Diagram

Of the three OEE factors, the percentage was lowest for the availability factor of the machine before using it with the longest time held when the machine stopped with 26.5% and the cleaning process with 36.7%. We can see in the following Pareto diagram (Figure 3) for each data losses in the machine filler which occurred in June 2015-October 2015. Incident when the machine stops can be tolerated because there is a lot of activity in which operators use a break or tidying up the room after production.

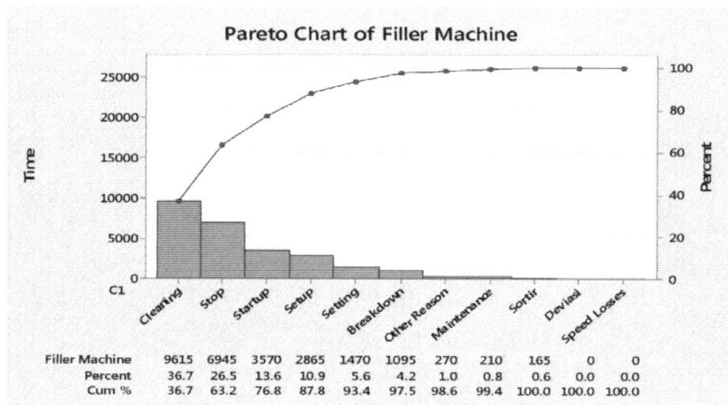

Figure 3. Pareto Chart of Filler Machine

To determine the factors influencing the decline in OEE and determine the causes and consequences of the length of the cleaning process, it can be used cause-effect diagram (Figure 4).

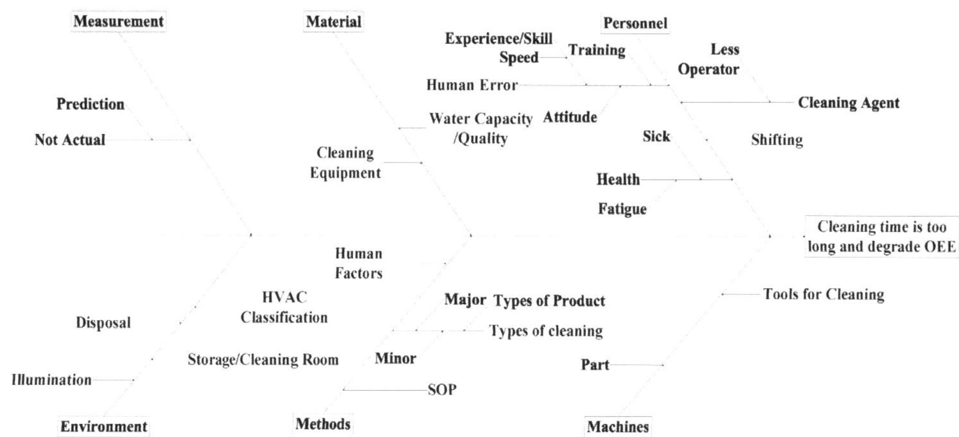

Figure 4. Cause-Effect Diagram

Based on the diagram, there are 6 main causes in the lengthy process of cleaning that is personnel, material, measurement, machines, methods, and also the environment. The causes are the ability/skill different operators, lack of operator which also interfere in the process of shifting work scheduling or operator. The different types of cleaning (major/minor) also affect the long or fast cleaning. Previous average time spent on cleaning is 2.5 hours for the major types. In addition to the Standard Operating Procedure (SOP) that does not explain the distribution of the cleaning process for the operator also

IOP Conf. Series: Earth and Environmental Science **101** (2017) 012039 doi:10.1088/1755-1315/101/1/012039

makes the process of being unstructured. The cause is not actually in the calculation of OEE can also occur because each room there is no clock or a stopwatch to see the exact time so that the operator will write time according to its own estimates. Then the problem of cleaning the room for those far from the machine room and less supportive tool for cleaning makes the cleaning process becomes longer do.

4.3. Critical Path Method

Critical Path Method (CPM) is applied to determine the division of tasks that can be carried by two operators. Designed a structured process based on the activities of the predecessor that had to be done before the next activity.

Table 3. Predecessor Activities on Cleaning Process

Activity	Time (Minutes)	Predecessor
A. Clean the bottle cap, the remaining drugs, etc.	8	-
B. Removing the spare part	12	-
C. Soak part (before)	7	B
D. Wash parts	26	C, F
E. Soak part (after) + dry	15	D
F. Wash holding tank	5	B
G. Cleaning machines	31	A, B
H. Preparation of the room and cleaning tools	3	G
I. Cleaning of walls	14	H
J. Clean the glass	17	I
K. Cleaning the floor	4	J
L. Dry floor	7	K
M. Record in log sheets	7	F, L

The result of the application of the critical path method can be seen in Figure 5.

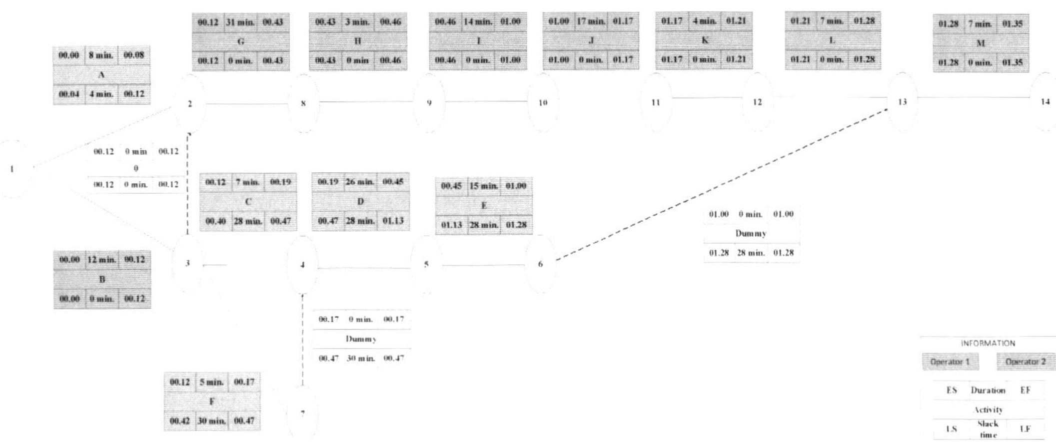

Figure 5. Critical Path Method

From CPM's optimal results, it was found that the process of cleaning the filler machine can be completed in a period of 1 hour 28 minutes (excluding record in the log sheet). When compared with the actual work time is 1 hour 53 minutes, then the application of this method can save time by 25 minutes with the work process has been structured.

The division of tasks to the two operators is determined based on the critical activities can be described in Table 4.

Table 4. Division of Tasks Based on CPM

Operator 1	Operator 2
Removing the engine part	Cleaning bottle, cap, remaining drug, etc.
Cleaning machines	Soak the parts (before)
	Wash holding tank
Preparing equipment for cleaning the room	wash parts
Cleaning of walls	Soak the parts (after) + dry
Clean the glass	Setting up the vacuum tool
Cleaning the floor	Record on *log sheet*
drying the floor	

4.4. Maynard Operation and Sequence Technique

In this study, the calculation of MOST is divided into three cleaning process, namely cleaning machine, cleaning the room, and cleaning parts. Application of MOST in this study is more focused on the cleaning process and is not applied when the operator cleans the bottle, cap, the rest of the drug, the release of spare parts, and also recording the activities on the log sheet. This is because the cleaning process has a more complex problem.

Time is produced in the form of Time Measurement Unit (TMU) and if converted into units of time, then the result is the normal time. Table 5 is a summary table of MOST calculation results in the form of time TMU and normal time.

Table 5. Maynard Operation and Sequence Technique

Cleaning Machine	A	B	G	P	M	X	I	S	C	TMU	*Total Time (minutes)*
Total Work Element	5190	34240	5230	580	460	140	100	2700	0	48640	29.184
Non-value added	3.114	20.544	3.138	0.348	0.276		0.06	1.62	0		29.1
Value added						0.084					0.084

Cleaning the Room	A	B	G	P	M	X	I	S	C	TMU	*Total Time (minutes)*
Total Work Element	9350	5330	630	680	2460	20	230	8780	0	27480	16.488
Non-value added	5.61	3.198	0.378	0.408	1.476		0.138	5.268	0		16.476
Value added						0.012					0.012

Cleaning Parts	A	B	G	P	M	X	I	S	C	TMU	*Total Time (minutes)*
Total Work Element	4560	20720	7530	9150	1100	870	370	1660	1560	47520	28.512
Non-value added	2.736	12.432	4.518	5.49	0.66		0.222	0.996	0.936		27.99
Value added						0.522					0.522

It can be seen in Table 5 that the process takes a lot more to be done is cleaning machines for ± 29 minutes and engine parts for ± 28 minutes. Cleaning the room normally can be done within ± 16 minutes. Thoroughness in cleaning machines and parts is needed thoroughness in the cleaning process because if there is an error then it can cause damage or loss. Besides, machine and parts have to be cleaned as well as possible because it will be reused for the production process afterwards. When viewed from the observation that has been done, cleaning the machine requires quite a long time because the equipment for cleanliness inadequate.

According to Table 5, the highest index parameter is B (Body Motion) which means that B has high motion waste. Several wastes in body motion are bending when taking water to rinse the machine, climbing a chair to clean high positioned window, bending to clean machine parts. In order to move effectively, the operators have to use the additional tools.

4.5. Allowance and Standard Time

To calculate the standard time it takes the data allowances each operator. Allowance is looseness interpreted as tolerance to rest to overcome fatigue or to factors that cannot be avoided. The value of the allowance can be seen from the ILO recommended allowances. Each cleaning process produces different allowances. Table 6 shows the difference between allowances and the calculation results of standard time at each cleaning process.

Table 6. The Result of Allowances and The Standard Time

No	Activity	Time (TMU)	Normal Time (minutes)	Allowances (%)	Standard Time (minutes)
1.	Cleaning Machine	48640	29.184	25%	38.912
2.	Cleaning the Room	27480	16.488	21%	20.857
3.	Cleaning Parts	47520	28.512	27%	39.057

MOST is used to set the standard time [13]. Based on Table 6, the standard time has been obtained. The total standard time for the cleaning process is 98.826 minutes or 1 hour 38.826 minutes. It can be a time reference for Company X in the process of cleaning.

If the cleaning process can be adapted to the standard time and the division of tasks in parallel to both operators, it is expected that the time for the cleaning process can be faster and increase the value of OEE through increased availability.

The results of this research shows that time study methods, such as MOST, is effective in reducing time in pharmaceutical industry setting. The result is accordance with other MOST research results [12, 13] and other time study method research results [14 - 16].

5. Conclusion and Suggestions

The findings show that CPM and MOST can decrease the cleaning process time of the filler machine. Based on the results of Critical Path Method for filler machine, the time to complete the cleaning process is reduced to 1 hour 28 minutes. It is 25 minutes faster than the current time of the cleaning process, which is 1 hour 53 minutes. With the Maynard Operation and Sequence Technique, the standard time for the cleaning machines is 38.912 minutes, 20.857 minutes for room, and 39.057 minutes for engine parts. The total standard time for the cleaning process is 1 hour 38.826 minutes. We suggest the company to make the work procedures specific to the division of tasks activity based on Critical Path Method along with the target time of the cleaning process in accordance with the standard time.

Wastes in body motion happens because the operators do not utilize the existing cleaning equipment properly. We suggest the company to pay attention to this problem by providing regular training to the operators to use the existing cleaning equipment properly, such as to use a hose to rinse the machine and extension stick for cleaning parts in high places.

References

[1] Frost & Sullivan's Industrial Automation Practice 2005 *Improving Plant Performance: Overall Equipment Effectiveness* (Frost & Sullivan)
[2] Karad AA, Waychale NK and Tidke NG 2016 *Int. J. of Eng. Res. & Gen. Sci.* **4** 657-62
[3] Sudri NM and Mareti A 2012 *J. Tek. & Ilm. Komp.* **1** 88-98
[4] Almeanazel OTR 2010 *Jord. J. of Mech. & Indust. Eng.* **4** 517-22
[5] Niebel BW and Freivalds A 2009 *Niebel's Methods, Standards, and Work Design* (New York: Mc-Graw Hill)
[6] Heizer J and Render B 2011 *Operations Management* (New Jersey: Prentice Hall)
[7] Sahid DSS 2012 *J. Tek. Inf. & Tel.* **5** 14-22
[8] Barnes RM 1980 *Motion and Time Study: Design and Measurement of Work* (New York: John Wiley & Sons)
[9] Mishra A, Agnihotri V and Mahindru DV 2014 *Glob. J. of Res. in Eng.* B **14** 1-7
[10] Sutalaksana IZ, Anggawisastra R and Tjakraatmadja JH 2006 *Teknik Perancangan Sistem Kerja*

(Bandung: Penerbit ITB)
[11] Jamil M, Gupta M, Saxena A and Agnihotri V 2013 *Glob. J. of Res. in Eng. Mech. & Mech. Eng.* **13** 45-56
[12] Puvanasvaran AP, Mei CZ and Alagendran VA 2013 *Proc. Eng.* **68** 271-7
[13] Pandey A, Deshpande VS and Gunjar S 2016 *Int. J. of Inno. in Eng. and Tech.* **6** 39-44
[14] Al-Saleh KS 2011 *J. of King Saud Univ. – Eng. Sci.* **23** 33-41
[15] Yusoff N, Jaffar A, Abbas NM and Saad NH 2012 *Proc. Eng.* **41** 1800-5
[16] Duran C, Cetindere A, Aksu YE 2015 *Proc. Econ. and Fin.* **26** 109-13

The Status of Implementation of Good Manufacturing Practices (GMP) Shredded Fish Production in UMKM Az-Zahrah, Makassar

Rindam Latif[1], Andi Dirpan[1], Sylvia Indriani[2]

[1] Department of Agricultural Technology, Hasanuddin University Makassar 90245, Indonesia
[2] Postgraduate Student at Department of Agricultural Technology, Hasanuddin University Makassar 90245, Indonesia

E-mail : rindamias04@yahoo.com

Abstract. Good manufacturing practices (GMP) describes the requirements that an industry should follow throughout the production process starting from sorting raw materials to handling final products. The purpose of this research is to disclose each phase of GMP in making shredded fish and to assess the status of the practices implemented by the UMKM Az-zahrah in producing its product. This descriptive study is conducted by doing interview with owner followed by observation and documentation of all activities related to the production process in order to assess the status of GMP applied by the UMKM Az-zahrah. In assessing the status, we referred to Permenperin RI No.75 / M-Ind / Per / 7/2010 and KBPOM regulation no. HK 03.1.23.04.12.22007 Year 2012. The findings indicate some deviation occurred when UMKM Az-zahrah implemented it, those are (1) there is a pet in production room (2) production employees are not yet wearing work uniform (3) production room has directly access to toilet (4) there is no health label about health and nutritional claims. The status of applied GMP in the UMKM Az-zahrah is at D level (less good) with rating IV.

1. Introduction

The existence of Micro, Small and Medium Enterprises (MSME) has a positive impact on the society's economy because MSMEs can grow and develop everywhere, from cities to villages. The increase of MSME can directly absorb labor so as to reducing unemployment rate; as the result, it will increase the economy of society. However, there is the fundamental problem often faced by MSME, such as there is limited capital support, have no skilled labor and have low standardized product. Another problem is there are many of MSME are labeled with no SNI standard since most of MSME entrepreneurs assume that standardization is not necessary. Not all MSME products in the market have been labeled with Indonesian National Standard (Standar Nasional Indonesia/SNI). Consequently, even though the products are qualified for export but they cannot compete with others because of no standard certification for MSME product.

In order to improve MSME competition with larger home industry in the food industry, they should increase consumer confidence by growing the awareness of producers and employees about the importance of hygienic food production process and responsible for consumer safety. Indonesian government through the Food and Drug Supervisory Agency sets out a guideline for Good Food

Production for Household Industries (CPPB-IRT), known as Good Manufacturing Practices (GMP), to improve the quality of produced food products. This guideline is not only ruling the nutritional ingredients but also the safety and viability of producing the products [1,2].

One of the processed food products that are popular is shredded fish produced by UMKM Az-Zahrah. The UMKM Az-Zahrah is one of the MSME producing processed food established since 2006 and it still exists until now. One of the problems faced by the MSME is having no SNI label for its product.

The problems faced by the MSME such as knowledge, technology, and cost, have caused the product can only be marketed locally, that is in South Sulawesi. Attempting to market its product in national and international scale (export) are limited because of not meeting with required quality requirements. Therefore, it important for UMKM Az-Zahrah to implement GMP status in producing shredded fish. This application is important for all parties, in particular, to protect MSME, and to guarantee that food is safe for the community or consumers.

To the best of the author's knowledge, this is the first study in UMKM Az-Zahrah to evaluate the GMP status. Therefore the objectives of this study are to identify the application of GMP to the production of shredded fish and to assess the application status of GMP in UMKM Az-Zahrah.

2. Material and Method

This research was conducted at UMKM Az-Zahrah located in Patingaloang Village, Ujung Tanah Subdistrict, Makassar. The research method used in this study is descriptive research method. Descriptive research means describing or illustrating the existing phenomena, both natural phenomena or human engineering [3].

Research Stages

After assessing CPPB-IRT application status, CPPB-IRT Rating document was obtained. At this stage, we conducted observation, interviews, and documentation of all activities related to the production process. An observation was conducted to observe closely the activity and behavior, environment, and general picture of the UMKM location. An in-depth interview is conducted with the owner. In assessing the status, we adopt Permenperin RI No.75 / M-Ind / Per / 7/2010 and KBPOM regulation no. HK 03.1.23.04.12.22007 Year 2012. The data obtained is processed by qualitative and quantitative description.

3. Results and discussion

3.1. Temperature UMKM Az-Zahrah Shredded fish Production

The UMKM *Az*-Zahrah is a local small business entity that is established to manage the processed food of shredded fish. The UMKM is located in Patingaloang Village, Ujung Tanah Sub district, Makassar, and it was incorporated in 2006. The production capacity of shredded fish on average is 500 kg per day. The number of employees is 25 and they are residing at around the industry and have the low-level education. Therefore it is important the UMKM conduct training about the processing procedures of shredded fish for its employees so that the procedures of GMP can be implemented maximally.

3.2. Status of Application Good Manufacturing Practices (GMP)

Following is the stages of the process production of shredded fish at UMKM Az-Zahrah: (1) sorting of raw materials, (2) weeding (3) boiling, (4) initial pressing, (5) early stage shredding, (6)) Frying, (8) second stage pressing, (9) second stage shredding, (10) refrigeration and (11) packaging. Based on the observations conducted in site (UMKM Az-Zahrah).

3.3. Location and Production Environment
The location of the UMKM Az-Zahrah is close to the fish auction. The site is also located in a densely populated residential area. As a result, household waste and poor sanitation around it may contaminate the product.

3.4. Building and Facilities
Factory building, including its facilities in the UMKM Az-Zahrah is made of materials that are easy to be maintained. The outside of the building, such as the street, parking lot and yard are well managed. The surface of the floor, wall, roof/ceiling, and ventilation is fine and non-slippery. it is also easy to wash the dirt and grime off the walls. The interior of walls is painted with light colour, easy to wash the dirt, durable and not flaking. Interior lighting design in every room is good since the rooms have enough lights from both sunlight and electricity. Doors and windows are made of stainless, waterproof material, smooth surface and easy to clean. Drainage and sewerage liquid system are sufficient and strong enough so that the liquid waste can flow smoothly. It is also easier to clean it so that rodents and other insects cannot have access to enter the room. The room is also equipped with a sink, hand soap, and dryers for employees to keep clean their hands. It has a bathroom that is directly related to the treatment room, however, it is easy to clean and enough wash water.

3.5. Production Equipment
All appliances such as plastic baskets, trays, buckets, and pans that are used in the process of production are clean. Before and after using the appliances, an employee will brush and wash them with soap and then dry them. Such actions may reduce contamination into the product. All appliances are placed in a room where processing and packing activities are done. The equipment used for production is conveniently designed based on its usefulness and it is regularly cleaned. The UMKM also uses an accurate scale that is commonly used to weigh food.

3.6. Water Supply or Water Supply Facility
Water used in the production process at the UMKM Az-Zahrah is pumped and flowed to water reservoirs through pipes/hoses. Thus, the water is precipitated first in the sterile container before using it into the production process. In addition, the water used is available in sufficient quantities for production activities.

3.7. Hygiene and Sanitation Facilities and Activities.
At the UMKM Az-Zahrah, cleanliness and sanitation of processing area are well maintained. Similarly, the practice of worker hygiene, clothing and hand washing is in good condition, although still in a simple condition. However, there is a toilet close to the production area.

3.8. Health and Employee Hygiene
Health and hygiene employees aim to ensure that employees involved in the processing process are not a source of contamination. At UMKM Az-Zahrah the condition of production employees is always clean and healthy, in addition to employees at UMKM is also always wash hands when starting to process, after handling raw materials and equipment and out of the toilet. Employees also work with a serious attitude (not eating, drinking or spitting indiscriminately). In the UMKM there is personnel responsible for health and hygiene. However, the UMKM has not yet provided the special uniforms imposed by employees.

3.9. Maintenance and Hygiene and Sanitation Program
Maintenance and hygiene and sanitation programs aim to maintain and apply sanitation to production facilities such as buildings, machinery or equipment, pest control, waste handling and others. If any pet is not justified to roam around the production room because it can cause contamination. Production space at UMKM Az-Zahrah there is a pet that is a kind of bird that is in a cage that is placed near the

wash site. The presence of waste in the production environment is immediately discarded as well as conducting a program of hygiene and health on a regular basis. In addition, the chemicals used in the production are used according to the rules and stored in containers labelled appropriately.

3.10. Storage
In order to keep foodstuffs such as raw materials, supplementary materials, and final products from damage, a manufacturer should have a good storage system. The UMKM Az-Zahrah has good storage system to keep all materials from damage. The materials have been stored in a clean room. Foodstuffs and packing materials are stored separately from the final product, in a clean and bright room [2].

3.11. Process Control
The purpose of controlling the production process on P-IRT is to produce quality food products and provide benefits for consumers so it is expected that all materials used should be good not to harm consumers. At the UMKM Az-Zahrah, has a record of raw materials that have been damaged, hazardous materials and food additives that meet the requirements. In the MSME using special food packaging materials and food additives used are given a special marker. In the MSME has a food production flow chart.

3.12. Food Labeling
At the UMKM Az-Zahrah, the label used includes the product name, ingredient composition, net weight, expiration date, production code, P-IRT number, and name and address of the UMKM. However, the label has not included health claims and nutritional claims.

3.13. Supervision by Responsible Person
Every industry should have someone who is responsible for the production process so that the quality of food products can be controlled properly. Similarly, there is a person in charge (PIC) that is responsible for that activity at the UMKM Az-Zahrah. A PIC has a certificate of food safety instructor who conducts internal supervision and correction action in MSME on a regular basis.

3.14. Product Withdrawal
A company can withdraw its product from selling if there is found any problem with the product. One reason for the withdrawal of food products is safety problem such as perhaps the product will cause illness or poisoned to consumers if they consume the product. As a consequence, it is very important to follow a guideline that the Indonesian Government has ruled about how to process the food product until it is distributed to society. Similarly, the UMKM Az-Zahrah will withdraw the shredded fish products if the products are not safe to be eaten.

3.15. Recording and Documentation
This activity is very helpful to owners to record what problem faced with regard to product quality. It also important to record production process, materials used, expiration date, and distribution system. There is a bookkeeper at the UMKM Az-Zahrah that is responsible to record production activities such as calculate materials used, the amount and date of production testing, storage and distribution. In addition, production documents are always updated, easy to store and easy to find.

3.16. Employee training
A P-IRT leader and employees must have the skills and knowledge of sanitation in the process of processing food products in order to avoid problems that could be harmful to both owners and consumers. There is a safety training program held at the UMKM Az-Zahrah to train employees about that.

3.17. Discussion

Based on findings found on site, the status of GMP application run by the UMKM Az-Zahrah that was assessed based on Permenperin RI No.75 / M-Ind / Per / 7/2010 [4] and KBPOM regulation no. HK 03.1.23.04.12.22007 Year 2012 is at D level with IV rating (less good). It is because there are some deviations that occur at the processing site. These deviations consist of two critical deviations, one serious deviation and one major deviation.

Minor irregularities are deviations in which if they are not corrected or disregarded continuously, they will potentially affect the quality of food. Major irregularities are irregularities in which if they are not corrected, they will potentially affect food security, and serious and critical irregularities are irregularities in which if they are not corrected, they will affect food security [5]

This study has shown that the implementation of GMP in processing shredded fish is very important in order to produce good quality of shredded fish products, guaranteed safety and fulfilled consumer expectations. Thus, in the future, the stages of the corrective actions of existing deviations are the most important action in the processing activity[1,6,7].

The results of this study provide some managerial implications, that is improvements toward critical deviations [8]. In this regard, prohibition all kind of pets such as dogs, cats, chickens, sheep, and others to enter production room because it can contaminate the product. In order to keep storage room from insect pests, the room must be regularly cleaned and the food must be placed in the container that does not have direct contact with floor, wall or ceiling of the room.

Pest control is the important thing that cannot be disregarded because pests such as birds carry biological contaminants that can reduce the quality and food security [9]. Based on research on the implementation of Good Manufacturing Practices by Ristyanandi & Darimiyya (2012), the number of pest/animal intruders in the production area should be of concern, because the animal faces or diseases that they usually have, may transmit to humans, then it can be harmful to the quality of the final product. In addition, according to [9], dust and soil is a source of microbial transmission and its spores. A room without adequate sanitation is a good growth medium for many types of pathogenic bacteria, which cause various diseases..

Additionally, the attachment of health labels such as health and nutritional claims is also important. Based on the Regulation of the Head of the POM Number HK.03.1.23.11.11.09909 Year 2011, the claim is any form of description stating, suggesting or indirectly stating the specific quality of processed food such as nutrient content, nature, production, processing or other quality factors. Claims on processed foods include nutritional claims and health claims. The improvement implication of serious deviation according to BPOM,[8] is related to the requirement of production employees. They should wear work clothes, gloves, masks, and work shoes. Employees should also wash their hands with hand soap after work and in particular after going to the toilet. Employee habits such as eating and drinking at work, smoking, spitting, sneezing in front of the food and using jewellery must be prohibited since these habits may contaminate the processed food products. According to Pupitas and Monika (2016), employees should wear standard clothing for a food industry, namely head, mouth and nose, and gloves covers [10].

While the improvement implication of major deviations according to BPOM [11] is the existence of the bathroom that is not directly related to the treatment room, enough ventilation, easy to clean and enough washing water.

4. Conclusion

The status of GMP application run by UMKM Az-Zahrah to produce shredded fish is at D level (less good) with rating IV. The UMKM Az-Zahrah needs to implement CPPB well and correctly in order to produce better products. It is important for the UMKM Az-Zahrah to take corrective action to reduce the existing deviations to improve its GMP application status.

References

[1] Costa Dias M A, Sant'Ana A S, Cruz A G, Faria J de A F, Fernandes de Oliveira C A and Bona E 2012 On the implementation of good manufacturing practices in a small processing unity of mozzarella cheese in Brazil *Food Control* **24** 199–205

[2] He T T, Ung C O L, Hu H and Wang Y T 2015 Good manufacturing practice (GMP) regulation of herbal medicine in comparative research: China GMP, cGMP, WHO-GMP, PIC/S and EU-GMP *Eur. J. Integr. Med.* **7** 55–66

[3] Sukmadinata S.N 2016 Penerapan Sistem HACCP dan GMP pada Proses Pembuatan Abon di Industri Abon45 di Kabupaten Semarang *Semin. Nas. Pengabdi. Kpd. Masy.*

[4] Peraturan Menteri Perindustrian Republik I 2010 Pedoman Cara Produksi Pangan Olahan yang Baik (Good Manufacturing Practices) *Jakarta : Kementrian Perindustrian.*

[5] Masrifah E, Noorachmat B P, Sukmawati A and Pertanian F T 2015 Kesesuaian Penerapan Manajemen Mutu Ikan Pindang Bandeng (Chanos chanos) Terhadap Standar Nasional Indonesia **10**

[6] Abou-El-Enein M, Römhild A, Kaiser D, Beier C, Bauer G, Volk H D and Reinke P 2013 Good Manufacturing Practices (GMP) manufacturing of advanced therapy medicinal products: A novel tailored model for optimizing performance and estimating costs *Cytotherapy* **15** 362–83

[7] Godinho Filho M, Marchesini A G, Riezebos J, Ganga G M D and Vandaele N 2017 The application of Quick Response Manufacturing practices in Brazil, Europe, and the USA: An exploratory study *Int. J. Prod. Econ.*

[8] Badan Pengawasan Obat dan Makanan 2012 Pemberian Sertifikat Produksi Pangan Industri Rumah Tangga. Badan Pengawasan Obat dan Makanan *Jakarta Badan Pengawas Obat dan Makanan.*

[9] Winarno, F.G. and Surono 2009 GMP Cara Pengelolaan Pangan yang Baik *Bogor M. Brio Press*

[10] Pupitas, S. V, Monika R. M R, T. J D, T. osephine D and Mayer T. S. & M M S 2016 Penerapan Sistem HACCP dan GMP pada Proses Pembuatan Abon di Industri Abon45 di Kabupaten Semarang *Semin. Nas. Pengabdi. Kpd. Masyarakat.*

[11] Badan Pengawasan Obat dan Makanan 2012 Cara Produksi Pangan yang Baik untuk Industri Rumah Tangga *Jakarta Badan Pengawas Obat dan Makanan.*